Topics in Biological Inorganic Chemistry

Volume 2

Springer

Berlin
Heidelberg
New York
Barcelona
Hong Kong
London
Milan
Paris
Singapore
Tokyo

Metallopharmaceuticals II

Diagnosis and Therapy

Editors: M.J. Clarke · P.J. Sadler

With contributions by
M. W. Brechbiel, C. L. Hill, J. F. Kronauge,
K. Kumar, J. H. McNeill, C. Orvig, A. Packard,
P. J. Sadler, R. Schinazi, C. F. Shaw III, H. Sun,
J. T. Rhule, K. H. Thompson, M. F. Tweedle,
Z. Zheng

Springer

Volume Editors:

Professor Michael J. Clarke
Department of Chemistry
Boston College
Merkert Center
Chestnut Hill, MA 02467
USA

Professor Peter J. Sadler
Department of Chemistry
University of Edinburgh
King's Buildings
West Mains Road
Edinburgh EH9 3JJ
Scotland, GB

ISSN 1437-7993
ISBN-13: 978-3-642-64239-5 e-ISBN-13: 978-3-642-60061-6
DOI: 10.1007/978-3-642-60061-6

Library of Congress Cataloging-in-Publication Data

Metallopharmaceuticals II ed.: M. J. Clarke; P. J. Sadler. - Berlin; Heidelberg; New York; Barcelona; Hong Kong; London; Milan; Paris; Singapore; Tokyo: Springer
2. Diagnosis and therapy/with contributions by M. W. Brechbiel... - 1999
(Topics in biological inorganic chemistry; Vol. 2)
ISBN 978-3-642-64239-5

Softcover reprint of the hardcover 1st edition 1999

Coverdesign: Friedhelm Steinen-Broo, Pau/Spain; MEDIO, Berlin
Typesetting: Scientific Publishing Services (P) Ltd, Madras
SPIN: 10551883 2/3020-5 4 3 2 1 0 - printed on acid-free paper

Editorial Board of the Series

Preface

Inorganic chemistry is beginning to have a major impact on medicine. It offers great potential for the design of novel therapeutic and diagnostic agents. ***Volume 1*** in this series was concerned with anticancer drugs, especially the successful platinum complexes which target particular sites on DNA. In ***Volume 2***, the wider scope of inorganic medicinal chemistry is illustrated.

About one quarter of all magnetic resonance imaging (MRI) scans in the clinic now involve administration of a contrast agent. The challenges involved in optimising the electronic relaxation properties of paramagnetic contrast agents through chemical design, their formulation and dosing are described by *Tweedle* and *Kumar*. Progress is being made with agents that can also probe biochemical functions and be targeted to specific organs and tissues.

Packard, *Kronauge* and *Brechbiel* describe recent advances in the targeting of radioactive compounds for diagnosis and therapy, which encompasses radionuclide production and processing, organic chemistry and coordination chemistry for radiopharmaceutical synthesis, as well as associated biochemistry and molecular pharmacology. The outstanding success of man-made ^{99m}Tc, with its rich variable-oxidation-state co-ordination chemistry, is evident.

The versatile chemistry of antiviral polyoxometallates with their variable charge distribution, shape, acidity, hydrolytic stability and redox potentials is described by *Rhule*, *Hill*, *Zheng* and *Schinazi*. They also speculate that the primary mode of action of fullerenes involves inhibition of human immunodeficiency virus protease. Future progress with improving the water solubility of fullerenes is important.

The potential of vanadium compounds as orally-administered insulin mimetics capable of lowering blood glucose and ameliorating other diabetic symptoms is described by *Orvig*, *McNeill* and *Thompson*. The main challenge is to control the toxicity of vanadium through the choice of oxidation state, types of chelated ligands, and amphiphilicity. A vanadium complex may well enter the clinic soon.

The chemistry and biochemistry of bismuth, the heaviest non-radioactive element in the periodic table, is poorly understood despite its use in medicine for several centuries. *Sun* and *Sadler* describe recent advances in understanding the structures of bismuth antiulcer drugs and their target sites on proteins.

Although gold drugs have been in widespread use for over 60 years for the treatment of rheumatoid arthritis (chrysotherapy), their chemistry and biochemistry are also poorly understood. *Shaw* describes how both injectable and oral gold drugs are biotransformed before they reach their biological target sites: they are pro-drugs.

Could it be that the metabolite gold(I) dicyanide is an active species? This and some other gold complexes also exhibit antiviral activity. The realisation that gold(I) can be oxidised to gold(III) in vivo, and that this has major effects on T-cell activation, is likely to lead to progress in understanding the toxic side-effects of gold drugs.

Overall this volume will provide chemists, biochemists, molecular biologists and pharmacologists with new insights into the mechanism of action of metallodrugs and diagnostic agents, and inspiration for the design of novel ones.

August 1999

Peter J. Sadler
Michael J. Clarke

Contents

Magnetic Resonance Imaging (MRI) Contrast Agents

Michael F. Tweedle, Krishan Kumar

Bracco Research USA, P.O. Box 5225, Princeton, NJ 08543-5225, USA

This chapter covers all current types of contrast agents (CA) for use in Magnetic Resonance Imaging (MRI). It is intended for learning rather than exhaustive review, presenting and discussing terms and sufficient theory to understand the original literature in the field. The emphasis is on the CA themselves as chemical entities, rather than on the images they generate, but sufficient examples of MRI are shown to demonstrate the observed effects. The chapter begins with an historical perspective setting MRI agents in the context of the older X ray and radiopharmaceutical agents, which bracket the MRI agents in tolerance and sensitivity to detection. Following a description of MRI, the mechanisms of image contrast generation with contrast agents are introduced, including proton water displacement, T1 enhancing agents, and T2 enhancing agents such as the iron oxides. Relaxivity is defined, and the mechanisms of inner sphere relaxivity pertinent to paramagnetic metal ions, particularly Gd chelates, are detailed, including the Solomon-Bloembergen Morgan theory. The next section d'eals with the most widely used class of MRI CA, the water soluble Gd chelates. Fundamental chemical and biological properties and their importance are described in detail, including chemical structures, dosing, formulations, relaxivity, colligative properties, in vitro and in vivo stability, tolerance, and the mechanism by which the agents enhance CNS abnormalities. A section on liver imaging agents follows including structures and MRI images of agents (water soluble) for the hepatobiliary and (particulate) for the reticuloendothelial systems. A short section follows on new agents for the near term for gastrointestinal and blood pool use (MR angiography), including recent images. The chapter ends with a detailed discussion of the possibilities for biochemically targeted MRI agents that would combine the exquisite spatial detail of MRI with the biologic specificity of the newest targeted radiopharmaceuticals.

Keywords. Magnetic resonance imaging (MRI), Contrast agents (CA), T1 and T2 enhancing agents, Inner sphere relaxivity, Gd chelates, CNS abnormalities, Liver imaging, MR angiography, Biochemically targeted MRI agents

1 Historical Perspective

1.1 X-ray Contrast Agents

The field of contrast agents in diagnostic medicine opened in 1895, [1] six weeks following the discovery of X rays by Röntgen [2]. These first contrast agents were solutions of heavy elements with greater differential absorption of X rays than tissue and thus they cast a dark shadow on the film following intravenous administration to cadavers. They were far from ideal, being bare metal ions, and hence quite toxic. The concentrations required were (and still are) on the order of mM. Achieving 1–3 mM in heavy atom concentration in vivo requires injections of tens of grams of a heavy atom. Iodine as the sodium salt was proposed in 1918 [3], but it was not until the late 1920s that intravenously administered organoiodine agents with acceptable tolerance were developed [4]. Three full generations of intravenous agents have evolved since the first commercial agents, with improvements in tolerance being the driving force. Hundreds of triiodinated benzenoids have been synthesized and tested by several large commercial R&D groups [5]. These agents are highly water soluble and contain hydrophilic moieties alternatively substituted with the iodine to mask the hydrophobic iodines. The well tolerated commercial examples are renally excreted to

>97%, even though the applications are throughout the body. Around the 1960s similar agents (lacking one hydrophilic moiety) were commercialized which were excreted by the liver rather than wholly by the kidneys. These were poorly tolerated relative to renally excreted agents, and are now seldom used [6]. Today the rate of serious (life threatening) complications resulting from contrast agent injections has dropped from frequent to about 1:100,000. Table 1 shows the historical progress of the development of the iodinated X-ray contrast agents, measured in terms of the primary factor limiting development: tolerance.

1.2 Radiopharmaceuticals

By contrast, Table 2 shows the development of Technetium (^{99m}Tc) radiopharmaceuticals. These diagnostic pharmaceuticals emit gamma photons of 140 k eV with a half-life of 6 h, which pass through tissue and are detected by an Anger (gamma) camera. The concentration required for imaging is of the order of nM or less. Hence tolerance is usually not a serious obstacle to development of new radiopharmaceuticals. The limiting factor in their development has been the

Table 1. Tolerance of some X-ray contrast agents in rodents

Name	Formula	LD_{50},[a] g Iodine/kg
Sodium Iodide	NaI	1.07
Iodopyracet	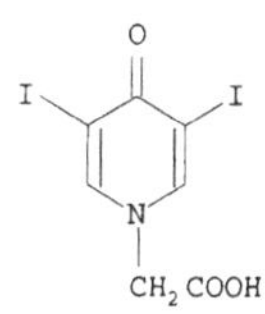	3.15
Acetrizoate	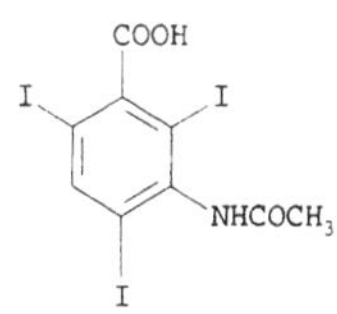	10.7
Ditrizoate	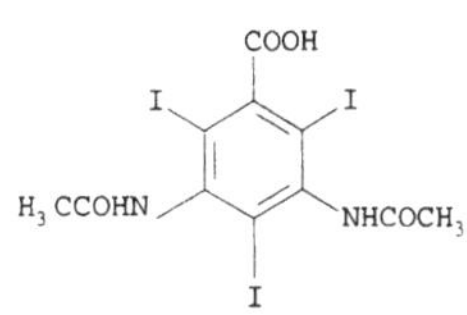	13.65
Iopamidol	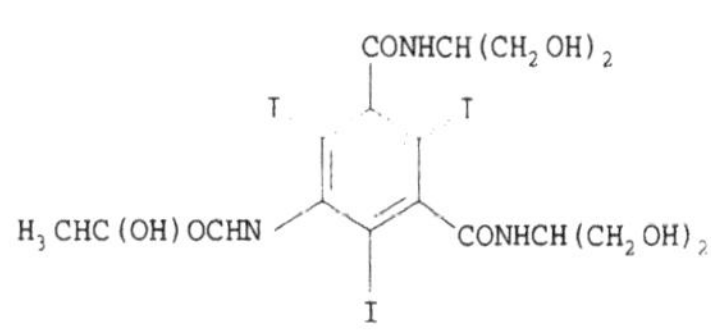	20[b]

[a] Ref. 1
[b] Sovak M (1986) In: Sovak M (ed) Radiocontrast Agents. Springer, Berlin Heidelberg New York

Table 2. List of potential ^{99m}Tc radiopharmaceuticals and MRI contrast agents for targetting organs[a]

Radiopharmaceutical	Clinical use	MRI agent	Development status
TcO_4-	Small size hydrophilic agent	Mn(DPDP) or Mn^{2+} (aq) ion	Clinical trials. Mn^{2+} liberated. Heart pancreas, and liver uptake. Tolerance (?)
Tc-albumin	Blood Pool imaging. Agent remains intravascular throughout imaging time	1. Gd-proteins and polymers 2. In vivo binders 3. Fe_3O_4 particles	1. Preclinical. Toxic so far 2. Clinical trials 3. Clinical trials
1. Tc-albumin (large aggregates) 2. Tc-S Colloids	Focal liver disease (e.g. liver metasteses)	1. None 2. Fe_3O_4 3. MnS	2. Clinical trials 3. Preclinical, probably toxic
$Tc(DTPA)^{2-}$	Blood Brain Barriers abnormalities	$Gd(DTPA)^{2-}$ $Gd(DOTA)^-$ Gd(DTPA-BMA) Gd(HP-DO3A)	Commercial products
Tc-HIDA	Hepatobiliary function. Focal liver diseases	Gd(BOPTA) Gd(EOB-DTPA)	Clinical trials
Tc-diphosphonates	Bone seekers for bone metastases. Myocardial infarcts.	Gd(Phos-DOTA) Gd(Phos-DTPA) analogs	Preclinical, probably toxic due to Ca binding
Tc-BATO Tc-MIBI	Myocardial perfusion	None	Possible (?)
Tc-MAG_3	Renal tubular secretion. Renal blood flow	None	Possible (?)
Tc-Nitroimidazole	Hypoxic tissue	None	Possible (?)
Tc-HMPAO	Cerebral perfusion	None	Possible (?)

[a] In(Octreotide) binds to Somatostatin receptors and is marketed in US and Europe for use in Oncology

chemistry required to direct the agents to specific targets in vivo following intravenous administration. Over the past thirty years in particular, a variety of novel Tc chelates have been developed which are used in imaging not only the excretory organs and those with high macrophage content (particle sequestering cells) such as liver and spleen, but also bones and calcified infarcts (dead tissue) [7]. In the past decade, more sophisticated complexes with the ability to enter endothelial cells (cells lining the capillary) and myocytes (heart cells) were developed [8]. Today, animal and clinical studies are testing the ability of Tc complexes to be retained in the tissue as a direct function of tissue oxygen concentration in an effort to use imaging to isolate and identify tissue at risk of further damage and worth salvaging via medical intervention [9]. New peptide-conjugated metal complexes are also emerging as imaging agents for specific targets [10], one of which is marketed in the US and Europe. This fascinating chemical and biological variety is made possible by the lack of toxicity manifested by such low administered doses. The primary barrier to development has been finding ligands that possess useful physiologic or biochemical activity. Unfortunately, targeting is always imperfect, and the imperfections – Tc atoms that disintegrate in organs adjacent to the target organ (e.g. blood) - are imaged as background noise. The weakness of the radiopharmaceutical technique is the poor spatial resolution (ability to resolve small structures) of the images - about 1 cm - relative to about 1 mm for X-ray imaging and Magnetic Resonance Imaging (MRI).

1.3 Magnetic Resonance Imaging (MRI) Contrast Agents

Magnetic Resonance Imaging (MRI) was developed 30 years after the initial discovery of NMR by Bloch and Purcell [11]. MRI was developed during the 1970s. Initially it required hours to obtain images which had spatial resolution on the order of a cm. Practicality required application of two dimensional NMR [12] techniques and the very fast computers and array processors of the 1980s led to dramatic improvements in speed and resolution. MRI scans can now be routinely obtained in a few minutes with the same spatial resolution as X-ray CT (computed tomography) scans, namely about 1 mm. Bloch also foresaw paramagnetic contrast agents for NMR, adding ferrous ion to his solutions to reduce relaxation times [13]. The first attempt at using contrast agents in MRI was published by Lauterbur, Mendoca Diaz, and Rudin in 1978 [14]. They used complexes of Mn^{2+}, an ion known to concentrate in myocardial (heart) tissue [15], and demonstrated the most fundamental principle of the art: exogenous paramagnetic substances reduce relaxation times sufficiently to be readily visualized in MR images of tissues [16,17].

The metal ions which are reasonably stable in aqueous media and which possess the highest spin only magnetic moments are Mn^{2+}, Fe^{3+} and Gd^{3+}. The aqua ions and simple salts are too toxic for human use. Mn^{2+}, although an excellent relaxation probe with the longest electron spin relaxation times among transition metals, produces cardiovascular toxicity due to calcium blockade at dose of 10 μmol/kg. Gd^{3+} and Fe^{3+} ions precipitate due to hydrolysis ($K_h^{Gd} = 10^{-8.3}$ and $K_h^{Fe} = 10^{-2.8}$) and reaction in vivo with phosphate and carbonate and due to the low solubility products of $GdPO_4$ and $GdCO_3$ ($K_{sp}^{GdPO_4} = 10^{-22.26}$, $K_{sp}^{GdCO_3} = 10^{-31}$). Thus, the first commercialized (1989) chelated MRI contrast agent was an *N*-methylglucamine salt of $Gd(DTPA)H_2O^{2-}$ (where DTPA = diethylenetriamine-pentaacetic acid) [18,19]. The effective in vivo concentration of Gd chelates is about 50–100 μM, or about an order of magnitude lower than for X-ray agents and several orders of magnitude greater than for radiopharmaceuticals. Thus MRI agents have about an order of magnitude inherent advantage over X-ray contrast agents in terms of effectiveness versus tolerance, without the loss in imaging spatial resolution inherent in radionuclide imaging. Even at these lower administered doses, tolerance, excretability, and in vivo stability are significant concerns. MRI agents therefore fall between the X-ray agents and the radiopharmaceuticals in terms of the barriers to further development of new chemical entities. There are already MRI agents marketed (a Mn chelate and a dextran coated Fe_3O_4 particle) and in clinical trials (two Gd chelates), for imaging liver [20–24] and macrophage organs [25], which are analogous to the Tc-based agents used for these purposes (Table 2).

2
Magnetic Resonance Imaging (MRI) and Contrast Mechanisms

2.1
Magnetic Resonance Imaging (MRI)

An MRI imager (Fig. 1) is essentially a large version of a modern multidimensional NMR spectrometer, with field gradients running in three perpendicular dimensions. The MRI image of human head in Fig. 2 is made up of 65,536 volume elements of

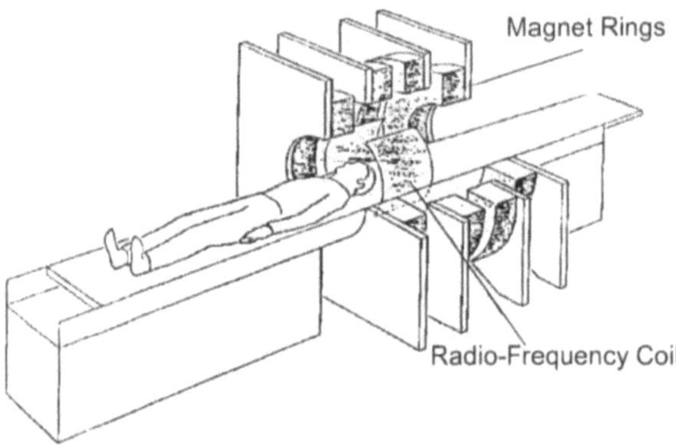

Fig. 1. Schematic diagram of an NMR imager

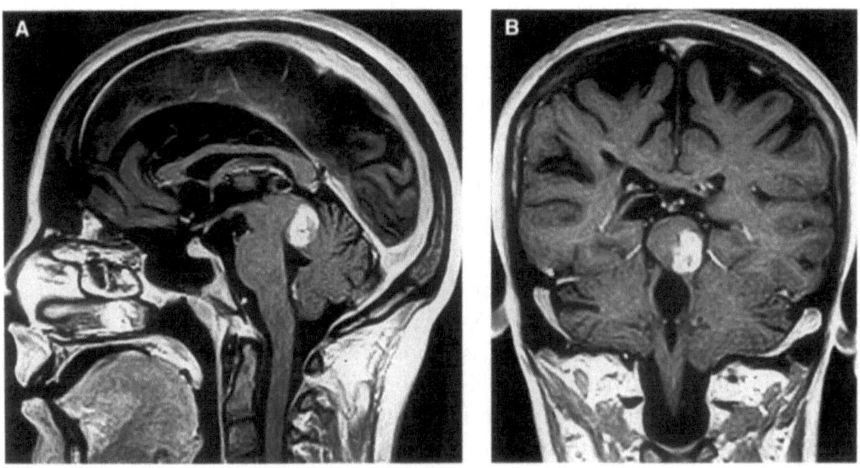

Fig. 2. A T_1 weighted image of the human head obtained following intravenous administration of 0.1 mmol/kg ProHance^R (Gadoteridol, (Gd(HP-DO3A)) shows a Tectal glioma. Abnormal tectal brightness is obvious as a centrally located bulbous mass (3 o'clock from center) (**A**) Sagittal (nose – on) image, and 5 o'clock dead center in (**B**) coronal (ear to ear) 1 mm reformatted images from post-contrast 3D MP-RAGE pulse sequences. Images courtesy of Dr. Val Runge, University of Kentucky

about, 1 mm × ~1 mm × 5 mm dimensions. The field gradients spatially located each element in single experiments (line by line of data) which required 3 minutes of total imaging time. The signal intensity (SI) in each volume element ranges from 0 (black) to 256 (white) and is governed approximately by the following equation:

$$SI = [H]H(v)\exp(-TE/T_2)\{1 - \exp(TR/T_1)\} \quad (1)$$

where TE (spin echo time) and TR (pulse repetition time) are operator-adjusted variable times, [H] is the concentration of water protons in the volume element, H(v) is a motion factor which in theory accounts for the potential of water protons to move into or out of the volume element in the time of the pulsed NMR experiment. The radiofrequency burst used to excite the protons in the magnetic field is of the order of a microsecond. T_1 (longitudinal) and T_2 (transverse) are the classical relaxation times of the excited water protons in the volume element. The reciprocal of these are rate constants governing return to equilibrium alignment with the static magnetic field. T_1, the spin lattice relaxation time, represents energy transfer to the surroundings, and T_2, the spin-spin relaxation time, represents the process of exchange of energy between excited state and ground state nuclei. The concentration of water in tissues is variable, but the high contrast in MR images derives primarily from the high heterogeneity of the relaxation times in tissues. For example T_1 relaxation times for different tissues determined at 1.4 T are reported in the range 500 to 200 ms [26]. In practice the operator adjustable parameters are used to generate a wide variety of pulse sequences with different weightings of the key parameters to give variously contrasting "spin echo" images.

Despite significant research efforts, most imaging with and without contrast agents is qualitative as far as signal intensity is concerned. Two factors limit quantitative analysis of signal intensities. The field inhomogeneities strongly affect T_2, reducing it in practice. Inhomogeneities are inherent in any NMR experiment, but are magnified by the size of the equipment needed for human-sized samples. In addition, the relaxation times are on the order of several tens to several hundreds of milliseconds, and water diffusion events in tissues and especially flowing blood are relatively rapid and heterogeneous among imaged tissues. Hence, the motion term interferes with absolute quantitation of relaxation times. Even so, some success is achievable [27].

2.2
Water Proton Displacement Agents

Conceptually the simplest way to generate contrast in an MR image is to displace the water protons with a substance lacking protons, or whose protons resonate outside of the imaging frequency band. Off-resonance absorptions have been used to enhance MR images in a technique called "phase-contrast imaging" [28]. But the imaging band is wide enough that the latter technique is, so far, impractical.

One might imagine deuterium substitution for imaged protons in tissue, since the image is primarily of water protons (approaching 110 M as the concentration of water is 55.5 M). However, the toxicity of deuterium oxide is limiting. In mice toxicity was determined to be at 30–40% of the total tissue water deuterated in chronic studies [29,30]. Use at this equilibrium concentration would probably be just visible. About

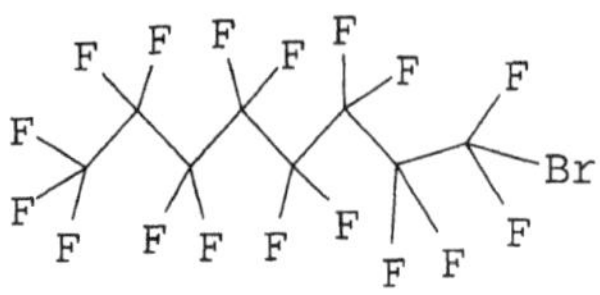

Fig. 3. Structure of Perfluorooctyl bromide

300 ml of D_2O would be needed for imaging work in humans which would cost about \$300. This high cost would be prohibitive. Currently MRI agents need to be competitive with X-ray and radiopharmaceutical agents in the range of about \$100 per dose or less, unless the agent is functionally unique and the medical information provided is essential. Intravenous use of Deuterium oxide has been seriously considered as a perfusion agent [31], however, where smaller doses are possible.

A practical attempt at near complete displacement of water protons was made in the gastrointestinal tract. The only commercialized agent to accomplish this is perfluoro- octylbromide (Imagent^{R}) (Fig. 3) [32]. This versatile substance was originally designed as a blood substitute. The mechanism of displacing protons carries the advantage of making the contrast agent effective on numerous pulse sequences. Unfortunately the quantity required to fill the entire gastrointestinal tract (hundreds of mL), the cost (several hundred dollars), and a high incidence of diarrhea were significant drawbacks, despite the excellent imaging characteristics.

2.3 Proton Relaxation Catalysis

2.3.1 *Relaxivity*

Dissolved or colloidal paramagnetic compounds catalyze proton relaxation of bulk water. The reduction of T_1 when a water molecule becomes bound to a paramagnetic metal ion is about a factor of 10^6. The term relaxivity is commonly used to refer to a second order rate constant, $r_{1,2}$ (where 1 and 2 refer to T_1 and T_2 reduction, respectively) which is determined by using Eq. 2.

$$(T_1^{-1})_P-(T_1^{-1})_O=r_{1,2}[P] \quad (2)$$

In Eq. 2, [P] is the concentration of the paramagnetic compound in mmol L^{-1} (mM) for solutions or mmol kg^{-1} for soft tissues, which are 50–90% water), $(T_1^{-1})_P$ and $(T_1^{-1})_O$ are reciprocals of relaxation times in the presence and absence of paramagnetic compound, respectively. A plot of $1/T_1$ vs [Gd], for dilute aqueous solutions of Gd(HP-DO3A) (0.1–10 mM), is shown in Fig. 4, r_1 being determined as the slope. Relaxivities are highly dependent on measurement frequency and temperature. It is therefore essential that these parameters be reported with relaxivity data. Viscosity also directly affects relaxivity [33], as demonstrated by a plot of relaxivity vs viscosity in Fig. 5. Hence, different biological fluids, water ($\eta = 1$), buffers, and blood ($\eta = 2.5$) may yield substantially different $r_{1,2}$ values due to viscosity differences alone. r_2 values may be measured in the same way.

X-ray and radiopharmaceutical agents affect the observed signal in the images with a straightforward linear dependence on agent concentration. In MRI the situ-

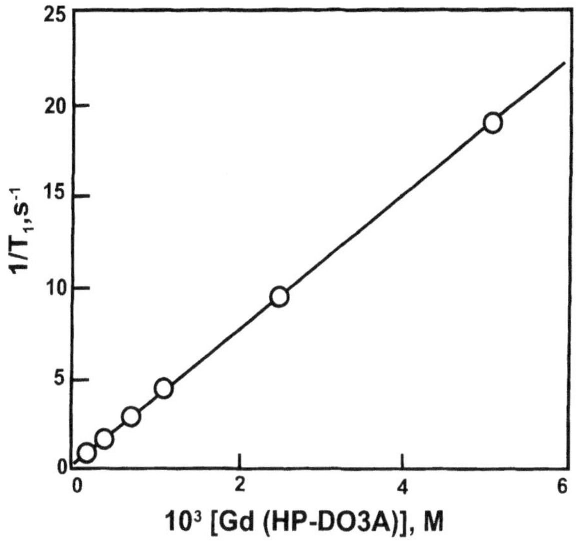

Fig. 4. Plot of $1/T_1$ vs [Gd(HP-DO3A)] to determine relaxivity of the chelate [45b]

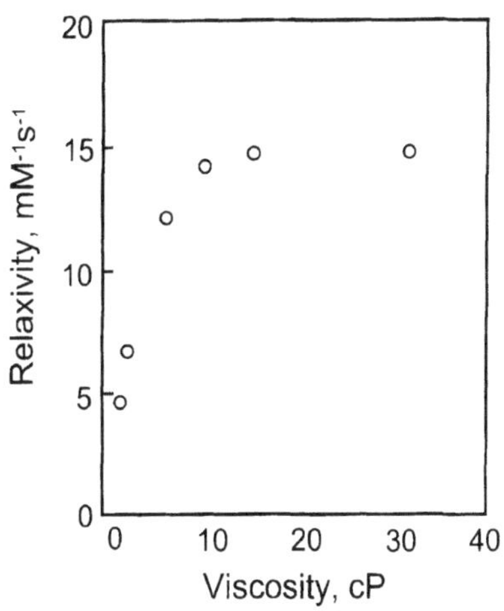

Fig. 5. Effect of increasing viscosity on the relaxivity of Gd(HP-DO3A) at 85 MHz

ation is more complex. Using Eqs. 1 and 2, for typical contrast agent and tissue and imaging parameters, calculated signal intensity vs [Gd] curves for four different contrast agents are shown in Fig. 6. The behavior is simple at relatively low concentrations, but becomes complex after about 0.5 mM, where T_2 reduction (first exponential in Eq. 1) begins to compete with T_1 reduction (second exponential in Eq. 1) and ultimately lowers the signal intensity. The effect has been observed in vivo in T_1 weighted images [34] and is the basis of some T_2-weighted imaging studies (vide infra). Given the instrument variables, TE and TR in Eq. 1 (note that there are numerous other radiofrequency pulse sequences in use with different parameters), the different tissue relaxation times and the fact that both T_1 and T_2 inevitably contribute in opposing directions, it is not surprising to see data from MRI experiments on live rats like that shown in Fig. 7 [35]. In this case, in addition to T_2

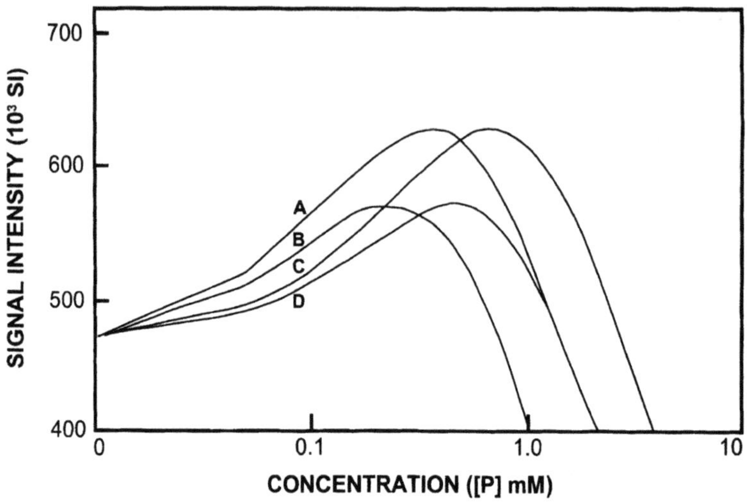

Fig. 6. Simulated intensity curves for four contrast agents with variable adjustable parameters. Before addition of the contrast agent the T_1 and T_2 values in a tissue were 0.5 s and 0.1 s, respectively. The pulse sequence used was a spin echo with TE = 0.030 s and TR = 0.5 s. The values of r_1 and r_2 (mM^{-1} s^{-1}) used were: (*A*) $r_1 = r_2 = 10$, (*B*) $r_1 = 10$ and $r_2 = 20$, (*C*) $r_1 = r_2 = 5$, and (*D*) $r_1 = 5$ and $r_2 = 10$

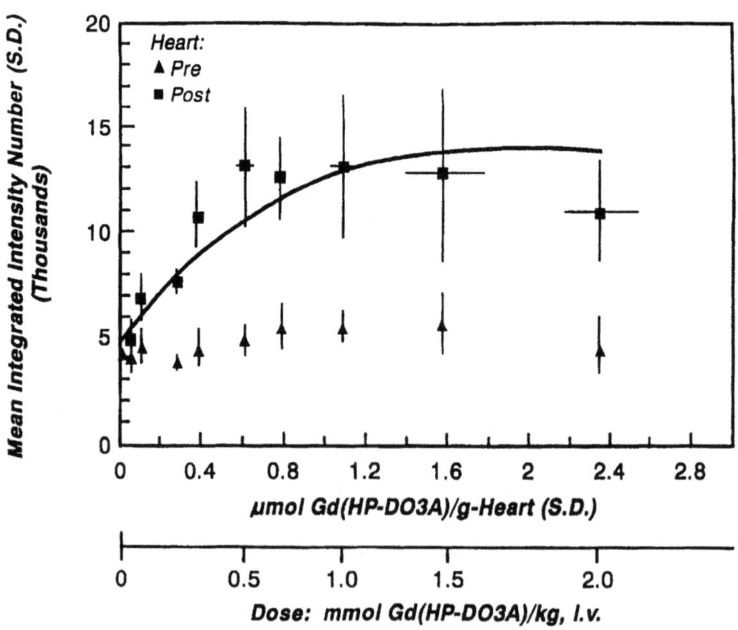

Fig. 7. Effect of concentration on signal intensity measured in vivo in rat heart muscle at 2.0 T [35]. For this combination of spin-echo pulse sequence and cardiac phase, maximum image enhancement was achieved at 0.5 mmol Gd(HP-DO3A)/kg. Rats were nephrectomized to prevent excretion. The solid line is calculated using Eq. 1, with measured $T_{1,2}$ values

effects, the authors observed nonlinear T_1 vs concentration data in the heart and postulated water compartmentation, where water relaxation times are reduced to a level that is competitive with the exchange of water among several regions containing different contrast agent concentrations [36]. The effect tends to place an upper limit on the concentrations of contrast agents that rely on reducing T_1, but it is a minor limitation because the signal enhancement generated by existing T_1 agents (up to 0.3 mmol/kg dose) is no more than half that at 0.5 mM, and still on the relatively linear portion of the signal vs concentration curve.

2.3.2
T_2 Agents

The average T_2 relaxation time of most tissues is only a fraction of their average T_1. For example T_1 of Putamen tissue in brain is 747 ± 33 ms, relative to 71 ± 4 ms for the T_2 relaxation time [26]. Because the concentration and $T_{1,2}$ are inversely related, it is therefore inherently more difficult to reduce T_2 with contrast agents. For agents with $r_1 \sim r_2$, the choice of operator selected imaging parameters is usually to emphasize T_1. On the other hand particles of ferrites (Fe_3O_4) in the 5–50 nm size range have extraordinarily high relaxivity values and large r_2/r_1 ratios [37]. For example the r_2/r_1 ratio for AMI 25 is as high as 4, as compared to 1.06 for $Gd(DTPA)^{2-}$ (vide infra). Other experimental particles with similar Fe cores but different coatings have relaxivities as high as r_2 = 350–450 $mM^{-1}\ s^{-1}$, with $r_2/r_1 > 40$ [137]. Ferrites co-crystallized with dextrans have proven safe enough to undergo clinical trials [38], and a product related to the AMI 25 particles is now marketed as a liver specific contrast agent, Ferridex^R (see http://www.Berlex.com). The primary limitation of the particles may be tolerance. While there are no adverse reactions reported for commercial particulate radiopharmaceuticals, these agents are administered in very small doses (~1 mg) compared to the Fe oxide based MRI agents (~120 mg). Both iron and dextrans are known to elicit hypersensitivity reactions, and large particulates sequestered by macrophages are retained in the body for far longer times than small water-soluble molecules. Acute, severe back pain requiring treatment was reported in some patients during infusion of the larger Fe-dextran particles.

The higher relaxivity parameters, combined with a favorable selection of instrumental parameters can easily outweigh the inherent advantage of T_1 reduction. Unlike the T_1 agents, which increase signal intensity, the T_2 agents tend to reduce (in practice erase) the signal intensity wherever they concentrate. For example AMI 25 (Ferridex) localizes in liver (Fig. 8) where there are abundant (2%) macrophages [39]. The T_2 agents have large magnetic susceptibility values relative to monomeric contrast agents. For example, the reported values of magnetic succeptibility are 25,000 × 10^{-6} cgs/gm for AMI 25 and 163 × 10^{-6} cgs/gm for $Gd(DTPA)^{2-}$ [40]. More recently, "magnetoferritins" have been reported to have a very high r_2 value, 175 $mM^{-1}\ s^{-1}$, with $r_2/r_1 = 22$ at 1.5 T and 37 °C [41].

Unlike particulates, small water soluble agents diffuse rapidly out of the blood stream into the interstitial space of tissues. Despite this, a sharp bolus of a Dy^{3+} complex may also create a situation where T_2 is effectively reduced. As long as the injected bolus (on the order of 0.5–1.0 M) remains concentrated and predominantly

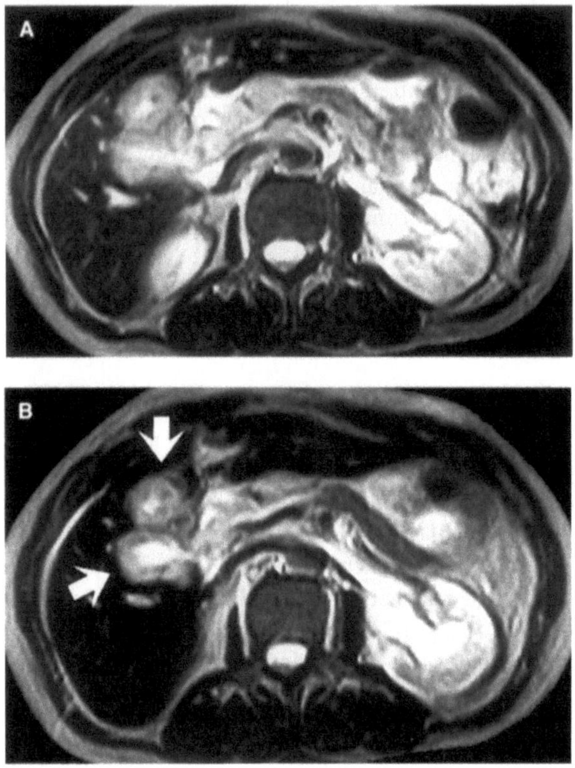

Fig. 8. Liver metastases (*arrows*) from ovarian carcinoma. Pre- (A) and 1 h post-contrast (B) T_2 weighted scans are presented. The contrast agent, AMI 25, Ferridex, was given at a dose of 10 μmol Fe/kg. The signal intensity of normal liver is markedly reduced post-contrast, improving lesion detectability. Taken from: Runge VM, PelsRijeken TH, Davidoff A, Wells JW, Stark DD (1994) J Mag Res Imag 4:28

inside the blood vessel, a field gradient is generated at the vessel wall proportional to the magnetic susceptibility of the contrast agent. Water molecules diffusing randomly through the space near the wall become dephased, causing a drop in T_2 visible as a signal decrease in the imager [42]. Dysprosium has a negligible T_1 effect because of its extremely short electron spin relaxation rate (vide infra). Owing to a large experimental magnetic moment (μ_{eff} = 10.5 BM) with a substantial orbital contribution (calculated spin only magnetic moment due to 5 unpaired electrons is 5.9 BM) it is an effective T_2 susceptibility agent. The only dysprosium complex investigated in humans is Dy(DTPA-BMA) (where DTPA-BMA the bis methyl amide of DTPA). This complex has the same pharmacokinetic behavior as Gd(DTPA-BMA) and $Gd(DTPA)^{2-}$. Also the toxicological profile of the compound appears to be similar to that observed for Gd(DTPA-BMA). The experimental diagnostic use of Dy(DTPA-BMA) was to differentiate qualitatively between ischemic and normally perfused tissues in brain, heart, kidney or liver [43]. The perceived advantage over the Fe particles of such agents are better tolerance and the ability to inject as a rapid bolus. The Fe particles have been restricted to slower injections, so far, which does not allow applications such as tissue perfusion. A potential flaw is that Gd chelates may also be used, as their susceptibility is about half that of the Dy chelates, and the Gd chelates require no further significant development costs.

2.3.3
T_1 Agents

"T_1 agents" are loosely defined as those where r_2 is no more than about twice the value of r_1. These can be any paramagnetic substance, including natural substances such as oxygen [44]. Examples of the most widely studied paramagnetic structure types are shown in Fig. 9. The initial design criteria has included water solubility, for safe excretion, and very strong chelation of metal ions to avoid toxic free metal ions or ligands. For effective catalysis of bulk water relaxation there must also be at least one coordination site for a labile water molecule. The four Gd chelates shown in Fig. 9 are the only commercially available paramagnetic extracellular MRI contrast agents at this writing [45].

2.3.4
Mechanism of Inner Sphere Relaxivity of T_1 Agents

The relaxivity of a paramagnetic complex depends on the magnitude of the dipole-dipole interaction between the electron spin on the metal ion and the proton spin on the water molecule coordinated to the metal. Relaxivity can be divided into inner and outer sphere contributions:

$$r_{1,2} = r_{1,2} \text{ outer} + r_{1,2} \text{ inner} \tag{3}$$

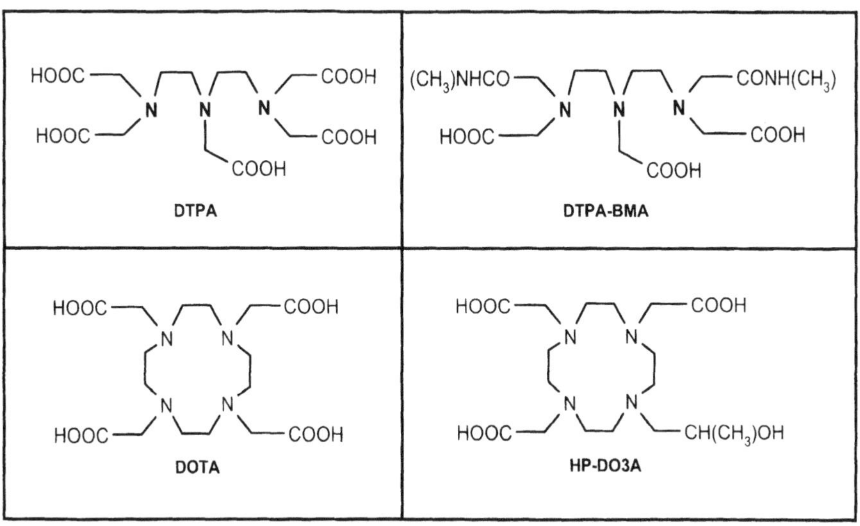

Fig. 9. Chemical structures of the ligands used to complex Gd^{3+} in commercial relaxation agents

Inner sphere contributions are those arising from relaxation catalysis when the water molecule is bound to the metal through its oxygen atom. Outer sphere contributions include those from hydrogen bonded water protons and contributions from translational diffusion. A further distinction between these two outer sphere contributions can be made based on the length of time the water protons reside close to the metal ion relative to the translational diffusion times of water and the complex. In principle, hydrogen-bonded water would be treated as "inner sphere" when its life time did not limit the observed water relaxation time (vide infra). The inner sphere equations in the next section apply to the long lived H-bonded case as well as to labile coordinated water.

2.3.5
Inner Sphere Relaxation: The SBM Equation

Efficient catalysis by a dilute metal complex in 55.5 M water requires that a water molecule reside at the metal no longer than necessary to be relaxed. Relaxivity may be defined in terms of exchange theory (dilute solution) $r_1 = \frac{q}{55.5}\left(T_{1p} + \tau_M\right)^{-1}$ where q is the number of water molecules bound to the metal ion; T_{1p} is the T_1 relaxation time of the protons on the water that is coordinated to the metal; and τ_M is the lifetime of the association. T_{1p} for protons on water coordinated to, for example, Gd^{3+} are on the order of microseconds. To the extent τ_M is long relative to T_{1p}, "slow exchange" conditions exist, and poor communication between the metal and the bulk water occurs. "Fast exchange" conditions exist when $\tau_M \ll T_{1p}$ and it is the nearly universal presence of labile water ($k_{ex} > 10^6\ s^{-1}$) for Gd^{3+} and Mn^{2+} that make these ions more desirable than the kinetically inert metal ions such as Cr^{3+} ($k_{ex} \sim 10^{-6}\ s^{-1}$).

A quantitative theoretical description of T_{1p} for Gd^{3+} and Mn^{2+} complexes would facilitate design of more effective contrast agents. Although this is not yet available, it is useful to be aware of the theory which has been developed for paramagnetic aqua ions, which should eventually be a special case of a broader theory. The dipole-dipole (through space) and scalar (through bonds) interaction between the protons on coordinated water and metal complexes is described by the Solomon-Bloembergen (SB) equations (Eq. 4–6) [46]. An extensive literature and a review of the theory exist [47]. We provide here a summary as it relates to NMR contrast agents.

$$\frac{1}{T_{1p}} = \underbrace{\frac{B}{r^6}\left[\frac{7\tau_c}{(1+\omega_S^2\tau_c^2)} - \frac{3\tau_c}{(1+\omega_I^2\tau_c^2)}\right]}_{\text{Dipolar Term}} - \underbrace{\frac{2S(S+1)A}{3h^2}\left[\frac{\tau_s}{(1+\omega_S^2\tau_s^2)}\right]}_{\text{Scalar Term}} \tag{4}$$

$$\frac{1}{\tau_c} = \frac{1}{\tau_r} + \frac{1}{\tau_M} + \frac{1}{T_{1s}} \tag{5}$$

$$B = (2/15)\gamma_I^2 g^2 S(S+1)\mu^2 \tag{6}$$

where γ_I is the proton gyromagnetic ratio, g is the electronic g factor, S is the total electron spin of the metal ion, μ is the magnetic moment in Bohr magneton, r is the distance from the center of the Gd ion to each of the protons, ω_s and ω_I are the electron and proton Larmar precession frequencies, respectively. A/h is the electron nuclear hyperfine coupling constant. The dipolar and scalar relaxation mechanisms are modulated by correlation times τ_c and T_{1s} as given by Eq. 5 and 6, where τ_c is the overall correlation time, τ_r is the rotational tumbling time of the complex, τ_m is the lifetime of coordinated water on the metal, and T_{1s} is the longitudinal electron spin relaxation time. For $Mn(H_2O)_6^{3+}$ ion, the SB equations are not adequate as these cannot describe magnetic field dependence of r_1 and r_2 relaxivities. Bloembergen and Morgan [48] developed a theory to express field dependence of T_{1s} (Eq. 7), in which T_{1s} has a complex functional form weighted by a parameter, B′, dependent on the magnitude of transient zero field splitting of electronic spin levels caused by solvent-solute collisions or other molecular motions. These molecular motions or collisions induce distortions of the symmetry of the metal ion water complex, which leads to zero field splitting of electronic spin levels.

$$\frac{1}{T_{1s}} = B\left[\frac{\tau_V}{(1+\omega_s^2\tau_V^2)} + \frac{4\tau_V}{(1+4\omega_s^2\tau_V^2)}\right] \tag{7}$$

In Eq. 7, τ_v is another correlation time characterizing these motions. The relationships are used to describe relaxation of protons on coordinated water in the inner sphere, and in the outer sphere when protons are long lived relative to diffusion. Addition of Eq. 7 to Eqs. 4–6 constitutes the well known Solomon-Bloembergen-Morgan (SBM) theory.

A caveat applies to the arguments in the next three sections. The SBM theory is useful in simulating the interplay of the multiple parameters involved in relaxation enhancement. But the theory, as it applies to complexes of $S > 1/2$ and symmetry lower than that of the hydrated ions, particularly with regard to the influence of the correlation times, is still evolving from the SBM theory as it applies to hydrated ions.

In Table 3 are collected data that bear on the relaxivity of some strong complexes of Gd^{3+}, Mn^{2+}, and Cr^{3+}. The metal complexes are chosen with $q = 1$. τ_c is dominated by τ_r which is usually about 10^{-10} s for small (500 amu) molecules [49]. Gd^{3+} has the largest magnetic moment, $\mu_{eff}^2 = g^2\,S(S+1) = 63$, and so with other variables approximately equal, we would expect the Gd^{3+} complexes to have the largest ${}^{20}r_1$ values.

Three features are essential to make highly effective contrast agents, as shown in Table 4. The first criterion for high relaxivity is that T_{1s} be long relative to molecular rotation (about 100 ps [49]). Otherwise the fluctuations of the electrons dominate the overall correlation time, τ_c, and nuclear dipole–dipole interaction is inherently suboptimal. Most metal ions have unsymmetrical electronic ground states and are thus excluded. The high electronic symmetry and thus high spin only magnetic moment of Gd^{3+} and Mn^{2+} ions and complexes make them highly preferred species. Fe^{3+} tends to have relatively slower water exchange rate, k_{ex}, than optimal, but exceptions probably exist. Cr^{3+} may be (usually) excluded based upon the non labile coordinated water, giving rise to low exchange rate constants, k_{ex}, although it is not actually necessary for the water to exchange, only the protons. Hence, a very rapid, probably catalyzed proton exchange could conceivably carry the bulk relaxation.

Table 3. Physical properties of $Gd(DTPA)^{2-}$, $Mn(EDTA)^{2-}$, and $Cr(EDTA)^{-}$

Parameter	Gd^{3+}	Mn^{2+}	Cr^{3+}
μ^2_{eff}	63	35	15
q	1[a]	1[d]	1[g]
r_{M-H_2O}	2.490[a]	2.155[d]	2.002[g]
k_{ex}, s^{-1}	4.1×10^6[b]	4.4×10^8[e]	1[h]
$^{20}r_1$, mM^{-1} s^{-1}	3.8[c]	2.9[f]	0.2[i]

[a] Gries H, Miklantz H (1984) Physiol Chem Phys Med NMR. 16:105
[b] Ref. 50
[c] Ref. 45
[d] Richards S, Berit P, Silberton J, Hoard JL (1964) Inorg Chem 3:27
[e] Zetler MS, Grant MW, Wood EJ, Dodgen HW, Hunt JP (1972) Inorg Chem 11:2701, and Margerum DW, Cayley GR, Weatherburn DC, Pagenkopf GK (1978) In: Martell AE (ed) Coordination Chemistry, vol 2. American Chemical Society, Washington DC, p 1
[f] Koenig SH, Baslin C, Brown RD, Brewer CF (1984) Mag Res Med 1:496
[g] Hoard JL, Kennard CHL, Smith GS (1963) Inorg Chem *2*:1316
[h] Ogino H, Watanabe T, Tanaka N (1975) Inorg Chem 15:31
[i] Tweedle MF, Gaughan GT, Hagan J, Wedeking PW, Sibley P, Wilson LJ, Lee DW (1988) Nucl Med Biol 15:31

Table 4. Rate of water exchange, k_{ex} at 25 °C, activation volume, enthalpy and entropy of activation for $Gd(H_2O)_8^{3+}$, and some Gd^{3+} complexes[a]

Complex	k_{ex}, s^{-1}	ΔV, cm^3 mol^{-1}	ΔH, kJ mol^{-1}	ΔS, JK^{-1} mol^{-1}
$[Gd(H_2O)_8]^{3+}$	$(8.3 \pm 0.95) \times 10^8$	-3.3 ± 0.2	14.9 ± 1.3	-24.1 ± 4.1
$[Gd(PDTA)(H_2O)]^-$	$(1.02 \pm 0.1) \times 10^8$	-1.5 ± 0.5	11.0 ± 1.4	-54.6 ± 4.6
$[Gd(DTPA)(H_2O)]^{2-}$	$(4.1 \pm 0.3) \times 10^6$	12.5 ± 0.2	52.0 ± 1.4	56.2 ± 5.0
$[Gd(DOTA)(H_2O)]^-$	$(4.8 \pm 0.4) \times 10^6$	10.5 ± 0.2	48.8 ± 1.6	46.6 ± 6.0

[a] Ref. 50

Gd^{3+} chelates with a *single* water coordinated actually seem to be just labile enough, the life time of coordinated water at the Gd^{3+} $\tau_m \sim 10^{-6}$–10^{-7} s, compared with the lability of $Gd(H_2O)_9^{3+}$, $\tau_m < 10^{-9}$ s. This surprising result stems from a mechanism change on going from multiple coordinated water systems to monocoordinated water systems. Merbach [50] found, based upon comparison of activation volumes, that the exchange mechanisms for $Gd(H_2O)_9^{3+}$ and $Gd(PDTA)(H_2O)_2^-$ (where PDTA = 1,2 diaminopropanetetraacetic acid) were associative while those of $Gd(DTPA)(H_2O)^{2-}$ (where DTPA = diethylenetriaminepentaacetic acid) and $Gd(DOTA)(H_2O)^-$ (where DOTA = 1,4,7,10-tetraazacyclododecane-N,N′,N″,N‴-tetraacetic acid) were dissociative (see Table 4). Similar results were obtained for another monocoordinated water complex, $Gd(DTPA\text{-}BMA)(H_2O)$ by Aime and coworkers [51]. Their k_{ex} value was 1.23×10^6 s^{-1}. Apparently, the donor atom type is important (vide infra).

For Mn^{2+} and Gd^{3+} the difference in magnetic moment is partially offset by a counterbalancing difference in the distance between the proton and the metal (Table 3), the metal-oxygen (water) distance. A fine point here involves consideration of the means by which water coordinates to the metal. Thinking of coordinated water as a tetrahedral structure-central oxygen with two protons and two lone pairs-an inert metal-oxygen bond through one of the two lone pairs would yield a shorter metal-proton distance than a labile structure in which the two oxygen lone pairs were exchanging, creating an averaged structure (Fig. 10). Some circumstantial evidence supports the former situation for lanthanides [52], but there seems to be no prior reason why the range of possibilities could not be expressed by the lanthanides if interactions of the coordinated water with the ligand played a role.

2.3.6
Correlation Times

To stimulate proton relaxation the unpaired electron spin must fluctuate at a frequency matched to that of the proton and electron Larmor frequencies. This is represented mathematically by the "7" and "3" terms of the dipolar part of Eq. 4. τ_c^{-1} is the overall rate of the fluctuations; τ_c is called the overall correlation time. It is easier to think of the correlation times in terms of their reciprocals, first order correlation rate constants characterizing correlation rates, because these rate constants are additive. The overall correlation rate is generally dominated by one of the three processes, rotational (τ_r), electron relaxation (τ_s or T_{1s}), and water or proton exchange (τ_m) represented by the terms in Eq. 5. Further processes undoubtly exist.

Figures 11 and 12 show simulated and experimental NMRD curves (Nuclear Magnetic Resonance Dispersion), pioneered by Koenig [53], which depict the effects of changing the dominant correlation time, τ_r. Relaxivity versus Larmor frequency values were calculated using Eqs. 2 and 4–6 over a range of frequency. (For most clinical proton imaging the range is actually only from about 1 to 60 MHz.) The parameters in Fig. 11 are assigned based on the best current knowledge for a Gd^{3+} chelate with q = 1 and sufficiently rapid exchange kinetics. At the τ_r values measured for monomeric Gd chelates in water (~0.1 ns) [49], the curve "disperses" at about 5 MHz due to the term containing $7\tau_c$. The $3\tau_c$ term disperses out of range at >1000 MHz. As τ_c is increased to 1 MHz, the T_{1e} becomes competitive with τ_r and a characteristic minimax behavior is generated by the lowering of correlation terms. The $7\tau_c$ dispersion drops to about 2–3 MHz and the $3\tau_c$ dispersion drops to about 100 MHz. The peak at around 20 MHz is due to increasing T_{1e}. After 10 MHz the T_{1e}

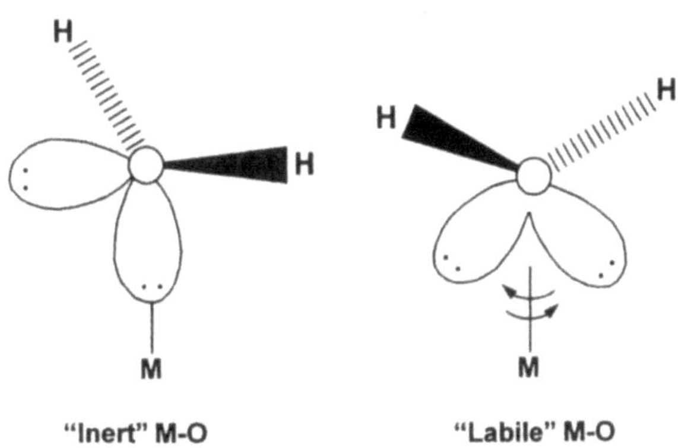

Fig. 10. Different coordination modes of water coordination to gadolinium

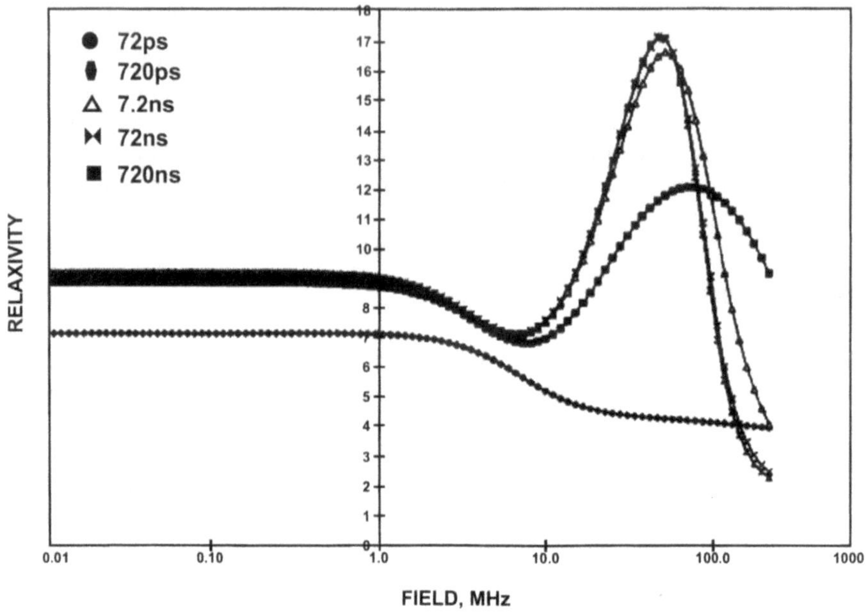

Fig. 11. The simulated NMRD curves for a Gd chelate

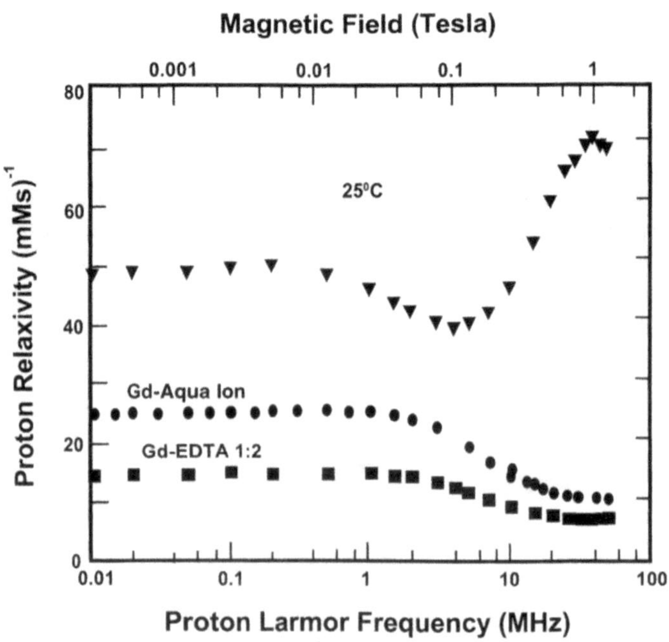

Fig. 12. Effect of τ_r variation on the actual NMRD curve for protein-bound Gd^{3+}. Data taken from [130]. $Gd(EDTA)^-$ and Gd^{3+} (aq) have ~2.5 and 8–9 coordinated water molecules

of the Gd ion dominates τ_c. Figure 12 shows actual data (collected on free and protein-bound Gd chelates) whose shape parallels those in Fig. 11.

The SBM theory does have inherent limitations. Most important are the use of the Redfield assumption and second order perturbation theory in its derivations (essentially, these assume that τ_r dominates and that correlation terms are not coupled). Also, the necessary assumption of a S = 1/2 ion has been made. Neither the S = 1/2 nor τ_r dominance are the case for Gd^{3+} bound to a protein. The data in Fig. 12 nevertheless strongly resemble those calculated in Fig. 11, both qualitatively and quantitatively. The SBM theory therefore provides us with a reasonable theoretical framework for understanding existing contrast agents and a point of departure for research aimed at increasing the relaxivity and refining our theoretical understanding of the inner sphere mechanism.

2.3.7
Outer Sphere Relaxivity

In the absence of inner sphere water, outer sphere relaxation can dominate the relaxivity, despite the larger inherent distance between the water and the electron spins. Outer sphere relaxation refers to the relaxation of noncoordinated water protons by paramagnetic nuclei. Mathematical descriptions for T_1 relaxation exist (Eq. 8) [54] which emphasize increasing relaxivity with increasing τ_d, the relative diffusional correlation time (Eq. 9).

$$\frac{1}{T_1} = \frac{C\Pi N_s\gamma_I^2\gamma_S^2 h^2 S(S+1)}{r^3\tau_D^2)}[7I(\omega_S\tau_D T_{1s}) + 3I(\omega_I\tau_D T_{1s}] \qquad (8)$$

$$\tau_d = r^2/3\,(D_H + D_P) \qquad (9)$$

where C is a constant, N_s is the number of metal ions per cc, r is a distance of closest approach between the electron spin on the metal and the proton spin on the coordinated water and D_H and D_P are the diffusion coefficients of water and the paramagnetic species, P, respectively. Diffusion coefficients can be estimated by describing the motion as diffusion of a rigid sphere in a medium of viscosity η as expressed by Eq. 10, where a is the molecular radius.

$$D = kT/6\Pi a\eta \qquad (10)$$

For free Gd aqua ion, the outer sphere portion of the relaxivity is calculated to be about 10% of the total relaxivity or ~1 mM^{-1} s^{-1} (20 MHz and 40 °C) [55]. However, for anionic and neutral Gd chelates with q = 1, the inner and outer sphere contributions are about 2 mM^{-1} s^{-1} each [56]. The relaxivity, at the low field limit, of a coordinately saturated cationic Mn^{2+} complex was approximately half of that observed for coordinately saturated anionic Mn^{2+} chelate such as $Mn(DTPA)^{3-}$ [57]. This was attributed to the opposite orientation of water protons toward the metal. Thus the distance of closest approach of water protons is about 1 Å longer for cations than for anions. The lower outer sphere contribution for the cationic Gd^{3+}_{aq}

ion relative to the chelates is therefore rational. The magnitude of the relaxivity for Gd^{3+} ion in slightly acidic conditions is about ~10 mM^{-1} s^{-1} compared to 4 mM^{-1} s^{-1} for a q = 1 Gd^{3+} chelate of about 600 MW [56]. Subtracting 10% and 50% for outer sphere contributions, respectively, leaves 1 mM^{-1} s^{-1} and 2 mM^{-1} s^{-1} (20 MHz and 40 °C) for inner sphere relaxivity per coordinated water for the aqua ion and the complex, respectively. The difference is probably due to the smaller size, and hence greater rotational correlation rate for the Gd^{3+} aqua ion.

3 Extracellular Agents with Renal Elimination for Imaging CNS Pathology

The first generation of MRI contrast agents are water soluble Gd chelates weighing about ~500–600 g mol^{-1} (Fig. 9). They are administered intravenously as 0.5 M aqueous solutions and have the same biological tissue distribution pattern as the currently used water soluble X-ray contrast agents (vide infra). The primary clinical indications are in the central nervous system, mainly in the brain (about 80%) [58,59], where they are frequently used to diagnose brain tumors and other pathologies, as shown in the example of use in a typical human brain tumor patient in Fig. 2.

3.1 Blood Brain Barrier

The neuronal activity of the brain places much stricter limits on the movement of the compound within and between its cells. Extracellular contrast agents move throughout the vascular space and leak out from here to the extracellular space. To protect the highly regulated tissue of the brain from blood born entities the capillaries feeding neural tissue are different from most other tissues. They protect the brain with a series of physiologic barriers to the entry of freely diffusing natural chemicals (other than very small molecules such as water and ethanol). The various barriers are collected in the schematic representation in Fig. 13, and are, collectively, referred to as the "blood brain barrier" (BBB) [60]. The spacing and composition (with fibrils) of the cell-cell junctions, and secondary barriers, including a continuous basal membrane, block passive diffusion. Pinocytosis is the tendency of the cell membrane to engulf contents of the blood plasma into a small subcellular package and allow it to release the contents by a reverse process at the other side of the cell. In neural tissue a higher mitochondria mass is necessary because active transport of nutrients replaces the lost diffusion, requiring energy. Also, the higher mitochondria mass supports additional enzyme which constitutes a biochemical BBB. The lack of these structures in the normal tissue allows much more facile access of diffusable chemicals, including small water soluble contrast agents. Brain tumors grow their blood supply by secreting angiogenesis factors which stimulate the growth of new capillaries from existing ones. However the new capillaries tend to lack the differentiating features of the normal neural capillaries (the BBB) despite the fact that they are clones (probably due to matrix effects). Water soluble contrast agents pass through the new capillaries and catalyze relaxation of the water in the brain tumors.

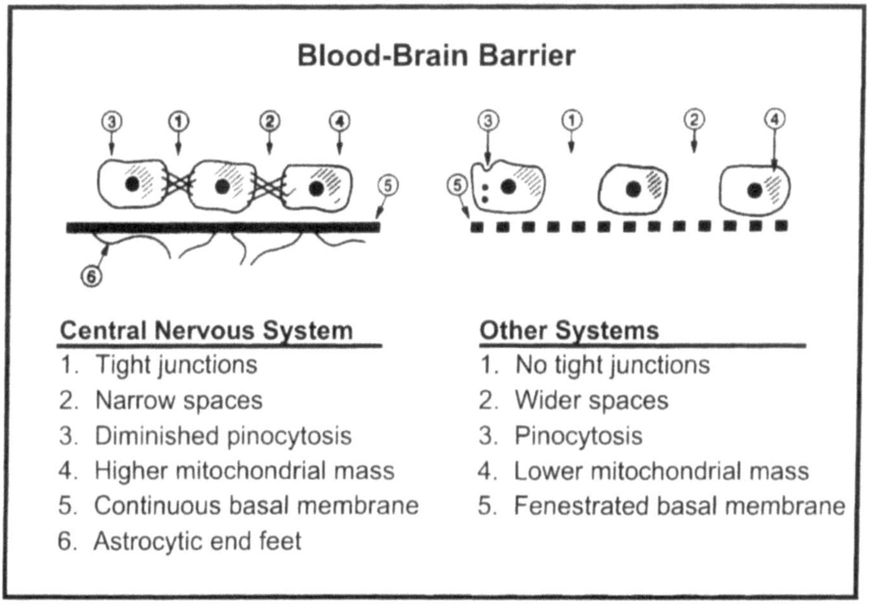

Fig. 13. Schemetic of blood brain barrier (*left*). The capillary endothelial cells and related structures represent a portion of the membrane in capillaries feeding blood to the CNS. Diffusion of water soluble contrast agents is stopped. Systems lacking the blood-brain barrier (*right*) allow diffusion of water soluble contrast agents

Other space filling diseases, such as infections, are less subtle than the cancers, but luckily tend to disrupt the BBB and so allow some contrast agent to enter.

Contrast agents of this type are less successful outside the brain because the normal vascular/tissue interface is already highly permeable to them. Indications in the abdomen exist, but the contrast agents are far less frequently used there (15%). Future indications that appear promising include the breast [61] where the absence of ionizing radiation offered by MRI is attractive, particularly in view of recent molecular genetics findings that appear to allow screening of patients for a gene that is associated with breast cancer [62]. In these patients, breast scanning may be indicated every six months, and MRI scanning would be expected to offer a significant radiation safety advantage over X-ray mammography. MRA (magnetic resonance angiography) is another new area of application which is beginning to compete with X-ray imaging, and which uses Gd chelates in a high percentage of procedures (vide infra).

3.2 Chemistry and Biology of Gadolinium Chelates

The four extracellular MRI contrast agents currently in use in humans, are shown in Fig. 10 [45]. The counter ion in $Gd(DTPA)^{2-}$ and $Gd(DOTA)^{-1}$ formulations is NMG^{+} (NMG = *N*-methylglucamine ion). These gadolinium chelates are delivered in

0.5 M aqueous solutions as a matter of convenience and to minimize the administration volume. A 70 kg person would receive 0.1–0.3 mmol/kg or about 7–21 mmol. This is 14–42 ml of 0.5 M solution or ~4–12 g of chelate without counterions, about 1/3 of which is Gd.

The four chelates have very similar relaxivity in water (Table 5) at 20 MHz and at 40 °C. This is due to similar τ_r (the rate controlling rotational correlation time) [49], q (the hydration number) [56], and Gd-H_2O bond distances [63–67]. The slight increase in q for Gd(HP-DO3A) is probably due to the presence of the hydroxypropyl group, which is expected to contribute somewhat to the Tb(OH) luminescence quenching from which q is determined [67], but not to the relaxivity of Gd(HP-DO3A) which requires a proton exchange rate constant ($k_{ex} > 10^6\ s^{-1}$) faster than that of the hydroxyl protons. $^{20}r_2$ values (not shown) are also very similar, and are unlikely to significantly effect MRI using T_1-weighted pulse sequences at concentrations envisioned for most clinical applications.

The charge status of the chelates is readily verified by measuring the molar conductivity in aqueous solutions which is shown in Table 6 [45]. When NMG_2[Gd(DTPA)] and NMG[Gd(DOTA)] are dissolved in distilled water, they ionize fully liberating NMG^+ and anionic $Gd(DTPA)^{2-}$ and $Gd(DOTA)^-$ to conduct electric charge through the solution. A near zero molar conductivity is diagnostic for nonionic substances such as Gd(DTPA-BMA) and Gd(HP-DO3A).

The relatively high osmolality and viscosity values for the ionic chelates (Table 6) are consequences of the magnetically inactive NMG^+ cations. Osmolality is a colligative property, and therefore depends on the total number of particles in solution, regardless of charge. For example $(NMG)_2$[Gd(DTPA)], contributes three particles

Table 5. Important physical parameters controlling relaxivity of Gd^{3+} complexes

Complex	q[a]	τ_r, ps[b]	Gd-OH_2 (Å)	$^{20}r_1$, $mM^{-1}\ s^{-1}$[g]
Gd(HP-DO3A)	1.3 ± 0.1	57	2.50[c]	3.7 ± 0.1
Gd(DTPA-BMA)	1.1 ± 0.1	53	2.42[d]	3.8 ± 0.1
$Gd(DTPA)^{2-}$	1.1 ± 0.1	55	2.49[e]	3.8 ± 0.1
$Gd(DOTA)^-$	1.1 ± 0.1	63	2.46[f]	3.5 ± 0.1

[a] Ref. 56
[b] >Ref. 49, determined from ^{13}C-NMR studies of corresponding diamagnetic Y complexes assuming C-H bond distance as 1.00 Å
[c] Ref. 63
[d] Ref. 64
[e] Ref. 65
[f] Ref. 66
[g] Ref. 45

Table 6. Molar conductivity, osmolality, and viscosity at 37 °C

Complex	Conductivity µS	Osmolality of 0.5 M, Osmol/kg	Osmolality of 1.0 M, Osmol/kg	Viscosity of 0.5 M, cP	Viscosity of 1.0 M, Cp
Gd(HP-DO3A)	1	0.63	1.91	1.3	3.9
Gd(DTPA-BMA)	5.5	0.65	1.90	1.4	3.9
$Gd(DTPA)^{2-}$	117	1.96	5.85	2.9	>30
$Gd(DOTA)^-$	54	1.35	4.02	2.0	11.3

per Gd, whereas the non ionic Gd(HP-DO3A) contributes only one particle per Gd. The higher viscosity for the ionic chelates is also due to the NMG^+ cations (*N*-methylglucamine), which contain multiple hydroxyl groups. The osmolality of blood and body fluids is about 0.3 mmol/kg. The global increase in osmolality after injection of 0.1 mmol/kg of an ionic compound in humans is probably insignificant (~2% increase). However, transient local effects, such as crenation (deformation) of erythrocytes and other cells, may be significant, especially if a bolus injection is administered. Pain and warmth at the site of injection, damage to the endothelial surfaces, and other adverse reactions are undesirable consequences of hyperosmolality. As doses increase, adverse reactions due to hyperosmolality are expected to increase proportional to the osmotic load, thus non ionic contrast agents are expected to be better tolerated at higher dose. In addition to the better tolerance of the non ionic chelates they appear to offer greater flexibility of formulation at higher concentration as evidenced by the data given in Table 6. Higher concentration formulations will result in smaller injection volumes when larger doses are administered. 1.0 M formulations also provide for sharper bolus injections, and greater Gd concentration at the temporal bolus peak. Elevating the height of the bolus would be an advantage for dynamic first-pass MRI studies [68].

In vitro equilibrium stability for Gd chelate-based MRI agents is usefully defined in terms of three constants. The thermodynamic stability constant, K_{eq}, is useful in determining the presence of free Gd^{3+} or the free ligand at equilibrium, both of which are toxic (Table 8). In addition to acute toxicity, Gd^{3+} can also have high affinity (>10 [10]) for calcium binding proteins [69]. A more physiologically appropriate method of comparison is through the conditional stability constant at pH 7.4, K′, which considers the protonation constants of the ligand, and describes the position of binding at pH 7.4. A third in vitro equilibrium test is the calculation of stability constant in the presence of endogenously available ions, e.g. Ca^{2+}, Cu^{2+}, Zn^{2+}, and Fe^{3+} and PO_4^{3-} etc. The latter can be demonstrated by carrying out experiments in the presence of these ions.

Table 7 contains binding constants and results of dissociation rate studies [45, 63, 70–72]. A comparison of the data suggests that negatively charged carboxylate donor atoms are more powerful donors for Gd^{3+} than the uncharged hydroxyl oxygen atom in HP-DO3A or the amide oxygen atoms in DTPA-BMA. This appears to be corre-

Table 7. Equilibrium constants and kinetic data for some Gd^{3+} complexes

Chelate	log K	log K′	$t_{1/2}$ in 0.1 M HCl[c]	% Reaction with Cu^{2+}[d]	% Reaction with Zn^{2+}[d]
Gd(HP-DO3A)	23.8[a]	17.1[a]	3 h	<1	<1
$Gd(DOTA)^-$	25.3[a]	18.6[a]	>1 mo	<1	<1
Gd(DTPA-BMA)	16.9[b]	14.9[b]	~30 s[e]	35	25
$Gd(DTPA)^{2-}$	22.2[a]	17.8[a]	10 min[e]	25	21

[a] Ref. 63
[b] Ref. 70
[c] Refs. 45,71
[d] Ref. 72: $[Cu^{2+}] = [Zn^{2+}] = 25.0$ mM, $[PO_4^{3-}] = 66$ mM, room temperature, pH 7.4, [Trizma] = 50 mM, and [sodium acetate] = 125 mM
[e] Ref. 152: Dependent on conditions, $t_{1/2}$ may be <2 s at pH 1

Table 8. LD_{50} values of active ingredients of commercial MRI contrast agents (rodents)

Chelate	LD_{50}, mmol/kg
Gd(HP-DO3A)	12^a
$Gd(DTPA)^{2-}$	6^b, 10^e
$Gd(DOTA)^-$	11^c
Gd(DTPA-BMA)	15^d
$Gd(EDTA)^-$	0.3^e
$GdCl_3$	0.5^e
Na_2H_3DTPA	0.1

[a] Runge VM, Gelblum DY, Jacobson S (1991) Mag Res Imag, 9:79
[b] Ref. 70
[c] Meyer D, Schaefer M, Doucet D (1990) Invest Radiol, 25:S53
[d] Ref. 70
[e] Ref. 19

lated with the sum of the basicity of the ligands. A good linear correlation between log K_{GdL} vs ΣpKa has been observed for a variety of ligands with a variable number of donor atoms (Fig. 14). The equilibrium K_{eq} and K′ values are each fairly high, however, even for the lowest binding constant.

In vivo proteins, surfaces, and a host of other metal cations (e.g. Ca^{2+}, Cu^{2+}, Zn^{2+}, and Fe^{3+}) compete with the Gd^{3+} ion for the ligand and a number of other anions (e.g. PO_4^{3-}, CO_3^{2-}, and OH^-) compete with the ligand for the Gd^{3+}. The experimental stress test data in the last two columns of Table 7 measure the free Gd released in 10 min at 22 °C (room temperature) after mixing Gd chelate (25 mM) with copper or zinc cations (25 mM) in the presence of phosphate (66 mM) at pH 7 [72]. Both thermodynamics and kinetics may play roles in this experiment, but the results are obvious. Overall, the most important chemical feature controlling relative stability and dissociation inertia is the presence of the macrocyclic framework. Mechanisms proposed from the above kinetics studies are consistent with rate limiting steps involving multiple bond breaks.

Equilibrium stability data are especially useful in formulation work, where aqueous solutions have years of shelf life to equilibrate. In vivo, relatively rapid excretion prevents attainment of equilibrium. Hence, the rate of excretion relative to the rate of chemical/biochemical dissociation of the Gd^{3+} ion will determine the degree to which Gd chelates release Gd^{3+} in vivo. A more important parameter for understanding relative in vivo dissociation is the rate of dissociation, reported as the $t_{1/2}$ at pH 1 in Table 7 [45,70,71].

A more biologically relevant stress test is the residual free Gd remaining in mice several weeks after intravenous administration of radiolabeled ^{153}Gd chelates [73]. Over this length of time the effect of in vivo competition may be appreciated because the excretion rate of free Gd is on the order of 1% per day in mice, compared to >95% per day for the Gd chelates. Any free Gd formed from Gd-chelate dissociation is detectable after 14 days. Figure 15 shows results of such an experiments plotted as log percentage injected dose vs time. $Gd(EDTA)^-$, used as a positive control, is as poorly tolerated as Gd^{3+} (Table 8), seemingly due to dissociation. The other chelates are excreted (into urine), with >95% excreted by 1 day. At 7 and 14 days, each chelate shows <1% residual free Gd^{3+}. However, it is obvious that the macrocyclic chelates have lowest residual free Gd^{3+}. Despite the faster acid-assisted dissociation

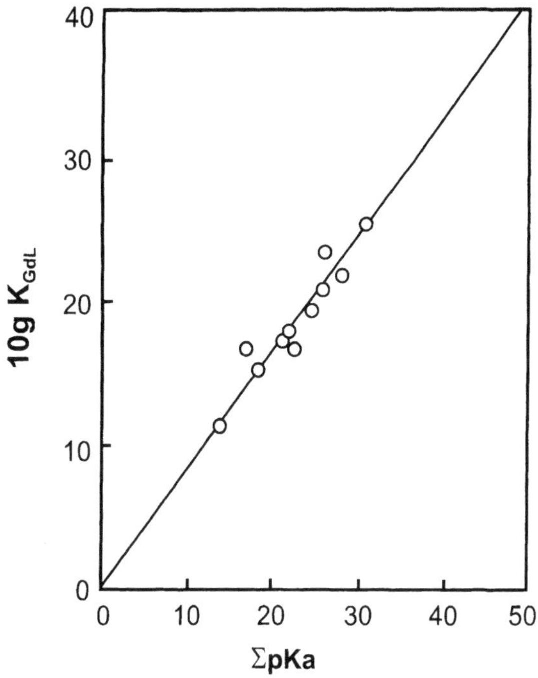

Fig. 14. Plot of log K_{GdL} vs ΣpKa for some Gd^{3+} polyamino-carboxylate chelates which form 5-membered chelate rings

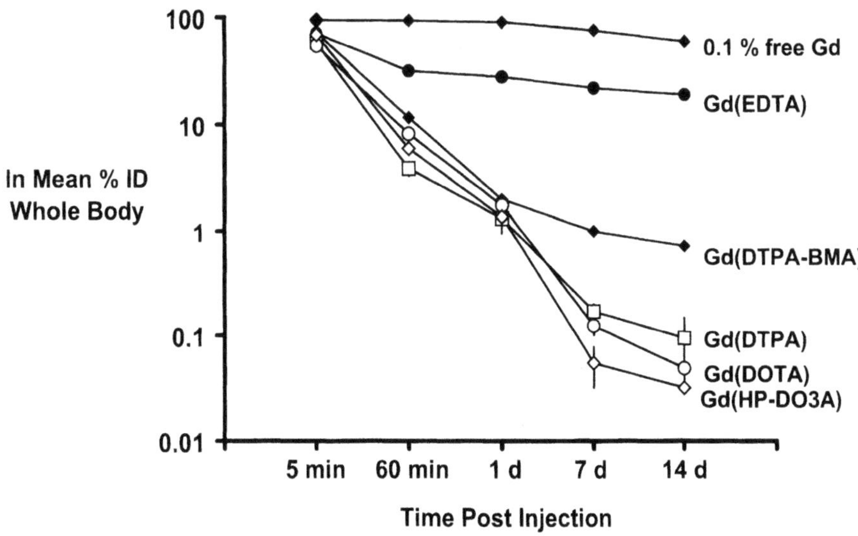

Fig. 15. Total whole body residual gadolinium in mice after intravenous administration of 0.4 mmol/kg of radiolabeled ^{153}Gd chelates

rate, $Gd(DTPA)^{2-}$ also appears to have relatively low residual Gd, although it is three times greater than Gd(HP-DO3A). Gd(DTPA-BMA), on the other hand, shows considerably greater residual Gd at 7 and 14 days.

In humans and animals, the pharmacokinetics of the four chelates (Fig. 9) are the same. Their apparent volume of distribution is in the range of 0.2 to 0.3 L/kg (20%–30% of body weight). The elimination half-life of all commercial MRI Gd chelates in normal healthy adults is approximately 1.5 hour, which is expected for a compound which is distributed in extracellular space and is excreted by glomerular filtration. The toxicity of these chelates is remarkably low as shown by the LD_{50} values in mice (Table 8). Adverse events in humans are very rare.

4
Hepatobiliary Agents for Liver Pathology Imaging

The natural course for contrast agent development will run from extracellular agents to those for imaging the liver (hepatocytes and reticuloendothelial cells), and then to agents for other reticuloendothelial cells. These are likely to include: lymph nodes, bone marrow, spleen and lung, and perhaps some body abnormalities such as infections, which encourage macrophage accumulation. The extracellular agents lend themselves to renal excretion, which generally is the safest excretion pathway. The next safest pathway is through the hepatobiliary system, followed by the macrophages, which generally decompose the agents, rather than excrete them. Here again the X-ray agents [74] and radiopharmaceuticals [75] offer some guiding precedents. The principles of distribution governing these agents are directly applicable to MRI contrast agents. The potential diagnostic utility of the hepatobiliary class of MRI contrast agents are: a) selective enhancement of normal functioning tissue to aid in detection of small lesions, such as metastatic tumors (focal liver disease), b) indication of liver function to detect diffuse liver disease such as cirrhosis, and c) high resolution visualization of bile ducts and the gallbladder.

4.1
Metal Chelates

Figure 16 shows structures of some clinical [20–22] and preclinical [76–80] hepatobiliary agents. Each agent has in common several characteristics. They are anionic and contain aromatic rings making them more lipophilic in some region of the molecule, relative to the extracellular agents. These features encourage plasma protein binding and hepatocyte uptake by the organic anion transport mechanism [80]. Molecules with greater lipophilicity tend to have greater hepatobiliary excretion, though this is far from a reliable rule. A balance between charge and lipophilicity is necessary for designing hepatobiliary MRI contrast agents. A very lipophilic agent may have longer retention time in reticuloendothelial cells in the liver and spleen, which will possibly lead to chronic toxicity [81].

The first ligand in this series, BOPTA (benzyloxypropionictetraacetic acid) [21], forms a nine coordinated Gd^{3+} chelate with eight sites occupied by the ligand and the ninth coordination position occupied by water [82]. The stability constant of the chelate was reported as 22.59 [82]. The T_1 and T_2 relaxivity values of the chelate at 20 MHz and 40 °C are measured as 4.4 and 5.6 $mmol^{-1}$ s^{-1}, respectively. The r_1 value

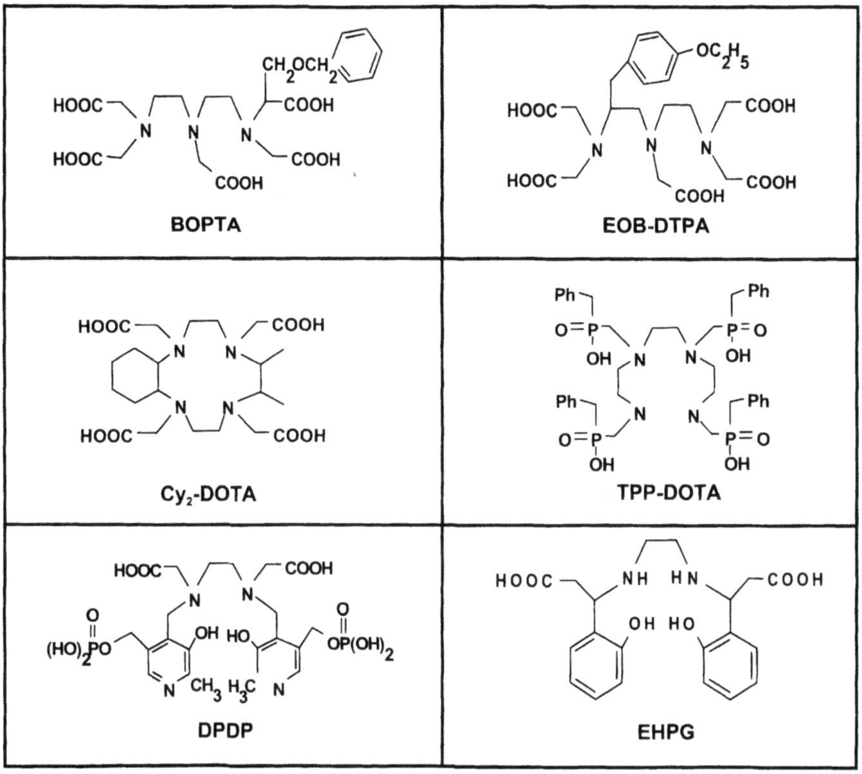

Fig. 16. Structures of some clinical and preclinical hepatobiliary MRI contrast agents

is slightly higher than $Gd(DTPA)^{2-}$, probably due to the larger size of the $Gd(BOPTA)^{2-}$ chelate. Markedly elevated relaxivity of the chelate was observed in liver homogenate and in liver tissues probably due to increases in volume of distribution, intracellular viscosity and slight protein binding. Pharmacological and imaging studies indicated that the chelate is useful for liver imaging [83]. The LD_{50} value of $Gd(BOPTA)^{2-}$ in mice after intravenous injection of 0.5 mmol/L solution is 5.8 mmol/kg, the same as for $Gd(DTPA)^{2-}$ [84]. Biodistribution studies of ^{153}Gd labeled $Gd(BOPTA)^{2-}$ demonstrated bimodal excretion of the chelate, i.e. renal and biliary. The excretion of hepatobiliary chelates is highly variable among species. For example, 52% in rats, 25% in rabbits, and 2% in humans at a dose of 0.2 mmol/kg. At a lower dose of 0.005 mmol/kg, hepatobiliary excretion of $Gd(BOPTA)^{2-}$ is increased to 7.2% in humans, as the uptake and transport processes are saturable.

$Gd(EOB\text{-}DTPA)^{2-}$ (EOB-DTPA = ethoxybenzyl DTPA) and $Gd(BOPTA)^{2-}$ are similar in structure and have very similar physical properties [20,85]. The LD_{50} value in mice was reported as 7.5 mmol/kg. $Gd(EOB\text{-}DTPA)^{2-}$ and $Gd(BOPTA)^{2-}$ may share the same transport mechanism. Both agents show the same liver signal enhancement within a few minutes of the injection. However, the enhancement of rat liver by $Gd(EOB\text{-}DTPA)^{2-}$ starts to decrease after about 20 min and is substantially eliminated within 1 h. In contrast, $Gd(BOPTA)^{2-}$ remains pronounced for a period of 2 h or more.

Figure 17 shows images of a human patient with liver metastases collected before and after intravenous administration of $Gd(BOPTA)^{2-}$. The contrast enhanced images are taken at longer times after administration than for the brain agents to give the blood concentration time to diminish. The hepatobiliary agents are distributed into extracellular interstitial spaces, as well as specifically taken up into hepatocytes, while being excreted by both renal and hepatic routes [84]. The liver metastases contain a blood supply and interstitial space but lack the hepatocyte character (having generally been derived from another cell type). Hence both the liver and the metastases are enhanced early after injection by means of their blood supply and interstitial spaces. Only after the contrast agent concentrations have dropped in interstitial spaces, through a combination of liver uptake and renal and liver excretion, can the tumors be reliably imaged as dark (unenhanced) lesions. This appears to require 20 to 60 min for the clinical Gd chelates.

Recently, new structure types have been reported. For example, one of these, Cy_2-DOTA (bis cyclohexyl derivative of DOTA) contains no aromatic group [76]. The mouse biodistribution data of this chelate suggested 20% liver uptake in 60 min. Hepatobiliary agent clearance vs dose in rats suggested 20% accumulation in rat bile over a period of 90 min (Fig. 18). The MLD (minimum lethal dose) was determined

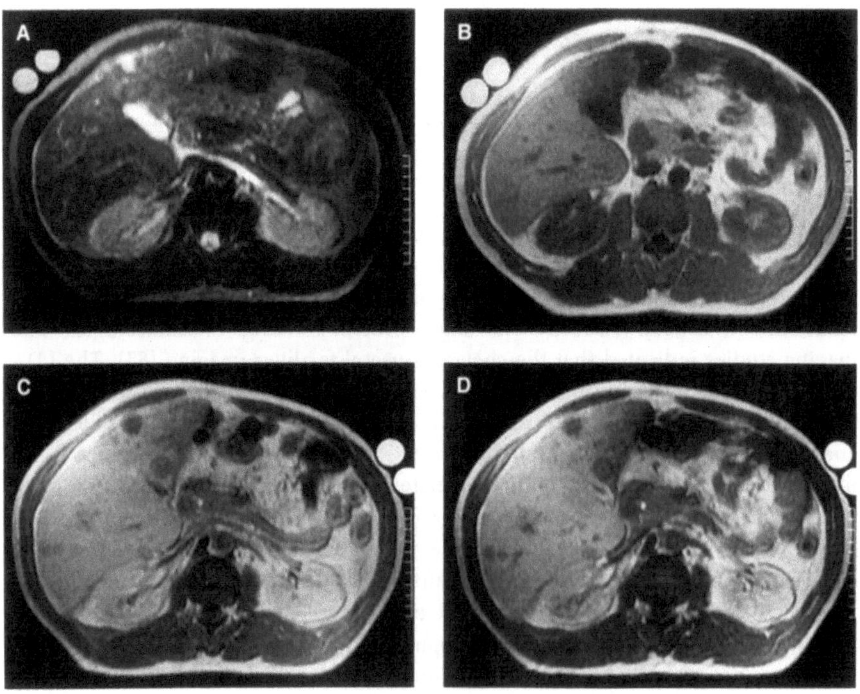

Fig. 17. T_1 weighted gradient echo MR images of liver metastases at 1.0 T. The dose of MultiHance, (Gadobenate dimeglumine, $NMG_2[Gd(BOPTA)]$, Bracco SpA) was 50 μmol/kg. **A** Precontrast T_2 spin echo (TR: 2050 ms; TE: 160 ms;)and (B-D) (T_1 gradient echo (TR: 110 ms, TE: 6 ms; FA: 70°), **B** Pre contrast, **C** 40–80 min, **D** 90–120 min. Courtesy of Prof. Christoph De Haen and Dr. Alberto Spinazzi, Bracco SPA

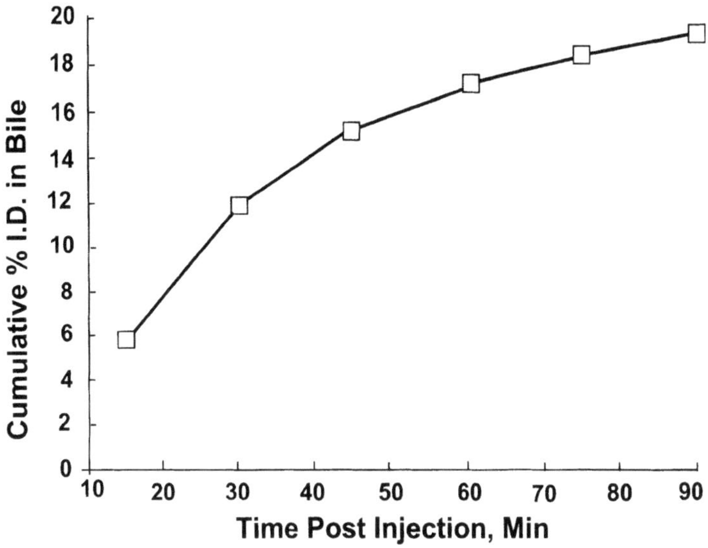

Fig. 18. Plot of cumulative %ID of $Gd(Cy_2DOTA)^-$ in rats with time

in mice as 7.5 mmol/kg. Parker et al. [77] investigated another structure type substituting the carboxylates of DOTA by methyl or phenyl phosphinic acids. The Gd^{3+} chelate (the ligand structures are given in Fig. 16) was found at 50% i.d. in rodent gut in 5 min after injection. Good images of the biliary system were obtained even at low doses (i.e. 5 μmol/kg). After six days, more than 99.99% of the complex was cleared from the body of the animal. Administration of a $^{153}Gd^{3+}$ labeled complex into tumor-bearing animals demonstrated that the chelate clears more slowly from the tumor than from the surrounding healthy tissue possibly due to albumin binding. Thus they were able to obtain images that defined the position of tumor sites up to 8 h after administration.

The chelating agent, DPDP (dipyridoxyl diphosphate) forms a distorted octahedral complex with Mn^{2+}. The metal ion coordinates with two tertiary amines, two carboxylate oxygens, and two phenolate oxygens, which deprotonate upon coordination [22]. Since the chelate does not have coordinated water, the aqueous T_1 relaxivity at 20 MHz and 40 °C is low, 2.8 $mM^{-1}\ s^{-1}$ with T_2 relaxivity also low, 3.7 $mM^{-1}\ s^{-1}$. However, it is higher than an outer sphere Mn^{2+} chelate [86], $Mn(DTPA)^{3-}$: $r_1 \sim 1\ mM^{-1}\ s^{-1}$. The stability constant of the chelate is lower than those of the Gd^{3+} chelates so far discussed by several orders of magnitude: log K = 15.1 and log K′ = 9.4 at physiological pH. The lability of Mn^{2+} chelates and the natural biologic sequestering mechanism tends to create free metal and the free ligand on injection of most Mn chelates, probably including this clinical example. A range of LD_{50} values for the chelate is reported as 1.9–5.0 mmol/kg [87].

Mn(DPDP) (now sold as Teslascan^R, http://www.amersham.com) dissociates quickly in vivo and provides liver enhancement through both Mn-chelate uptake and significant levels of free Mn due to dissociation, with subsequent hepatic and pancreatic uptake of Mn^{2+} by the natural sequestering mechanisms. The chelating agent protects the patient from most of the acute Mn toxicity if the doses are low

(10 μmol/kg) and the injections are administered as slow drip infusions (3 mL/min) of 10–50 mM solutions. Under these conditions side effects are limited to facial flushing and a sensation of facial warmth [78]. Clinical studies indicated useful enhancement of liver and some, though lesser, enhancement of pancreas. The pharmaceutical formulation includes substantial quantities of ascorbic acid, presumably as an antioxidant to inhibit Mn(IV) formation [88].

Another hepatobiliary chelate, the Fe^{3+} complex of EHPG (*N,N′*-ethylenebis(o-hydroxyphenylglycine) contains coordinated tertiary amines, carboxylates, and phenolate oxygens in an octahedral coordination environment [79]. Since Fe(EHPG) is coordinatively saturated, it relaxes water protons by an outer sphere mechanism with an r_1 of $\sim$1 mM^{-1} s^{-1}. The complex showed biliary clearance of the intact compound in rats. Bilirubin binding studies with HSA showed that the chelate binds with the bilirubin site [79]. However, the compound was considered unattractive because of its low relaxivity and less than ideal solubility.

4.2 Particulates

Water insoluble particulate based radiopharmaceuticals target organs rich in phagocytic cells, which engulf circulating foreign bodies in the nm to micron range. Protein aggregates labeled with ^{99m}Tc and colloidal ^{99m}Tc are used commercially. ^{99m}Tc agents targeted to asialoglycoprotein receptors in the liver have been tested clinically and is marketed in Japan [89]. As MRI contrast agents, Fe oxides coated with dextrans [38], arabinogalactan (a galactose-terminal polysaccharide) asiologlycoprotein [90], Gd_2O_3 suspensions [91], and Mn sulfides [92] have been reported, but only the Fe particulates have been examined clinically due to concerns over the long term retention of the metals. The iron particles coated with arabinogalactan are taken up by the hepatocytes in a process mediated by the asialoglycoprotein (ASG) receptor. AG-USPIO is an example of such a preparation stabilized with arbinogalactan. Biodistribution studies of these particles suggested that they accumulate in liver. Mn-sulfide colloids appear to be concentrated exclusively by reticuloendothelial (RES) phagocytosis. Similarly, Gd_2O_3 is also taken up by the RES, however, hepatotoxic effects are observed. Several other paramagnetic compounds have been encapsulated by liposomes and tested [93,94].

The use of superparamagnetic iron oxide particles for targeting liver and spleen [95] involves coating the otherwise toxic particulates with different materials, e.g. starch [96], albumin [97], dextran [95], or polyethylene glycol [98]. They are found to localize in liver. Biodistribution studies of iron oxide particles, based on starch and dextrans, have shown >80% liver uptake [96,99]. The physical properties of iron particles include very high r_2/r_1 ratios and r_2 values much greater than those of the Gd chelates (Table 9). These properties determine that the imaging mode is primarily one which emphasizes T_2, producing signal loss in regions of contrast accumulation. For example, the images in Fig. 8 show a human patient imaged with AMI 25 (Ferridex^R, vide supra) before and after intravenous infusion of 10 μmol Fe/kg [39]. The images through the liver show pronounced black out of the signal in liver parenchyma due to uptake into the reticuloendothelial cells. The metastases do

Table 9. Physicochemical properties of AMI 227, AMI 25, and $Gd(DTPA)^{2-}$

Material	$10^6\ \chi_g$, cgs/gm	r_1, $mM^{-1}s^{-1}$	r_2, $mM^{-1}s^{-1}$	r_2/r_1
AMI 227		7.1[b]	81.1[b]	11.4[b]
AMI 25	25,000	40[a]	160[a]	4.0[a]
$Gd(DTPA)^{2-}$	163	4.8[b]	5.1[b]	1.06[b]

[a] at 2.0 T
[b] at 0.5 T

not have the reticulocytes contained in the normal liver, and so are unaffected by the contrast agent, and are easily seen in most images.

The first particle listed in Table 9 differs in size, and so acquires the useful property of longer blood residence time and lymph node uptake. AMI 227 is undergoing clinical trials for liver [90c,100] and lymph node enhancement [38,101,102]. Results, particularly in head and neck lymph nodes, are very encouraging from the point of view of efficacy. The particulates have, so far, been administered as slow infusions to keep adverse reactions to a tolerable level (<10%). This is a disadvantage from the point of view of imaging time in the case of liver enhancement, but it is the same type of disadvantage as have the Gd chelates for liver imaging. While the tolerance of these agents is probably lower than that of the Gd chelates, measured in terms of the frequency of minor adverse events, it seems likely that the efficacy in some cases will outweigh the lower tolerance.

5
Blood Pool Agents

Blood pool enhancement is a potential clinical use of contrast agents of both Fe particulates and water soluble gadolinium chelates, and is an active area of clinical and preclinical research. Blood pool enhancers must remain in the blood during the imaging protocol, and then, ideally, be excreted rapidly thereafter. There are actually two uses imagined for clinical blood pool agents. Magnetic Resonance Angiography (MRA) [103], analogous in purpose and image appearance to X-ray angiography (imaging larger blood vessels) and perfusion imaging (imaging capillary blood flow).

In MRA the image consists only of the blood vessels, and relies upon the motion of the blood to differentiate it from unmoving tissues (Fig. 19). Reducing relaxation times in the blood using contrast agents increases the signal in blood and further differentiates blood from tissues, increasing signal to noise levels. This type of imaging requires considerable time (>10 min) to perform and thus an agent which maintains its signal disproportionately in the blood, rather than in extravascular spaces, is required. To achieve this goal by restricting access to the extracellular space, the molecular size of the agent must exceed approximately 25–50 kD. Gadolinium chelates attached to macromolecules can serve this purpose [104–106]. Several macromolecules such as albumins, dextrans, liposomes, and polylysines have been used [104–106]. Studies in animals have demonstrated that the macromolecular complexes distribute in the blood pool and enhance tissue signal [104–106]. Properties of some of these paramagnetically labeled macromolecules are given in

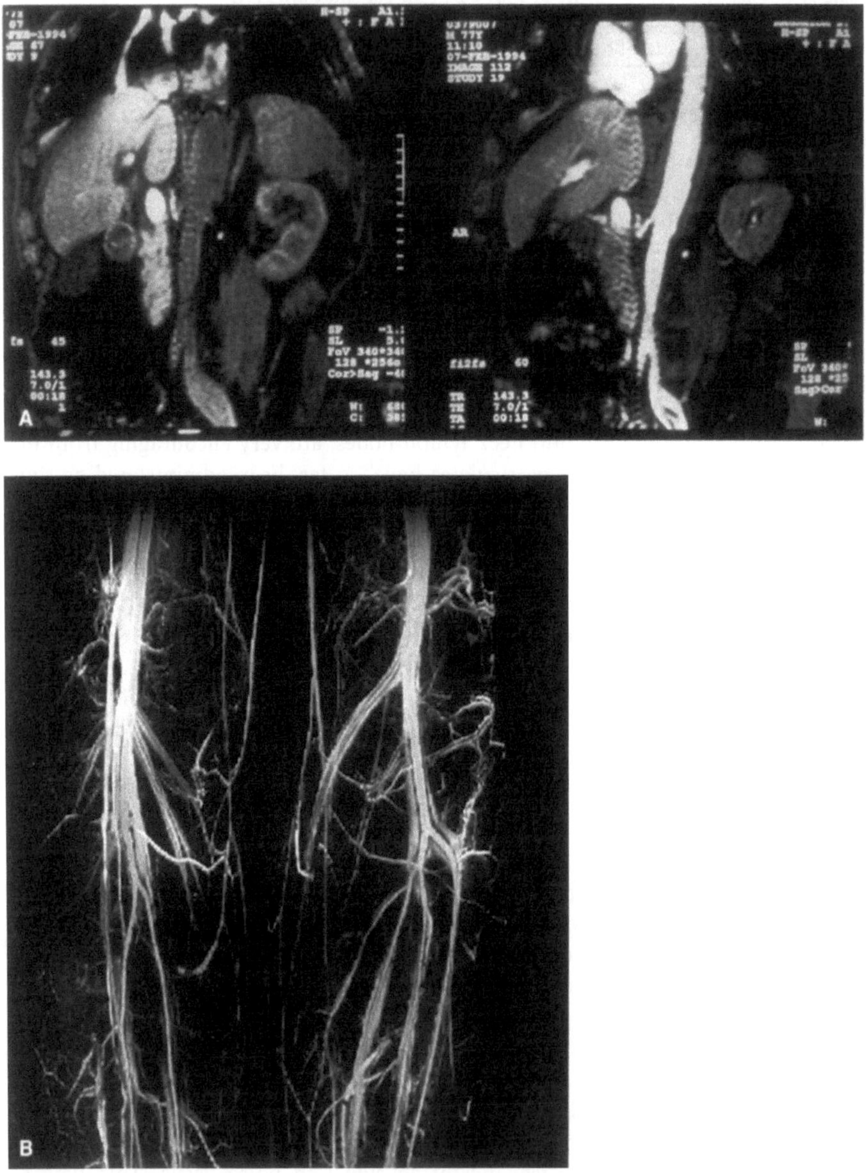

Fig. 19. Examples of Magnetic Resonance Angiography (MRA). **A** T_1 weighted MRA images using AMI 227. *Left*: unenhanced. *Right*: after administration of AMI 227. Courtesy of Dr. S. McLachlan and Ms. M. Morris, Advanced Magnetics, Inc. **B** Maximum intensity projection (T_1 weighted) images from a steady state image of the human calves of the lower limbs using MS 325, a derivatized Gd(R-DTPA) chelate (0.05 mmol Gd/kg) that binds human serum albumin on injection, rendering greater relaxivity in blood than in extracellular space . From Grist TM, Korosec FR, Peters DC, Witte S, Walovitch RC, Dolan RP, Bridson WE, Yucel EK, Mistretta CA , Radiology, 1998, 207:539–544. Courtesy of Thomas M. Grist, Univ. of Wisconsin-Madison

Table 10 [106]. Unfortunately, long blood retention tends to give long body retention and the long term retention of Gd as well as immunologic reactivity of polymers are seen as serious concerns [107]. Also, the indefinite organic ligand structure of large synthetic polymers represents a significant challenge to pharmaceutical development. Attempts to circumvent this problem have resulted in cascade polymers, e.g. dendrimers [108,109] which are claimed to be of definite structure, and these have been labeled with Gd. The G6 (G stands for generation) starburst dendrimer conjugates showed relaxivity as high as 34 mM^{-1} s^{-1} at 25 MHz and at 20 °C. Interestingly, a T_2 agent based on a Dy labeled polylysine has been tested [110]. Smaller dendritic structures have also shown some promise in animals [138], where a balance is achieved between extravasation and renal excretion which favors the latter, probably due to larger pores in the endothelial cell lining in the kidney than exist in most other tissues. This could allow renal excretion, along with enough blood retention to allow the imaging time needed for MRA.

Another blood pool imaging strategy is a variant of the Gd chelates covalently bound to serum albumin. MS 325 [139] is Gd chelate containing a lipophilic diphenylcyclohexyl group, which binds serum albumin noncovalently to about 85–90% at about 0.5 mM. On injection, the chelate binds albumin to the extent possible in an initial environment excess in chelate ([albumin] ~0.6 mM, [Gd] ~250 mM in the injectate), and also starts to fill the extravascular spaces, diffusing as a small chelate, probably binding to albumin there as well. The relaxivity when bound to albumin is tenfold higher than in the unbound state, so the relaxivity generated is heavily weighted to the space with the most albumin, the blood. Clinical trials on this compound are in progress.

Obtaining images whose signal is proportional to the amount of blood perfusing the tissue is a highly desirable clinical goal. Radiopharmaceutical tracers such as $^{201}Tl^{+}$ (a K^{+} mimic taken up into heart) and several ^{99m}Tc agents are routinely used for this purpose, but MRI holds the promise of achieving much greater spatial resolution if the right agents can be devised. The theory behind perfusion tracers is well known [111]. Ideally the perfusion agent must be >90% retained in blood, or they must be 90% extracted from blood in one pass through the tissue to accurately trace blood perfusion. The radiopharmaceuticals are of the extracted type, and from clinical experience it appears that >60% extraction is acceptable. Most of the research activity on development of perfusion agents is in Nuclear Medicine [112]. Some progress has also been made in MRI [113]. In the latter case, rapid injection

Table 10. Properties of some selected paramagnetically labeled macromolecules [106]

Properties	$Gd(DTPA)^{2-}$	Albu-Gd(DTPA)	Dex-Gd(DTPA)	Poly-Gd(DTPA)
Gd^{3+}/molecule	1	30	15	60
Mwt.	538	92,000	75,000	50,000
r_1, mM^{-1} s^{-1}/Gd in water	3.7[a]	14.4[b]	10.5[b]	13.1[a]
r_1, mM^{-1} s^{-1}/mol in water	3.7[a]	432[b]	157	786[a]
Half-life in rat plasma, h	0.3	>3.6	>1.5	>1
LD_{50} in mice, mmol/kg	7.5	–	–	>17
V_d, l/kg	0.6	–	–	0.25

[a] at 20 MHz and 39 °C
[b] at 10 MHz and 25 °C

and very fast imaging are used to map the transit of Dy^{3+} analogs of extracellular Gd^{3+} chelates [114], using the signal loss generated by the magnetic susceptibility of the Dy ions. The theory is that agents will serve, on the first pass through tissue, as perfusion agents (blood pool enhances), even though they rapidly diffuse out of the blood pool into interstitial spaces (they are only about 30% extracted). Here the safety advantage of the water soluble drugs allows them to be delivered as a sharp bolus injection. It is in this potential application that the nonionic chelates formulated at 1 M or higher concentrations are potentially useful.

Theoretically, if they could be bolus injected, fully non extractable agents are also useful as perfusion agents [115]. Ultrasmall iron oxides [116], for example, may be useful as perfusion agents [116] because they will remain in the capillary system (extraction <10%) and circulate for a long time (>1 h) before being trapped by the RES of the liver, spleen, and bone marrow [117]. Thus far, the difficulty has been the lack of bolus injectable compounds or formulations.

6 Gastrointestinal Agents

Numerous paramagnetic ions and chelates, ferromagnetic and ferrimagnetic particulates, and diamagnetic agents have been tested in animals and in humans. Among the agents tested were food supplements, e.g. Geritol (containing ferric ammonium sulfate), $Gd(DTPA)^{2-}$ in mannitol, and oil emulsions. The proton replacement agent, PFOB, was discussed above. This agent, Imagent^R, has shown to be an effective GI agent but many patients reported adverse reactions such as diarrhea in clinical trials. Some other examples include OMP (Oral Magnetic Particles), AMI 121 (Fe oxide), gas, barium sulfate, and clays [118–120]. The agents are all used as markers to highlight the bowel and so allow the examiner to better distinguish bowel from surrounding tissues and pathology.

A variety of foodstuff agents, e.g. baby and nutritional support formulas have been used as bowel markers for MRI [121,122], In these agents, T_1 shortening is attributed to the presence of trace amounts of paramagnetic substances. In general gastrointestinal (GI) agents are divided into three classes: a) positive or T_1 agents, b) negative or T_2 agents, and c) biphasic agents, which affect T_1 and T_2.

$Gd(DTPA)^{2-}$ in mannitol formulation was proposed as a positive oral MRI contrast agent. The agent was also administered rectally. Although phase I–III clinical trials were completed, approximately 30% of patients reported adverse reactions, e.g. diarrhea [119]. A formulation of $MnCl_2$ (LumenHance^R), is undergoing clinical trials as a biphasic agent. The formulation enhances signal intensity on T_1 weighted images and on T_2 weighted images signal intensity decreases [123]. Figure 20 shows an image. The r_1 and r_2 relaxivity of the product are: 17.3 and 30.1 mM^{-1} s^{-1}, respectively at 10 MHz and at 37 °C. This agent is very well tolerated.

Oral magnetic particles (OMP) containing superparamagnetic ferrite crystals incorporated into a polymer is being tested as a negative MRI contrast agent. OMP is generally well tolerated with ~5% adverse reactions [124]. AMI-121 (Gastromark^R) is another example of a superparamagnetic particle agent.

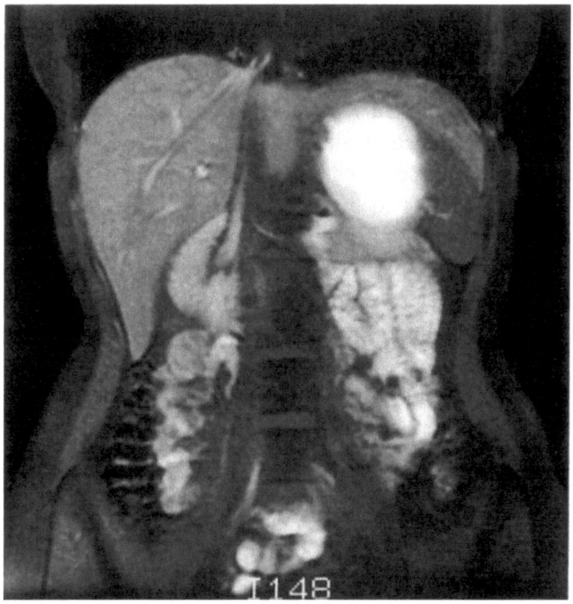

Fig. 20. Image of human GI using LumenHance[R]. This is a coronal fast gradient echo sequence that is T_1 weighted. Note that the bowel is filled with LumenHance[R] white bowel marking. The dose of the contrast agent was 40 mg of Mn^{2+} per liter. Courtesy of Ms. Roberta Muse and Bracco Diagnostics, Inc

Balkus et al. [125] found that microporus metal oxides such as zeolite molecular sieves and clays modified with paramagnetic metal complexes are effective MRI contrast agents for the GI. Zeolites are crystalline almunosilicates having well defined pore and channel systems of molecular dimensions. Gd^{3+} can be incorporated by an ion exchange method. The zeolites are thermally resistant and can tolerate the conditions in the stomach (pH < 2) and GI tract. More recently $Gd(DOTA)^-$ [126] and Gd(HP-DO3A) [127], based on their sluggish acid-asisted dissociation kinetics, have been proposed as a GI contrast agents for MRI.

7 Future Directions

Tissue targeted MRI agents are the most obvious future area of current research. A recent review outlined the prospects [145]. The conclusions were that while difficulty exists, there are no absolute physical constraints on using existing molecular structures to image receptors. There are biological constraints which add a greater degree of difficulty to the task. The biological problems are delivery of the contrast agents to the receptors in sufficient quantity and the biological implications of saturating those receptors. Biomolecules specific for organs, tumors, and other cellular targets labeled with Gd^{3+} and iron oxides are frequently proposed, but reductions in the biological constraints will be needed before practical agents are developed. The use of high turnover, internalizing receptors was proposed, as well as using metal polymers together with targeting receptors inside the blood vessels.

There are numerous reports in Nuclear Medicine on radiolabeled biomolecules, e.g. peptides for imaging infection and inflammation [10,140], nitroimidazoles for imaging hypoxic tissues [9], somatostatin and somatostatin receptor specific peptides [128], and carbohydrates to bind E-selectin/ELAM-1 receptors expressed on activated endothelium of imaging or inflammation sites [129]. Targeting of MRI agents has included mainly Fe particles due to their high relaxivity [90], Gd chelates have been reported to target receptors: for example, (folate)-targeted dendrimers [147], and antibody-targeted Gd-liposomes [148]. These are probably impractically expensive given their marginal efficacy, but, provided adequate controls and reproducibility verify the results, these early studies do demonstrate the principle. Recently, a novel Gd chelate with an enzymatically activated switch was reported [149]. An organic arm, R, containing a β-galactopyranosylethoxy group covered the water coordinating position of a Gd(R-DO3A) chelate. B-galactosidase enzymatically removed the galactopyranose from the chelate, causing a permanent gain in relaxivity as the q value increased. Targeted Gd chelates are also of some interest in microscopy, and bifunctional chelates have been made which include fluorescent markers [150].

The most important restriction on the paramagnetics, which does not exist for the radionuclide chelates, is the requirement for tissue concentrations ~50 μM in the target site [141]. This will dramatically restrict new developments, especially in gadolinium chelates. However, this problem can be addressed, in principle, by developing chelates with greater relaxivity, and creating oligomers of these [130].

The work toward optimizing relaxivity is currently aimed at small improvements in the existing gadolinium chelates as relaxation agents, but the larger goal is to endow the field of MRI with a set of injectable pharmaceuticals that will serve the same function that ^{99m}Tc pharmaceuticals serve in nuclear medicine. This requires that the effective concentration gap between Tc (nM) and Gd (>μM) be bridged. The extent to which this is achieved will govern the degree to which the gadolinium agents can be made tissue and biochemically specific.

There are tantalizing 100 fold gains possible if we eventually optimize the relaxivity of the gadolinium chelates. Existing theory, the Solomon-Bloembergen-Morgan (SBM) theory [46,48], predicts an optimum relaxivity on the order of 200 mM^{-1} s^{-1} per Gd-H_2O obtainable through optimization of molecular rotation (τ_r), electron relaxation (τ_s) and water exchange (τ_m) rates (Eq. 5). While the hydration number and distance might be altered chemically, a practical barrier has so far been reached at $q = 1$. This is probably a question of stability that may eventually be overcome. In existing Gd chelates with multiple inner sphere waters, the water molecules are coordinated adjacent to one another in the inner sphere. This arrangement allows undesireable biological chelation. For example, the Gd(DO3A)$(H_2O)_2$ binds carbonate ion (DO3A is a derivative of DOTA with one of the acetate arms absent, leaving a secondary amine) [66a]. Gd(EDTA)$(H_2O)_{2,3}$ has also been shown to bind ternary endogenous ligands [144].

Most work has centered on slowing the rotational rate, increasing τ_r, by which as much as an order of magnitude relaxivity enhancement is obtainable [131,132]. The limiting relaxivity in protein bound situations is often <20 mM^{-1} s^{-1} due to either a loose attachment of the chelate (and generation of intramolecular motions which limit τ_r [142]), or to the participation of electron relaxation or slower water exchange. It is apparent that slow water exchange will be a limiting factor in current chelates bound to protein. Using equation 3, with a relaxivity of 45 mM^{-1} s^{-1} gives $(T_{1P} + \tau_m)$ = 450 ns compared to the measured τ_m for $Gd(DTPA)H_2O^{2-}$ of 300 ns (Table IV). This means that the fast exchange regime requirement, $T_{1P} \gg \tau_m$ is not satisfied. When carboxylates on the DTPA and DOTA frames are substituted with amides and hydroxyls, the τ_m is, unfortunately, increased to between 600 ns and over 1000 ns, depending on the number of substitutions [143]. A relaxivity of 100 gives $(T_{1P} + \tau_m)$ = 180 ns, and requires an even lower τ_m, a practical number probably being 30–50 ns. Recently, it has been found that substitution of the acetate arms of DOTA with methyl groups on the second carbon, making the ligand, DOTMA, leads to a Gd-OH_2 distance increase of 0.126 angstrom, and an decrease in τ_m to about 35 ns [146].

The electronic limitations are considered to be the greatest challenge. The chemical problem will be to understand the relationship between electron relaxation rate and chemical structure in gadolinium chelates. Electron relaxation will be especially difficult to control because the unpaired *f* electrons in Gd^{3+} are masked by the filled 5*d* shell. Aside from noting that symmetry will play a role, little has been accomplished so far. $Gd(DOTA)^-$ has been reported to have a considerably longer electron relaxation time than the less symmetric $Gd(DTPA)^{2-}$ and Gd(DOTA-PA) (Fig. 21) [133]. Pulsed EPR studies at cryogenic temperatures (18 K) showed a factor of two longer T_{1e} (T_1 of the unpaired electron) for $Gd(DOTMA)^-$ and $Gd(CY_2\text{-}DOTA)^-$ (Fig. 16) compared to $Gd(DTPA)^{2-}$ and $Gd(DOTA)^-$ [134]. Apparently more rigid chelates may have longer τ_s values, as well as shorter τ_m values, both of which are theoretically desirable. Rigidity itself is now under study [151].

In addition, there is room for improvements in the theory, because the SBM theory assumes at the beginning that $\tau_r \ll \tau_s$, while our path toward greater relaxivity should take us to and beyond that limit. Research groups are currently using classical NMR and spectroscopic techniques to directly probe the parameters that appear in the SBM formulations, rather than attempting only to fit them simultaneously to NMRD curves. The most recent work in this area has been toward establishing precise values for the rotational, electronic, and water exchange correlation time for chelates. La^{3+} and Y^{3+} chelates have been synthesized and τ_r values extracted from the ^{13}C spectra arising from relatively rigid carbon atoms on the ligand [49,135]. The atoms used must not exchange on the time scale of the τ_r. The data obtained depend to about the sixth power on the C-H bond distance, however, and this distance has yet to be established by neutron diffraction in the appropriate environment. Measurements of τ_m are being made on Gd^{3+} chelates using ^{17}O NMR [50], and these combined with τ_v from EPR studies. Pulsed EPR has been used to determine electron relaxation times directly [134]. Unfortunately, each physical technique suffers limitations. The ^{13}C-NMR techniques require a knowledge of the C-H distance, the ^{17}O-NMR techniques use a fitting routine for multiple parameters and require assignment of q and the τ_s, and the pulsed EPR spectra must be measured in frozen glasses at cryogenic temperatures. Nevertheless, with enough data from multiple techniques there is every reason to believe that progress will

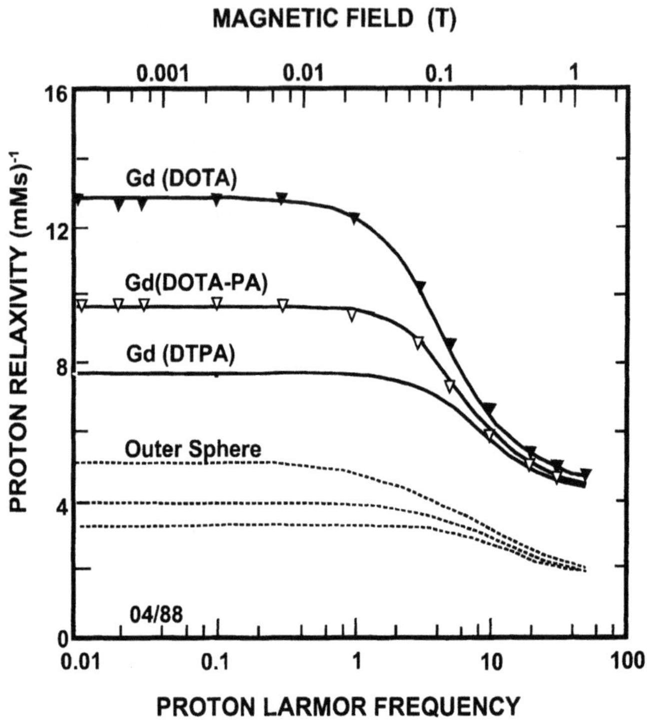

Fig. 21. $1/T_1$ NMRD profiles [133] of $Gd(DOTA)^-$ and $Gd(DTPA)^{2-}$

continue to be made both toward understanding of the theory and the realization of much greater relaxivities.

Acknowledgements. The authors would like to thank Drs. Val Runge, Stuart McLachlan, Christoph DeHaen, and Ms. Roberta Muse for supplying MR images and Dr. Rajesh Shukla for the NMRD simulation and graph shown in Fig. 11.

References

1. For Historical development see: a) Strain WH (1976) In: Knoefel PK (ed) International Encyclopedia of Pharmacology and Therapeutics: Radiocontrast Agents, vol 1, sect 76. Pergamon, Oxford, p 1; b) Grainger R (1982) Br J Radiol 55:1
2. Röntgen WC (1898) Sber Phys-Med Ges Wurzb 137:132
3. Cameron DF (1918) J Am Med Assoc 70:754
4. Binz A (1931) J Urol 25:297
5. Hoey GB, Wiegert PE, Rands RD (1976) In: Knoefel PK (ed) International Encyclopedia of Pharmacology and Therapeutics: Radiocontrast Agents, vol 1, sect 76. Pergamon, Oxford, p 23
6. For a review on the subject see: Sperber I, Serber G (1976) In: Knoefel PK (ed) International Encyclopedia of Pharmacology and Therapeutics: Radiocontrast Agents, vol 1, sect 76. Pergamon, Oxford, p 165
7. Fritzberg AR (1986) Radiopharmaceuticals Progress and Clinical Perspectives, vol 1, 2. CRC, Boca Raton, FL
8. Nowotnik DP, Nunn AD (1992) Drug News Perspect 5:174

9. Nunn A, Linder K, Strauss HW (1995) Eur J Nucl Med 22:265
10. a) Breeman WAP, Hofland LJ, Bakker WH et al. (1993) Eur J Nucl Med 20:1089; b) Fishman AJ, Picke MC, Kroon D, Fucello AJ, Rexinger D, Ten Kate C, Wilkinson A, Rubin RH, Strauss HW (1991) J Nucl Med 32:483
11. a) Bloch F (1946) Phys Rev 70:460; b) Purcell EM, Torrey HG, Pound TV (1946) Phys Rev 69:37
12. Martin GE, Zekter AS (1988) Two Dimensional Methods Establishing Molecular Connectivity. VCH, New York, NY
13. Bloch F, Hansen W, Packard M (1948) Phys Rev 70:474
14. Lauterbur PC, Mendoca-Dias MH, Rudin AM (1978) In: Dutton PL, Leigh LS, Scarpa A (eds) Froenters of Biological Energetics. Academic Press, New York, p 752
15. Chauncey DM, Schelbert HR, Halpern SE, Delano F, Kegney ML, Ashburn WL, Hogan PL (1977) J Nucl Med 18:933
16. Brady TJ, Goldman MR, Pykett IC, Byananno FS, Kistler JP, Newhouse JH, Burt CT, Hinsaw WS, Pohost GM (1982) Radiology 144:343
17. Goldman MR, Brady TJ, Pykett IL, Burt CT, Byananno FS, Kistler JP, Newhouse JH, Hinsaw WS, Pohost GM (1982) Circulation 66:1012
18. a) Laniado M, Weinman H-J, Schorner W, Felix R, Speck U (1984) Phiolog Chem Phys Med NMR 10:157
19. Weinman H-J, Brasch RC, Press WR, Wesbey GE (1984) Am J Roent 142:619
20. Weinman H-J, Sculma-Giampieri G, Svhmidt-Willich H, Vogler H, Frenzel H, Gries H (1991) Mag Res Med 22:233
21. Cavagna F, Dapra M, Maggioni C, DeHaen C, Felder E (1991) Mag Res Med 22:333
22. Rocklage SM, Cacheris WP, Quay SC, Raymond KN (1989) Inorg Chem 28:477
23. Baba Y, Lerch MM, Tanimoto A, Kraft BP, Saluja AK, Zhao L, Chen J, Steer KL, Stark DD (1994) Invest Radiology 29:S300
24. Young SW, Sidhu MK, Qing F, Muller HH, Neuder M, Zanassi G, Modi TD, Hemmi G, Dow W, Mutch JD, Sessler JL, Miller RA (1994) Invest Radiology 29:330
25. Rogers J, Lewis J, Josephson L (1994) Invest Radiol 29:S81
26. Wehrli FW (1988) In: Stark DD, Bradley WG (eds) Magnetic Resonance Imaging. The CV Mosby, St. Louis, 1988, p 1
27. Davies PL, Wolf GL, Gillen JS (1989) Invest Radiol 24:400
28. Stark DD, Wittenbery J, Middleton MS, Ferrucci JT (1986) Radiology 158:327
29. Katz JJ, Crespi HL, Czajka OM, Finkel AJ (1962) Am J Physiol 203:907
30. Evelhock JL, McDouall JBL, Mattiello J, Simpson NE (1992) Mag Res Med 24:42
31. a) Ackerman JJH, Ewy CS, Kim S-G, Shalwitz RA (1987) Annals New York Acad Sci 508:89; b) Mitchell MD, Osbakken M (1991) Mag Res Mag 9:545
32. a) Freeman DM, Muller HM, Hurd RE, Young SW (1988) Mag Res Mag 6:61; b) Lee HK, Nalcioglu O, Buxton RB (1991) Mag Res Med 21:21; c) Lee HK, Nalcioglu O, Buxton RB (1992) Mag Res Med 23:254; d) Ordemiral B, Nalcioglu O (1992) Mag Res Med 23:324
33. Zhang X, Tweedle MF, Unpublished Work
34. Laniado M, Weinman H-J, Schorner W, Felix R, Speck U (1984) Physiol Chem Phys Med 16:157
35. Wedeking P, Sotak CH, Telser J, Kumar K, Chang CA, Tweedle MF (1992) Mag Res Mag 10:97
36. Mauss V, Grucker D, Fornasiero D, Chambron J (1985) Mag Res Med 2:187 and Koenig SH, Brown RD III (1985) Invest Radiol 20:297
37. a) Saini S et al. (1967) Radiology 162:211; b) Rogers J, Lewis J, Josephon L (1994) Investigat Radiol 27:S81; c) Josephon L, Lewis J, Jacobs P, Hahn PF, Stark DD (1988) Mag Res Mag 6:647
38. Anzai Y, McLachlan S, Morris M, Saxton R, Lufkin RB (1994) Am J Neuroradiol 15:87
39. Stark DD, Weissledger R, Elizondo G, Hahn PF, Saini S, Todd LE, Wittenber J, Ferrucci JT (1988) Radiology 168:297
40. Josephon L, Bider J, White D (1991) Mag Res Med 22:204
41. Bulte JWM, Douglas T, Mann S, Frankel RB, Moskowitz BM, Brooks RA, Baumgarner CD, Vymazal J, Strub MP, Frank JA (1994) J Mag Res Mag 4:497
42. a) Villringerm A, Rosen BR, Belliveau JW, Ackerman JL, Lauffer RB, Buxton RB, Chao Y-S, Wedeen VJ, Brady TJ (1988) Mag Res Mag 6:164; b) Moseley KE, Kucharczyk, Mintorovitch J (1990) AJNR 11:423
43. Kucharczyk J, Mintorovitch J, Asgari H (1991) Mag Res Med 19:311
44. Grucker D, Chambron J (1993) Mag Res Mag 11:692

45. a) Tweedle MF (1992) Investigative Radiology 27:S2; b) Kumar K, Tweedle MF (1993) Pure and Appl Chem 65:515; c) Kumar K, Tweedle MF (1995) Analytical Profiles of Drug Substances. Wiley, New York
46. Solomon I (1955) Phys Rev 99:559
47. Bloembergan N, Morgan LO (1961) J Chem Phys 34:842
48. Lauffer R (1987) Chem Rev 87:901; b) Kowalewski J, Nordenskiold L, Benetis N, Westland PO (1985) Prog Nucl Mag Res Spectrosc 17:141
49. Shukla R, Zhang X, Tweedle MF (1991) Invest Radiology 26:S224
50. a) Micskei K, Powell DH, Helm L, Brucher E, Merbach AE (1993) Mag Res Chem 31:1011; b) Micskei K, Helm L, Brucher E, Merback AE (1993) Inorg Chem 32:3844
51. Aime S, Botta M, Fasano M, Paoletti S, Anelli PL, Uggeri F, Virtuani M (1994) Inorg Chem 33:4707
52. a) Harten AH, Hahn RL (1982) Science 217:1249; b) Giovanni F, Franchini-Angela M (1972) Acta Cryst 328:3572
53. a) Brown RD III, Brewer CF, Koenig SH (1977) Biochemistry 16:3883; b) Koenig SH, Brown RD (1987) In: Gupta RK (ed) NMR Spectroscopy of Cells and Organisms, vol 2. CRC, Boca Raton, FL, p 75
54. a) Pfeifer H (1961) Ann Phys (Leipzig) 8:1; b) Freed JH (1978) J Chem Phys 68:4034
55. Koenig SH, Epstein M (1975) J Chem Phys 63:2279
56. Zhang X, Chang CA, Brittain GG, Garrison JM, Telser J, Tweedle M (1992) Inorg Chem 31:5597
57. Jackels SC, Durham MM, Newton JE, Henninger TC (1992) Inorg Chem 31:234
58. Nelson KL, Runge VM (1992) In: Runge VM (ed) Contrast Media in Magnetic Resonance Imaging: Clinical Approach, p 57, Lippincott, Philadelphia, PA
59. Gaughan GT (1992) In: Runge VM (ed) Contrast Media in Magnetic Resonance Imaging: Clinical Approach. Lippincott, Philadelphia, PA, p 105
60. Rappoport SI (1976) Blood Brain Barrier in Physiology and Medicine. Raven, New York, NY
61. Weinreb JC, Newstad G (1994) Mag Res Quart 10:67
62. Leutwyler K (1994) Scientific American 271:26
63. Kumar K, Chang CA, Francesconi LC, Dischino DD, Malley MF, Gougoutas JZ, Tweedle MF (1994) Inorg Chem 33:3567
64. Konings MS, Dow WC, Love DB, Raymond KN, Quay SC, Rocklage SM (1990) Inorg Chem 29:1488
65. Gries H, Miklautz H (1984) Physiol Chem Phys Med NMR 16:105
66. a) Chang CA, Francesconi LC, Malley MF, Kumar K, Gougoutas JZ, Tweedle MF, Lee DW, Wilson LJ (1993) Inorg Chem 32:3501; b) Dubost JP, Legar JM, Langiois MH, Meyer D, Schaeffer M (1991) CR Acad Sci, Paris, Ser 2, 312:349
67. Horrucks WO, Sudnick DR (1979) J Am Chem Soc 101:334
68. Lassen NA, Cerb J (1984) Blood Flow Meta 4:633
69. Corson DC, Williams TC, Sykes BD (1983) BioChem 22:5882
70. Cacheris WP, Quay SC, Rocklage SM (1990) Mag Res Mag: 8:467
71. a) Kumar K, Chang CA, Tweedle MF (1993) Inorg Chem 32:587; b) Wang X, Tianzhu J, Comblin V, Lopez-Mut A, Mercini E, Desreux JF (1992) IC 30:1095
72. Tweedle MF, Hagan J, Kumar K, Mantha S, Chang CA (1991) Mag Res Mag 9:409
73. Wedeking P, Kumar K, Tweedle MF (1992) Mag Res Mag 10:641
74. a) Koehler RE, Stanley RJ, Evans RG (1979) Radiology 132:115; b) Moss AA (1983) In: Moss AA, Gravsu G, Genaut HK (eds) Computed Tomography of the Body. Saunders, Philadelphia, p 615
75. a) Heindel ND, Burns HD, Honda T, Brady LW (1978) In: The Chemistry of Radiopharmaceuticals. Masson, New York; b) Chervu LR, Nunn AD, Loberg MD (1984) Semin Nucl Med 12:5; c) Loberg MD, Nunn AD, Porter DW (1981) Nucl Med Annual 1
76. Kumar K, Comblin V, Desreux J, DiMareo J, Gidgoul D, Gougoutas J, Marinelli E, Neubeck R, Ranganathan R, Tweedle MF, Ugwuneri Z, Wagler T, Wan X, Wedeking P (1994) Int Symp on Macrocyclic Chemistry, Lawrence KS
77. Parker D et al. (1993) Mag Res Mag 11:761
78. a) Hamm B, Vogi TJ (1992) Radiology 1872:167; b) Gehl HB, Urhahn R, Bohndrof K (1993) Radiology 186:795
79. a) Lauffer RB (1990) Mag Res Quart 6:65; b) Lauffer RB, Vincent AC, Padmnabhan S, Mead TJ (1987) J Am Chem Soc 109:2216

80. a) Berke PD, Stremmel W (1985) In: Popper H, Schaffner F (eds) Progress in Liver Disease, vol 8. Grune and Stratton, Orlando, FL; b) Okuda H, Nunes H, Vallabhajosula S, Strashun A, Goldsmith SJ, Berke PD (1986) J Hepatol 3:251
81. Klaasen CD, Watkins JB (1984) Pharmacol Rev 36:1
82. Uggeri F, Aime S, Anelli PL, Botta M, Brocchetta M, de Haen CH, Ermondi G, Grandi M, Paoli P (1995) Inorg Chem 34:633
83. a) Vittadini G, Felder E, Musu C, Tirone P (1990) Invest Radiol 25:559; b) Cavagna F, Dapva M, Maggioni F, deHaen C, Felder E (1991) Mag Res Med 22:329
84. deHaen C, Gozzini L (1993) J Mag Res Img 3:179
85. a) Schuhmann-Giampieri G, Schmitt-Willich H, Press WR, Negishi C, Weinman H-J, Speck U (1992) Radiology 183:59; b) Clement O, Muhler A, Vexler U, Berthezene Y, Brasch RC (1992) Invest Radiol 27:612
86. Geraldes CFGC, Sherry AD, Brown RD, Koenig SH (1986) Mag Res Med 3:242
87. a) Rocklage SM, van Wagoner M (1990) In: Ferrucci JT, Stark DD (eds) Liver Imaging: Current Trends and New Techniques. Andover Medical, Boston, p 374; b) Watson AD, Rocklage SM, Carvlin MJ (1992) In: Stark DD and Bradley WG (eds) Magnetic Resonance Imaging, 2nd edn. Mosby Year Book. St. Louis, MO, p 372
88. Elizondo G, Fretz CJ, Stark DD et al. (1991) Radiology 178:73
89. Pimstone NR, Stadalnik RC, Vera DS, Hutak DP, Trudeau WL (1994) Hepatology 20:917
90. a) Groman EV, Enriqucz PM, Chu C, Josephon L (1994) Biocon J Chem 5:547; b) Weissleder R, Reimer P, Lee AS (1990) AJR 155:1161; c) Josephon L, Groman EV, Menz E, Lewis JM, Bengele H (1990) Mag Res Mag 8:637
91. Burnett KR, Wolf GL, Schumaker HR, Goldstein EJ (1985) Mag Res Mag 3:65
92. Chilton HM, Jackels SC, Hinson HM, Ekstrand KE (1984) J Nucl Med 25:604
93. a) Caride VJ, Sostman HD, Winchell RJ, Gore JC (1984) Mag Res Mag 2:17; b) Magin RL, Wright SM, Niesman MR (1986) Mag Res Med 3:440
94. a) Unger EC, Fritz TA, Shen DK (1992) J Nucl Med 2(P):54; b) Unger EC, Shen DK, Fritz TA (1993) J Mag Res Mag 3:195
95. Ferrucci JT, Stark DD (1990) AJR 155:943
96. Fahlvik AK, Holz E, Leander P et al. (1990) Invest Radiol 25:113; b) Fahlvik AK, Holz E, Schroder U, Klaveness J (1990) Invest Radiol 25:793
97. Widder DJ, Greif WL, Widder KJ (1987) AJR 148:399
98. Maas R, Spielman R, Bonacker M et al. (1991) Abstracts of SMRM Meeting at Berkeley, CA, p 505
99. Weissleder R, Stark DD, Engelstad BL (1989) AJR 152:167
100. White DL, Aicher KP, Aria T et al. (1992) Mag Res Med 24:14
101. Rogers JM, Lewis J, Josephon L (1992) Mag Res Imag 12:1161
102 Bengele HM, Palmacci S, Rogers J, Jung CW, Crenshaw J, Josephon L (1994) Mag Res Mag 12:433
103. a) Brasch RC (1992) Radiology 183:1; b) Masaryk TJ, Lewin JS, Laub G (1992) In: Stark DD, Bradley WG (eds) Magetic Resonance Imaging. Mosby-Year Book, p 299
104. a) Meyer D, Schaefer D, Chambon C, Beute S (1994) Invest Radiol 29:S90; b) Meyer D, Schaefer D, Bouillot A, Beute S, Chambon C (1994) Invest Radiol 26:S50; c) Niemi P, Reisto T, Kormano M (1991) Invest Radiol 26:S48; d) Dubost D, Urvoas E, Karvtzoff R, Chaubon C, Schrimann-Claeys E, Frija G, Ropras C (1991) Invest Radiol 26:S46; e) Brasch RC, Berthezene Y, Vexler VS, Mosley M, Clemant O, Muehler A, Price D, Jerome H (1991) Invest Radiol 26:S42
105. a) Wang SC, Wilkstrom MG, White DL et al. (1990) Radiology 175:483; b) Tilcock C, (1994) J Liposome Res 4:909; c) Sieving PF, Watson AD, Rocklage SM (1990) Bioconjugate Chemistry 1:65
106. Brasch RC (1991) Mag Res Med 22:282
107. Duncan JR, Franano FN, Edwards WB, Welch MJ (1994) Invest Radiol 29:S58
108. Adam G, Neuerburg J, Spuntrup E, Muhler A, Scherer K (1994) Mag Res Med 32:622
109. Weiner EC, Brechbiel MW, Bothers H, Magin RL, Gansow OA, Tomalia DA, Lauterbur PC, (1994) Mag Res Med 31:1
110. Wendland MF, Saeed M, Yu K (1992) Abstract of Society of Mag Res in Med 1116
111. English RJ, Kozlowski J, Tumeh SS, Leonard BL (1987) J Nucl Med Technol 15:138
112. a) Jurison S, Berning D, Jia W, Ma D (1993) Chem Rev 93:1137; b) Nowotnik DP (1992) Drug Pharm Sci 55:37

113. Watson AD (1994) J Alloys Comp 14:207–208
114. a) Saeed M, Wendland WF, Yu KK, Higgins CB (1992) J Am College Cardiol 20:1634; b) Mosley ME, Vexler Z, Asgari HS, Mintorovitch J, Drugin N, Rocklage S, Kucharczyk J (1991) Mag Res Med 22:259
115. Axel L (1980) Neuroradiology 137:679
116. a) Kent TA, Quest MJ, Kaplan BJ (1990) Mag Res Med 13:434; b) White DL, Aicher KP, Aria T (1992) Mag Res Med 24:14
117. Weissleder R, Elizondo G, Wittenberg J, Lee A, Josephon L, Brady TJ (1990) Radiology 175:494
118. a) Mattrey AF, Trambert MA, Brown JN et al. (1991) Invest Radiol 26:S65; b) Hahn PF, Stark DD, Lewis JM (1990) Radiology 175:695
119. Kaminski S, Laniado M, Gogoll M (1991) Radiology 178:503
120. Rinck PA, Smevik O, Nilson G (1991) Radiology 178:775
121. Balzarini L, Aime S, Barbero L, Ceglia E, Petrillo R, Reyner J, Tesoro Tess JD, Musumeci R (1992) Eur J Radiol 15:171
122. Bisset GS (1989) Pediatric Radiology 20:61
123. Bernardino ME, Weinreb JC, Mitchell DG (1993) Radiology 189(P), 203
124. Oksendal AN, Jacobsen TF, Gundersen HG, Rinck PK, Rammeny E (1991) Invest Radiol 26:S67
125. a) Balkus K, Bresinska I, Kavalak S, Young SW (1991) Material Res Sympos Proc 233:255; b) Balkus K, Bresinska I (1994) J Alloy Comp 207/208:75; c) Brensinska I, Balkus K (1994) J Phys Chem 98:12989
126. Schwizer W, Fraser R, Maecke H, Siebold K, Funck R, Fried M (1994) Mag Res Med 31:388
127. Wan X, Wedeking P, Tweedle M (1994) Mag Res Mag 12:1009
128. a) Breeman WAP, Hofland LJ, Bakker WH et al. (1993) Eur J Nucl Med 20:1089; b) Kwekkeboom DJ, Krenning EP, Bakker WH (1993) Eur J Nucl Med 20:283
129. a) Bevilacqua et al. (1989) Science 243:1160; b) Paulson JC, Perez MS, Gaeta EC, Ratcliff RM WO 91/19501
130. Tweedle M (1992) J Alloys Compounds 180:317
131. Koenig SH, Baglin C, Brown RD, Brewer CF (1984) Mag Res Med 1:496
132. Lauffer RB, Brady TJ, Brown RD, Baglin C, Koenig SH (1986) Mag Res Med 3:541
133. Sherry AD, Brown RD, Geraldes CFGC, Koenig SH, Kuan KT, Spiller M (1989) Inorg Chem 28:620
134. Tweedle M, Shukla R, Weber R, Zhang X (1997) Abstracts of the Int Soc Magn Reson Med, 5th Annual Meeting, p 1575
135. Aime S, Nano R (1988) Invest Radiol 23:S264
136. Powell DH, Merbach AE, Gonzalez G, Brucher E, Micskei K, Ottaviani MF, Kohler K, von Zelewski A, Grinberg OY, Lebedev YS (1993) Helv Chim Acta 76:2129
137. Pochon S, Hyacinthe R, Terrattaz J, Robert F, Schneider M, Tournier H (1997) Acta Radiol Suppl 412:69
138. Clarke S, Weinmann H, Lucas A, Dai E, Rutt B (1998) Abstracts of the 6th Int Soc Magn Reson Med, abstract 215
139. Parmelee DJ, Walovitch RC, Ouellet HS, Lauffer RB (1997) Invest Radiol 32:741
140. Peers S, Tran L, Erikson S, Ballinger J, Goodbody A (1995) J Nucl Med 36:114P
141. Wedeking P, Shukla R, Nunn A, Tweedle M (submitted)
142. Shukla R, Fernandez M, Pillai R, Ranganathan R, Ratsep P, Zhang X, Tweedle M (1996) Mag Reson in Med 35:928
143. Toth E, Pubanz D, Vauthey S, Helm L, Merbach A (1996) Chem Eur J 2:1607
144. Burai L, Hietapelto V, Kiraly R, Toth E, Brucher E (1997) Magn Reson Med 38:146
145. Nunn A, Linder K, Tweedle M (1997) Quart J Nucl Med 41:155
146. Tweedle M, Kumar K, Shukla R, Aime S, Botta M, DiMarco J, Gougoutas JZ, (submitted)
147. Wiener EC, Konda S, Shadron A, Brechbiel M, Gansow O (1997) Invest Radiol 32:748; b) Wiener EC, Patrick TA, Roy EJ, Kranz DM, Lauterbur PC (1995) Abstracts of the Soc Magn Reson Med Meeting, p 48
148. Sipkins D, Gijbels K, Tropper FD, Steinmant L, Bednarski MD, Li KCP (1995) Abstracts of the Soc Magn Reson Med Meeting, p 1139; b) Storrs RW, Tropper FD, Li HY, Song CK, Sipkins DA, Kuniyoshi JK, Bednarski MD, Strauss HW, Li KC (1995) J Magn Reson Imaging 5:719
149. Moats RA, Fraser SE, Meade TJ (1997) Angew Chem Int Ed Engl 36:726

150. Huber MM, Staubli AS, Kustedjo K, Gray MHB, Shih J, Fraser SE, Jacobs RE, Meade TJ (1998) Bioconjugate Chem 9:242
151. Shukla RB, Nunn AD, and Tweedle MF (submitted)
152. Tweedle MF, Eur Radiol 7 (suppl. 5) S225

Metalloradiopharmaceuticals

Alan B. Packard[1], James F. Kronauge[2], Martin W. Brechbiel[3]

[1] Division of Nuclear Medicine, Children's Hospital, Harvard Medical School, Boston, MA 02115, USA (*E-mail: packard@a1.tch.harvard.edu*)
[2] Department of Radiology, Brigham and Women's Hospital, Harvard Medical School, Boston, MA 02115, USA (*E-mail: jfkronauge@bics.bwh.harvard.edu*)
[3] Radiation Oncology, National Cancer Institute, National Institutes of Health, Bethesda, MD 20892, USA (*E-mail: martinwb@box-m.nih.gov*)

Metalloradiopharmaceuticals are used in more than 80% of the diagnostic procedures performed in nuclear medicine each year and represent an increasing fraction of the therapeutic procedures. The early development of technetium-based radiopharmaceuticals was, to a large extent, anecdotal, and many of the agents developed during this period, of which several are still in use, are chemically uncharacterized. In the past 10 to 15 years, however, a transition has occurred, and essentially all of the metalloradiopharmaceuticals that were introduced in this period are based on rational design of metal-ligand systems that target specific biological systems. An overview of these developments is presented with an effort to describe the interrelationship between the chemical and biological properties of these compounds.

Keywords. Radiopharmaceuticals, Technetium, Nuclear medicine, Radioimmunoconjugates, Metalloradiopharmaceuticals

List of Abbreviations

ATSM	diacetyl-bis(N^4-methylthiosemicarbazone)
BAT-TECH	bis(aminoethanethiol)tetraethylcyclohexane
BATO	boron adducts of technetium oximes
BATO-2MP	chloro[bis[2,3-butanedionedioxime(1-)-0][2,3-butane-dionedioximato (2-)-*N,N',N'',N''',N'''',N'''''*](2-methylpropylborato(2-))technetium]
cpi	2-carboxymethylpropylisocyanide
CAMI	carboxymethylisocyanide
$CDOH_2$	cyclohexanedione dioxime
β-CIT	2β-carbomethoxy-3β-(4-iodophenyl)-*N*-methyltropane
CFT	2β-carbomethoxy-3β-(4-fluorophenyl)-*N*-methyltropane
CHX-DTPA	*trans*-cyclohexyl-DTPA
CPTA	4-[(1,4,8,11-tetraazacyclotetradec-1-yl)methyl]benzoic acid
CT	computerized tomography
Cy-EDTA	*trans*-cyclohexyl-EDTA
DADT	diaminedithiol
diars	*o*-phenylenebisdimethylarsine
dmpe	bisdimethylphosphinoethane
DMSA	*meso*-2,3-dimercaptosuccinic acid
DOTA	1,4,7,10-tetraazacyclododecane-*N,N',N'',N'''*-tetraacetic acid
DTPA	diethylenetriaminepentaacetic acid
DTTA	diethylenetriaminetetraacetic acid
DOTMP	1,4,7,10-tetraazacyclododecane-1,4,7,10-tetra(met-hylenephosphonic acid)
EC	L,L-ethylenedicysteine
ECD	ethyl cysteinate dimer
EDTA	ethylenediaminetetraacetic acid
EDTMP	ethylenediaminetetramethylenephosphonate
FDG	2-fluorodeoxyglucose
GSA	galactosyl human serum albumin
HBED	*N,N'*-bis(*o*-hydroxylbenzyl)ethylenediamine-*N,N'*-diacetic acid
HEDP	hydroxyethylenediphosphonate
HMPAO	hexamethylpropyleneamine oxime
IPT	*N*-(3-iodopropen-2-yl)-2β-carbomethoxy-3β-(4-chlorophe-nyl)tropane

MAG_3	mercaptoacetylglycylglycylglycine
MDP	methylenediphosphonate
$(4,6\text{-}MeO_2sal)_3BAPEN$	bis(4,6-dimethoxy)salicylaldimine)-*N,N′*-bis(3-aminopropyl)ethylenediamine)
mibi	2-methoxyisobutylisocyanide
NGA	galactosyl neoglycoalbumin
NOTA	1,4,7-triazacyclononane-*N,N′,N″*-triacetic acid
OIH	*o*-iodohippuran
PET	Positron-emission tomography
PnAO	propylene amine oxime
pompom	bis(dimethoxyphosphino)ethane
PTSM	pyruvaldehyde bis(N^4-methylthiosemicarbazone)
SPECT	single-photon emission-computed tomography
tbi	*tert*-butyl isocyanide
TETA	1,4,8,11-tetraazacyclotetradecane-*N,N′,N″,N‴*-tetraacetic acid
tmp	trimethylphosphine
tu	thiourea

1 Introduction

1.1 What is Nuclear Medicine?

Nuclear medicine is the use of internally administered radioactive materials (radiopharmaceuticals) for the diagnosis and treatment of disease. It is fundamentally different from other radiological procedures in that the source of the imaging or therapeutic emissions is located within the patient's body. In contrast, conventional radiography is performed using an external X-ray beam, and radiation therapy is usually performed using an external γ-ray source. A second important difference is that the distribution of the radiopharmaceuticals within the body typically is reflective of function rather than anatomy. Because of this, radiopharmaceuticals are most often used to evaluate a biological process, such as renal function or myocardial perfusion, rather than anatomy. The Society of Nuclear Medicine estimates that more than 10 million nuclear medicine procedures are performed each year in the U.S.

A radiopharmaceutical may be broadly thought of as any radioactive drug administered to a patient. The route of administration is most often intravenous, but also includes, for example, inhalation for lung ventilation studies or oral for gastric emptying studies.

From a chemist's point of view, radiopharmaceutical development is a fascinating confluence of several different areas of chemical research ranging from nuclear chemistry for radionuclide production, radiochemistry in target processing, organic and coordination chemistry in the synthesis of radiopharmaceuticals, and biochemistry in understanding the processes that affect the biokinetics and biodistribution of radiopharmaceuticals. The situation is also complicated by the need to understand the way that diseases and non-radioactive pharmaceuticals can alter the biodistribution of the tracers. The "perfect" radiolabeled compound is of little use as a radiopharmaceutical if: (1) it requires a radionuclide that is not readily available; (2) the synthesis is so complicated that it cannot be prepared in a time frame compatible with the half-life of the radionuclide; (3) it is metabolized in the blood before it reaches the target; (4) the target-to-non-target ratio is not high enough to allow visualization of the target tissue; or (5) the change in its biodistribution does not accurately reflect the status of the disease for which it is targeted.

For ^{99m}Tc, which is used in more than 80% of the nuclear medicine procedures carried out in the United States each year, the situation is further complicated. Technetium is unique among the transition metals in that, for all practical purposes, it does not occur naturally, and there are no stable isotopes. It was first reported by Perrier and Segre' in 1937 [1,2], and the only isotope available in gram quantities is ^{99}Tc ($T_{1/2} = 2.12 \times 10^5$ y), a β^- emitter that is a fission by-product. Consequently, the coordination chemistry of the element is less well developed than that of its neighbors in the periodic table. As recently as 1972, there were "no uses for technetium" [3], and the development of ^{99m}Tc radiopharmaceuticals did not begin in earnest until the development by Eckelman and Richards in 1971 of the "instant kit" for the synthesis of ^{99m}Tc radiopharmaceuticals [4].

1.2 Radionuclide Production

There are three primary sources of radionuclides: nuclear reactors, particle accelerators, and radionuclide generators. Nuclear reactors were the primary source of the radionuclides used in early nuclear medicine procedures and are today the source of ^{99}Mo, the parent radionuclide in the ^{99}Mo/^{99m}Tc generator, which is a product of the fission of ^{235}U. The primary advantage of nuclear reactors is that there are quite a few around and, for research purposes, it is relatively simple to irradiate a target with neutrons to produce modest quantities of an interesting radionuclide. For example, ^{64}Cu can be produced using the reaction ^{63}Cu(n,γ)^{64}Cu [5]. The disadvantage of this production route is that the specific activity (mCi/mg) of the material is typically relatively low because it is contaminated with the "cold" (i.e., non-radioactive) starting material.

Charged-particle accelerators are used for the production of ^{123}I, ^{201}Tl, ^{82}Sr, and the radionuclides most frequently used in positron-emission tomography (PET) (^{11}C, ^{13}N, ^{15}O, and ^{18}F). Iodine-123, ^{201}Tl, and ^{82}Sr are produced using high-energy cyclotrons or linear accelerators (linacs), and ^{11}C, ^{13}N, ^{15}O, and ^{18}F are usually produced using low-energy biomedical cyclotrons. With ^{11}C, ^{13}N, ^{15}O, and ^{18}F, it is possible to prepare radioactive analogs of known drugs or receptor substrates by isotopic substitution. These compounds, by definition, have the same biological properties as their non-radioactive counterparts. One important property of cyclotron-produced radionuclides is that they are typically carrier-free because charged-particle reactions, for example, ^{18}O(p,n)^{18}F, transmute the target element (increase Z)

rather than add a neutron (increase *N*). The product of a (p,n) reaction can be separated from the target relatively easily. This high specific activity is essential for the study of biological receptors. The principal limitation of biomedical cyclotrons is that they are expensive to purchase, maintain and operate.

Biomedical cyclotrons are not, however, limited to the production of ^{11}C, ^{13}N, ^{15}O, and ^{18}F. Welch et al. have recently described the production of several copper radionuclides using a biomedical cyclotron [6,7], and Nickles has summarized 100 radionuclides that can be produced with a biomedical cyclotron [8]. Higher energy cyclotrons and linear accelerators are used to produce other medical isotopes such as ^{123}I [9] and ^{82}Sr [10].

For daily clinical use, generators provide the most convenient and economical radionuclide source. A radionuclide generator is fundamentally a simple chromatographic system in which a parent radionuclide is adsorbed onto an ion-exchange resin from which the daughter radionuclide is selectively eluted in high yield. The constraint is that this separation must be achieved using an eluent that is either itself directly injectable into a patient or that can be quickly rendered injectable. The separation must also be performed in a sterile and pyrogen-free environment. The growth of nuclear medicine in the 1960s and -70s through today is directly attributable to the development of the $^{99}Mo/^{99m}Tc$ generator by Richards [11]. In the $^{99}Mo/^{99m}Tc$ generator, $^{99}MoO_4^{2-}$ is adsorbed on an acidic alumina column from which $^{99m}TcO_4^-$ is eluted in high yield with 0.9% (0.154 M) saline (Fig. 1).

The amount of ^{99}Mo that is co-eluted with the ^{99m}Tc is less than 0.1% of the total amount loaded onto the column. The specific activity of the ^{99m}Tc that is obtained

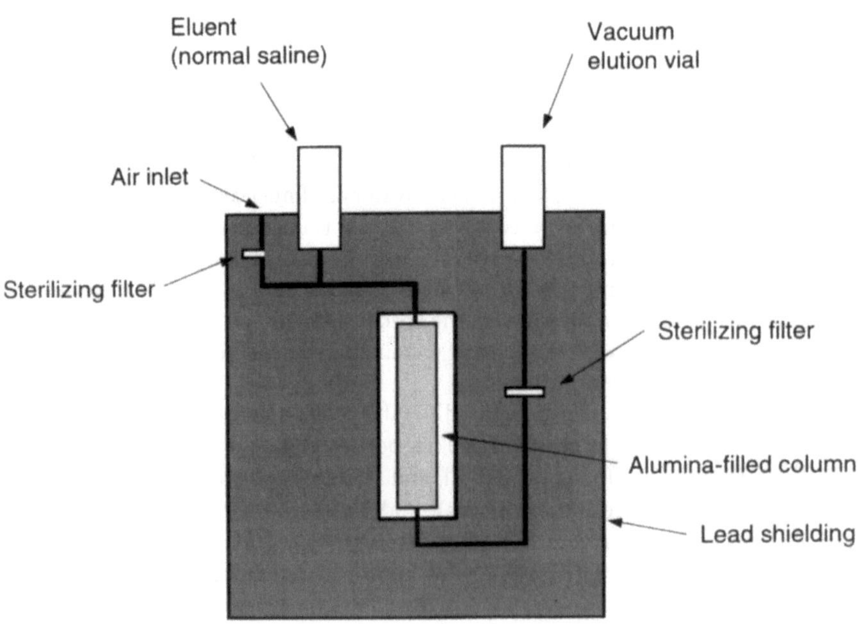

Fig. 1. Schematic of a $^{99}Mo/^{99m}Tc$ generator [11]. In the $^{99}Mo/^{99m}Tc$ generator, $^{99}MoO_4^{2-}$ is adsorbed on an acidic alumina column from which $^{99m}TcO_4^-$ is eluted in high yield with 0.9% (0.154 M) saline

from the generator is very high, but it is not carrier-free because decay of ^{99m}Tc on the column between elutions produces ^{99}Tc. The concentration of ^{99m}Tc in the eluate of a $^{99}Mo/^{99m}Tc$ generator is on the order of 10^{-9} M, which gives some context to the problems associated with the synthesis and characterization of new radiopharmaceuticals.

Given the superior convenience of this process compared to either reactor or cyclotron radionuclide production, it is not surprising that development of new generators is a major research focus in nuclear medicine. The properties of several generators are summarized in Table 1. Three of these generators produce radionuclides that are β^+ emitters and can, therefore, be used for PET. These generators offer the possibility of separating the PET-imaging process from the need for an on-site cyclotron to produce the radionuclides.

Table 1. Decay properties of several radionuclide generators

Parent	$T_{1/2}$	Decay	Daughter	$T_{1/2}$	Energy (keV)
^{99}Mo	6 d	β^-	^{99m}Tc	6 h	140
^{68}Ge	288 d	ec	^{68}Ga	68 min	511
^{82}Sr	25 d	ec	^{82}Rb	75 s	511
^{191}Os	15.4 d	β^-	^{191m}Ir	5 s	129
^{62}Zn	9 h	β^+, ec	^{62}Cu	10 min	511

1.3 Radiopharmaceutical Synthesis

The synthesis of radiopharmaceuticals for daily clinical use presents a challenge unlike that typically encountered in a chemical research laboratory or a standard pharmacy. For these compounds to be safe and effective, it must be possible to prepare them in a sterile, non-pyrogenic environment in high purity and high yield with minimal manipulation. Further, the synthesis must be performed rapidly (because the radionuclide is decaying during the synthesis) by technical staff, not trained chemists. This challenge is met in several different ways. Radiopharmaceuticals that are labeled with ^{111}In are usually prepared simply by the addition of a solution of the radionuclide to a solution of the chelating-agent-modified biomolecule. In cases where the synthesis is relatively complicated (e.g. [^{18}F]FDG), the synthesis is carried out at a central radiopharmacy and the product is delivered to the clinic as a "unit dose" ready for injection. The synthesis of ^{99m}Tc radiopharmaceuticals, however, presents a different challenge.

Technetium-99m is obtained from the $^{99}Mo/^{99m}Tc$ generator as a solution of $^{99m}TcO_4^-$ in normal (0.9%, 0.154 M) saline. As pertechnetate, it can be used to image the thyroid, gastric mucosa, and blood pool, but little else. To be incorporated into a molecule that will accumulate in other tissue, it is necessary to reduce the pertechnetate anion and incorporate ligands that will alter the biodistribution. The major breakthrough in the synthesis of ^{99m}Tc radiopharmaceuticals came in 1972 when Eckelman and Richards introduced the "Instant Kit" [4]. The instant kit, in most cases, consists of a sterile pyrogen-free vial containing a freeze-dried mixture of stannous chloride as a reducing agent and the ligand. When generator eluate containing $^{99m}TcO_4^-$ is added to the kit, the technetium is reduced by the stannous chloride and complexed by the ligand. Typically, these kits produce a single ^{99m}Tc complex, and the yield of the reactions is >90%. Recently, multi-vial kits have been developed for the preparation of radiopharmaceuticals that are not accessible via a single-step synthesis, but one of the vials is usually very similar to that described above.

Technetium radiopharmaceutical development is made more interesting by the fact that Tc can exist in nine different oxidation states (+7 through −1), and at least six of these can exist in aqueous solution (+7 through +1) [12]. This is one reason that several of the older radiopharmaceuticals that are still routinely used in the clinic, including the ^{99m}Tc-DTPA complex formed in the first kit [4], remain uncharacterized.

The development of ^{99m}Tc radiopharmaceuticals is now at a considerably more sophisticated stage than it was 15–20 years ago. At that time, most new agents were the result of "shake and shoot" experiments where a potential ligand, stannous chloride, and $^{99m}TcO_4^-$ were shaken together and injected (shot) into an animal to see where the product localized. Many commonly used ^{99m}Tc radiopharmaceuticals (e.g. ^{99m}Tc-DPTA) were developed in this way. In contrast, essentially all of the ^{99m}Tc radiopharmaceuticals introduced in the last 10 years are the result of chemical investigations where a specific tissue or function was targeted, a ligand set was designed that would provide a specific biological property associated with the target tissue or function, the complex was synthesized and characterized with ^{99}Tc or as a Re analog, and the biodistribution of the ^{99m}Tc complex determined using an animal model.

1.4 Instrumentation

There are two basic types of imaging devices used in nuclear medicine: single photon and dual photon. Single-photon cameras are internally similar to a standard well-type gamma counter in that they consist of a NaI(Tl) scintillation crystal coupled to a photomultiplier (PM) tube and an electronics package that converts the light emitted by the crystal when struck by a high-energy photon into an electronic signal. The intensity of this signal is proportional to the energy of the scintillation event and frequency is proportional to the number of events. In the case of a gamma camera, however, the crystal is a large thin disk rather than a cylinder containing a well, and a large number of PM tubes are attached to the crystal. Spatial information about the location of the scintillation event in the crystal is derived by interpolation of the relative intensities of the signals from the PM tubes adjacent to the event. Directional information is provided by placing a lead collimator in front of the NaI(Tl) crystal. This design was pioneered by Anger in the late 1950s [13]. Imaging devices have also been developed that rely on a large number of smaller crystals coupled via light pipes to an array of PM tubes [13] and gas-filled proportional counters rather than crystals [14].

The Anger camera provides a two-dimensional image of the distribution of tracer within the body. More recently, however, cameras have been developed that allow three-dimensional images to be obtained. This process, known as single photon emission computerized tomography (SPECT), calculates a three-dimensional image based on a computerized reconstruction of a series of planar images obtained as the gamma camera rotates around the patient. A single-head device can be used to obtain this data, but two- or three-head cameras are more frequently used to reduce the time required for data collection. The spatial resolution of current SPECT imaging devices is typically 7–10 mm.

PET radiopharmaceuticals contain positron (β^+) rather than single-photon emitting radionuclides. When the emitted positron encounters an electron, both are annihilated and two 511 keV photons are produced. Because angular momentum must be conserved, these photons are emitted at an angle of 180° to each other. If the patient who has been injected with the β^+-emitting radiopharmaceutical is surrounded by a ring of scintillation detectors, the two photons will impinge on the two opposing detectors at virtually the same time. By adding a logic circuit to the detector system that only counts those events that occur at the same time (coincidence), the event is known to have occurred along the direct path between the two detectors. This also means that, in contrast to Anger-type cameras, no collimator is required which increases the detection efficiency. The resolution of current PET devices is typically 4–8 mm.

The distinction between the two types of devices is becoming less sharp as multi-headed SPECT cameras are now available with coincidence circuitry that allows them to be used with either single-photon or positron-emitting radiopharmaceuticals [15].

1.5 Scope

The focus of this review is metalloradiopharmaceuticals in which the coordination chemistry of the metal ion plays an important role. Thus, tracers such as ^{201}Tl, which is used for myocardial perfusion imaging, are discussed only in comparison to coordination compounds such as ^{99m}Tc-MIBI ($^{99m}Tc(MIBI)_6$). This also excludes the majority of PET radiopharmaceuticals which are based on ^{11}C, ^{13}N, ^{15}O and ^{18}F. We have also generally excluded compounds for which there is little or no chemical characterization. Finally, we have chosen to focus on those radiopharmaceuticals that are now approved for human use (in the USA or elsewhere), may be approved in the near future, provide a background for the discussion of approved agents, or point to interesting new directions in radiopharmaceutical research. This review focuses on material from the late 1980s through early 1998.

2 Heart

2.1 Cationic Myocardial Perfusion Agents

The first widely used radiopharmaceutical for demonstrating patency of myocardial blood flow was ^{201}Tl [16]. This cyclotron-produced isotope ($T_{1/2}$ = 73 h, 74 keV photons) was developed in the mid-1970s to exploit its chemical similarity to K^+. The ionic radius of Tl^+ is only slightly larger than that of K^+, and $^{201}Tl^+$ was found to be taken up by myocytes, at least partially, by the ATP-driven Na^+/K^+ pump [17]. The active transport of Tl^+ into heart tissue produces a very high extraction efficiency from the blood (85%) on the first pass through the coronary arteries. Thallium-201 is cleared rapidly from the blood producing good heart/background ratios and has been used extensively to detect the permanent defects associated with coronary-artery obstruction and transient-flow defects that occur in myocardial ischemia [18]. In the case of ischemia, the release of potassium/thallium from the myocytes that occurs with each contraction can produce a rapid redistribution of the radioactivity that complicates the interpretation of the extent of arterial blockage. This redistribution phenomenon along with the less than optimal photon energy and long half-life provided the impetus for the development of a technetium-based myocardial perfusion agent.

The initial breakthrough in the development of a technetium-based myocardial perfusion agent was the synthesis and animal testing by Deutsch et al. [19] of the cationic ^{99m}Tc-arsine and -phosphine complexes first characterized with ^{99}Tc by Nyholm and Fergusson [20–22]. The technetium(III) complexes $[TcBr_2(diars)_2]^+$ (1) and $[TcCl_2(dmpe)_2]^+$ (2) accumulated in the hearts of dogs, rats and rabbits [23,24]. Although neither compound accumulates sufficiently in human myocardium to afford a commercial myocardial perfusion agent, this work spurred on the synthesis of a wide variety of cationic technetium compounds [25].

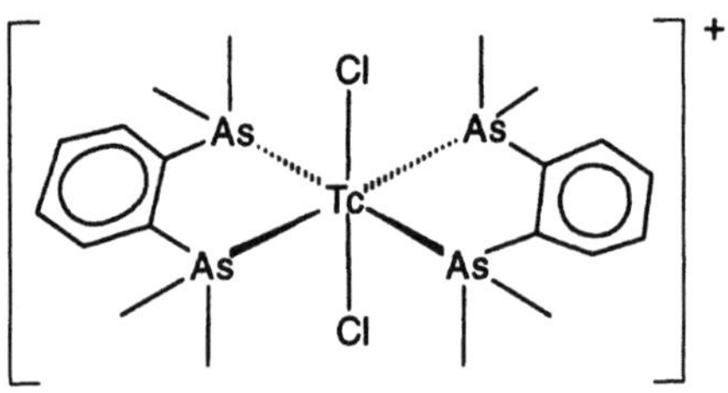

1

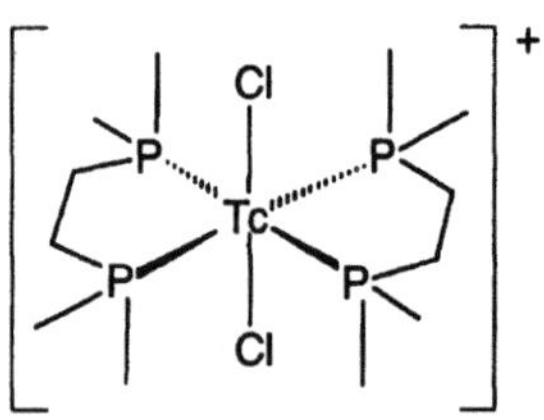

2

A second result of the technetium-phosphine studies was the recognition that oxophilic technetium, which tends to form polymeric oxo-bridged Tc(IV) $[TcO_2]$ species in aqueous conditions at macroscopic concentrations [26,27], would form stable, monomeric, octahedral, low-spin complexes under the dilute carrier-free conditions present in radiopharmaceutical kits [28]. During the characterization of the products obtained with phosphine ligands, it was observed that starting with pertechnetate a sequential two-electron reduction of the technetium core could be stabilized with the appropriate ligands. Reduction of TcO_4^- with $SnCl_2$ or excess phosphine ligand at 25 °C removed two of the tetrahedrally bonded oxygens and reduced the oxidation state of the central technetium from +7 to +5 producing *trans*-$[TcO_2(dmpe)_2]^+$ [29]. Heating this aqueous mixture for about 15 min. at 100 °C in the presence of chloride, which is present in the eluate of the ^{99}Mo/^{99m}Tc generator, produced the Tc(III) cationic species *trans*-$[TcCl_2(dmpe)_2]^+$ [30]. Finally, heating for longer periods or with excess $SnCl_2$ removed the chloro ligands with subsequent reduction to Tc(I) and formation of the *tris*-cationic species $[Tc(dmpe)_3]^+$ (**3**) [31–33]. Using these three basic cores, numerous functionalized phosphine and arsine ligands were tested to study the effects on in vivo distribution in animals [34].

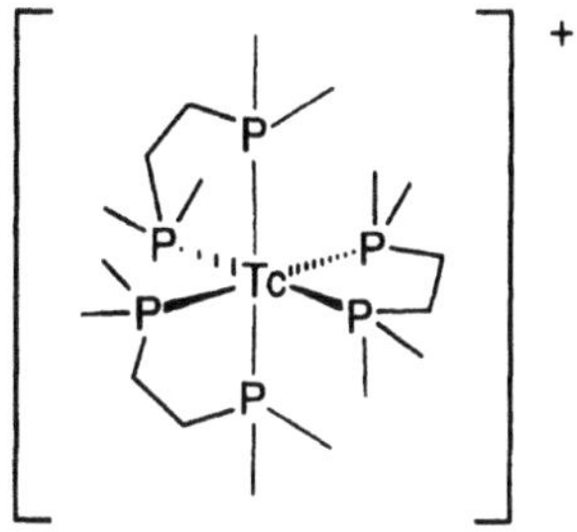

3

Cationic complexes of ^{99m}Tc in the +1 oxidation state were first isolated and tested by Linder et al. [33] and shortly afterwards by others [35]. The low-spin d_6 configuration of the near-perfect octahedral $[Tc(dmpe)_3]^+$ proved to be surprisingly thermodynamically stable. The ease of synthesis and stability in aqueous conditions of these complexes opened a new oxidation state for ^{99m}Tc radiopharmaceutical development. $[Tc(dmpe)_3]^+$ was the most lipophilic of the Tc(V), (III), and (I) dmpe series and, although it exhibited excellent myocardial localization in animals, prolonged blood retention in humans made it unsuitable as a myocardial perfusion agent [35]. For the evaluation of exercised-induced ischemia, rapid blood clearance is an essential requirement of the radiolabeled agent. Several other Tc(I) lipophilic cationic compounds were synthesized and tested including $[Tc(arene)_2]^+$ (**4**) [36], $[Tc(pompom)_3]^+$ (**5**) [37] and $[Tc(tmp)_6]^+$ (**6**) [38], but all exhibited high protein-binding that slowed their clearance from the blood [39].

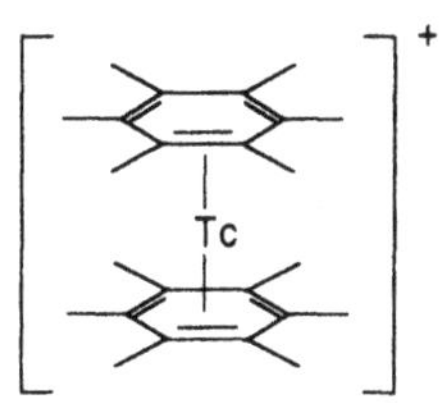

4

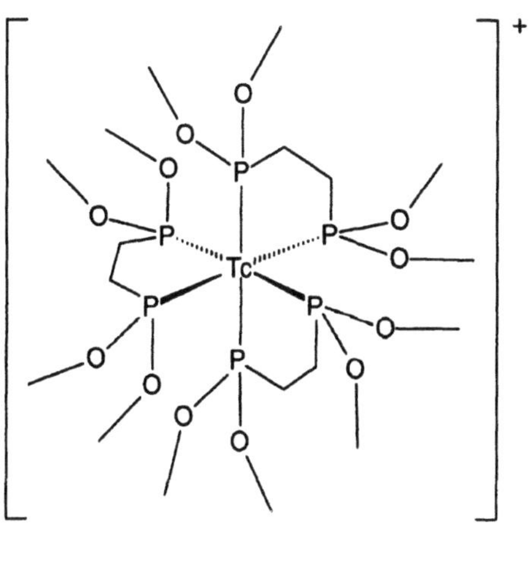

5

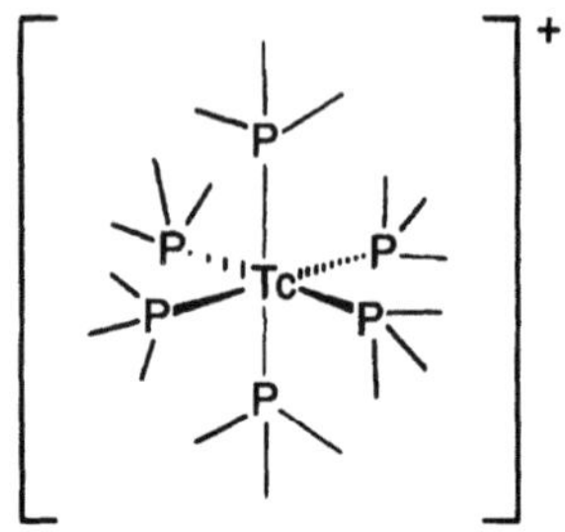

6

Numerous technetium cores in the I, III, and V oxidation states have now been studied as templates from which to develop myocardial perfusion agents. Several representative technetium cores with various types of ligands are shown in 7 [40–42]. The ligands' functional groups have been modified to optimize the lipophilicity and protein-binding characteristics of the stabilized monomeric technetium complexes [43,44].

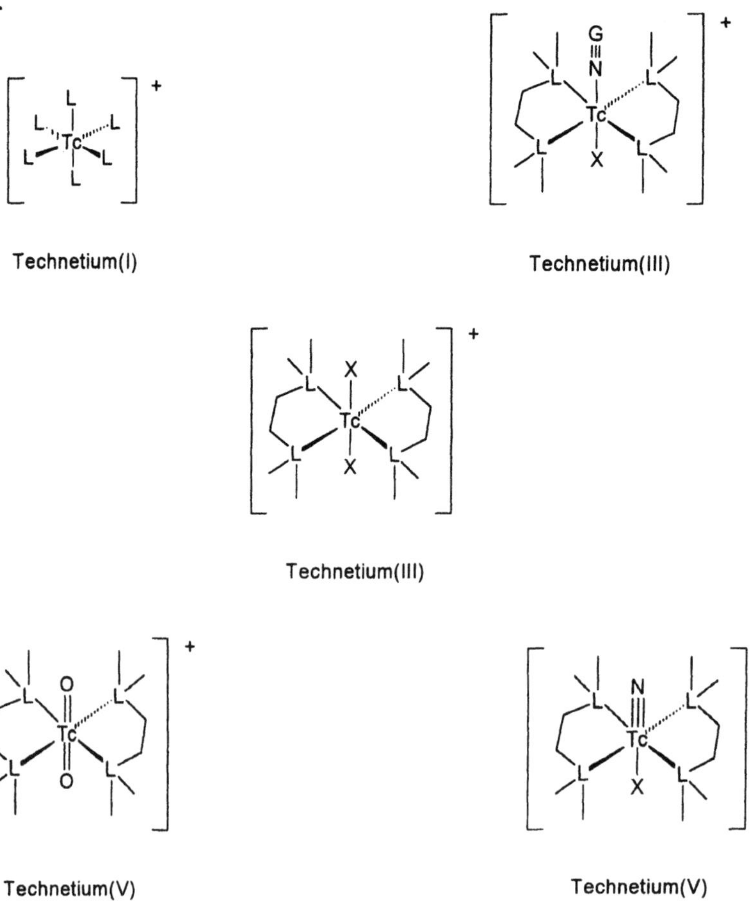

7

2.1.1
Technetium-99m Isonitriles

The first technetium complex to give consistently high extraction and retention in myocardial tissue in animals and humans was the technetium(I) cationic complex $[Tc(tbi)_6]^+$, developed by Davison et al. [45]. The compound was first prepared by Abrams from $[Tc(thiourea)_6]Cl_3$ using excess isocyanide ligand and sodium dithionite to reach the Tc(I) state [45]. The monodentate, highly back-bonding, isonitrile ligand allowed the complex to form an unstrained octahedral geometry with very strong Tc–C bonds as demonstrated in the crystal structure and shown in **8** [46].

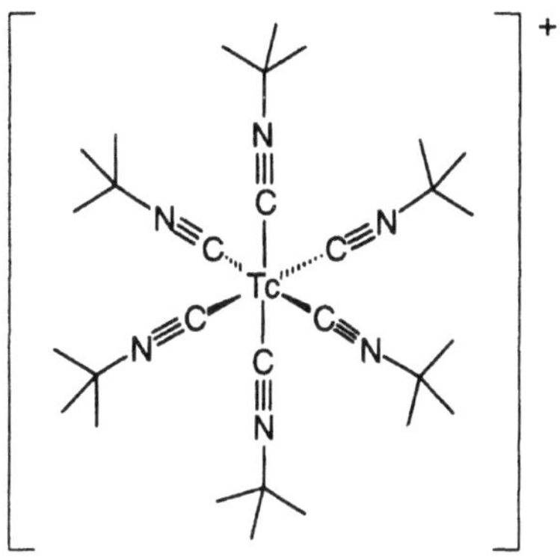

8

This water- and air-stable compound was found to be quite robust and formed readily by simply heating $^{99m}TcO_4^-$ with excess isonitrile and $SnCl_2$ as the reducing agent. Many isonitrile ligands are liquids at room temperature, so it was necessary to develop a ligand-exchange reaction to allow formulation of these agents as radiopharmaceutical kits [47]. The stability and shelf-life of both the ligands and the reducing agents are much greater in the solid state than in aqueous solution. Using a pre-formed Zn(II) or Cu(I) isonitrile complex, the ligand can be added as a sterile solution to the vial, freeze-dried along with the reducing agent, and sealed under an inert atmosphere. This procedure allows for rapid single-step reconstitution of a dry kit with generator eluate to form the desired Tc-isonitrile complex in a sterile environment with no need for further purification.

Although $[^{99m}Tc(tbi)_6]^+$ was a efficient myocardial perfusion agent, the substantial lipophilic character of the six coordinated *tert*-butyl isocyanide ligands resulted in prolonged retention in the lungs following intravenous injection [48]. Consequently, even though extraction of $[^{99m}Tc(tbi)_6]^+$ from the blood stream to the heart and liver was very high, a delay of several hours was necessary for the background activity to clear before high-quality images could be obtained. In addition, the gradual washout of the initial activity in the lungs back into the bloodstream produced a prolonged infusion of $[^{99m}Tc(tbi)_6]^+$ into the coronary arteries. The prolonged blood activity made this agent less than ideal for imaging the transient ischemic flow changes that occur during an exercise or pharmacological stress test. A high percentage of the administered dose also accumulated in the liver (60%) with prolonged retention that interfered with visualization of the apex of the heart even at 4–8 h after injection [48]. Efforts to reduce the activity in the liver led to the development of functionalized isonitriles that could undergo hydrolysis in vivo [49]. The most widely studied of the technetium ester isonitrile compounds for heart imaging was $[Tc(cpi)_6]^+$ (**9**).

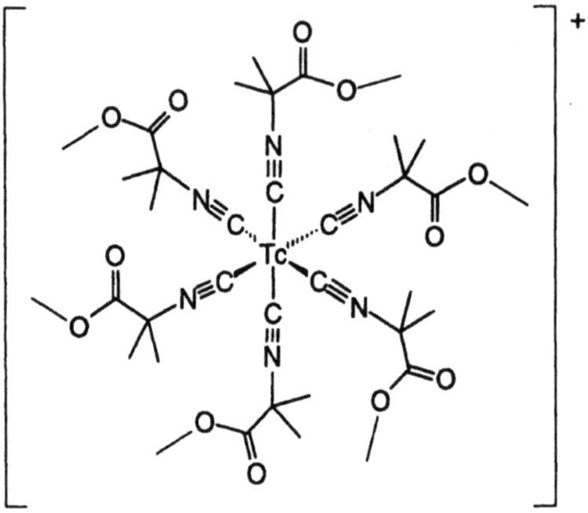

9

This hexakis methyl ester isonitrile was shown to undergo a species-dependent enzymatic hydrolysis of the terminal esters with no alteration of the central Tc(I) octahedral core [50]. Sequential hydrolysis of the neutral ester-isonitrile ligands resulted in decreasing lipophilicity and charge [51]. In humans, enzymatic hydrolysis occurred primarily in the liver, resulting in hepatobiliary clearance of the ^{99m}Tc, with much slower hydrolysis occurring in the blood and heart [52,53]. The mono-hydrolyzed neutral species was isolated and injected into animals but was not retained in heart tissue confirming the necessity of the cationic charge for myocardial retention [54].

Continued efforts to improve the heart to non-target organ ratios focused on adding various terminal functional groups on the isonitrile ligands. Literally hundreds of derivatives were prepared and tested leading to the clinical agent $[Tc(mibi)_6]^+$ (**10**) [55].

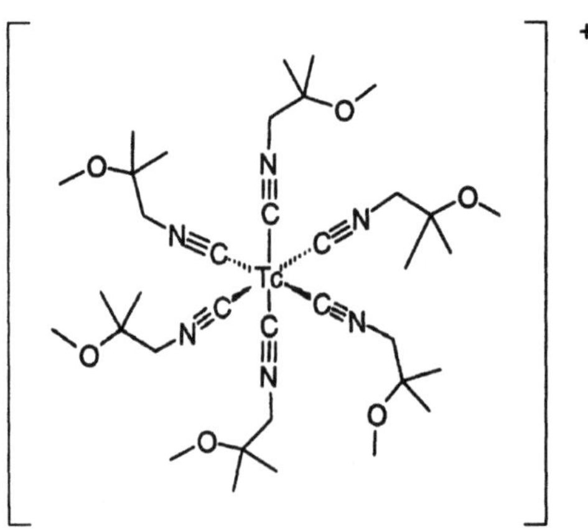

10

The freeze-dried kit formulation to produce this complex was approved by the FDA in 1990 (Cardiolite, Tc-SESTAMIBI, DuPont). The 2-methylether-*n*-propylisocyanide ligand is lipophilic but also possesses a large dipole moment because of the ether substituent. The large dipole moment increases the water solubility, decreases lung and liver accumulation, and increases kidney uptake [56]. As was the case with the ester derivatives, the ether-functionalized complexes were cleared from the liver through the hepatobilary system. Analysis of bile samples from guinea pigs by high performance liquid chromatography (HPLC) showed significant, but not complete, metabolism of the terminal methyl-ether groups to alcohols [57]. Human studies showed about 1.5% of the injected dose accumulated in the heart at rest one hour after injection with a very gradual washout to about 1% by 4 hours [58]. Injection immediately following exercise showed increased activity in the heart and peripheral muscles and reduced accumulation in the liver, consistent with the known changes in regional blood flow [58].

Mechanistic studies in cell cultures, isolated perfused rat hearts, and animals show this compound has flow-dependent diffusion-mediated accumulation in myocardial tissue [59]. The lipophilic nature allows free passage across the sarcolemmal membrane with intracellular retention facilitated by the large negative transmembrane electrochemical gradient [60]. Upon prolonged loading, the volume adjusted ratios of ^{99m}Tc in the mitochondria, compared with the cytosol, approach 50:1, correlating with the well-established more negative mitochondria membrane potential [61]. Technetium-NMR studies of $[^{99}Tc(mibi)_6]^+$ in isolated perfused rat hearts showed this compound to be freely solvated in an aqueous environment and not bound to a membrane or cytosolic protein [62]. These studies and others provide an explanation for the accumulation and retention of the wide variety of lipophilic cationic compounds that have shown uptake in myocardial tissue [63,64].

2.1.2
Technetium-99m-Tetrofosmin

Another lipophilic, cationic, perfusion agent approved for use in the US in 1996 is Tc-Tetrofosmin (Myoview, Nycomed-Amersham) [65]. The compound is a Tc(V) *trans*-dioxo bisdiphosphine complex based on the $[^{99m}TcO_2(dmpe)_2]^+$ core with two terminal ethoxyethyl groups on each of the four phosphines [66]. The crystal structure of the ^{99}Tc complex shows a linear *trans*-oxo core with the four phosphorus atoms from the two ethylene-bridged bidentate diphosphine chelates forming an equatorial planar array as shown in **11**.

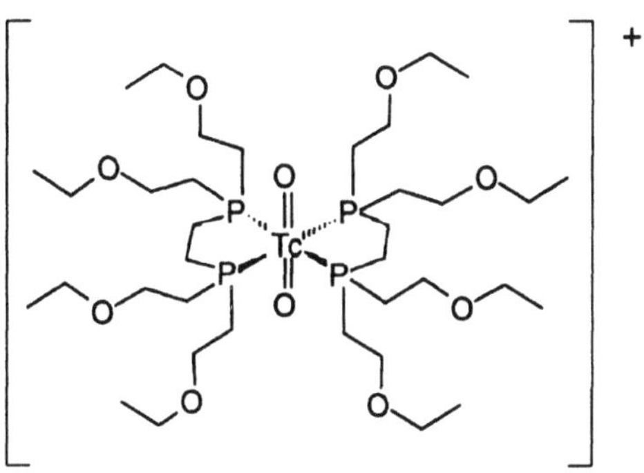

11

The ^{99m}Tc compound is formed in >90% yield from $^{99m}TcO_4^-$ after 15 min incubation at room temperature with the tetrofosmin ligand and $SnCl_2$. The Tc(V) complex is protected from further reduction to Tc(III) or Tc(I) species by not heating the reaction mixture and keeping the stannous and tetrofosmin concentrations low. The biodistribution properties are strongly influenced by the eight ethoxyethyl groups. Again, the combination of lipophilicity and dipole moment facilitates solubility in serum with low protein binding and permeability through the cytosolic membrane of heart cells [67]. Like Tc-SESTAMIBI, this compound exhibits membrane-potential-dependent retention in myocytes with a total heart accumulation at rest in the human of about 1.5% [68]. The compound is rapidly cleared from the blood stream by the liver and kidneys with excretion into the biliary and urinary systems, respectively [69].

2.1.3
Technetium-99m "Q" Complexes

Another potential myocardial perfusion agent is the lipophilic cationic compound Tc-Q12 [70]. This molecule is one of a series of octahedral Tc(III) complexes with a tetradentate Schiff base occupying the equatorial plane and tertiary phosphines in the two axial positions. These compounds are made in two steps by the reduction of $^{99m}TcO_4^-$ with $SnCl_2$ in the presence of a Schiff base ligand, L, to form a $[^{99m}Tc^{V}O(L)]^+$ intermediate which, upon addition of the phosphine, is reduced to Tc(III) with loss of the oxo group to form the desired complex [70]. In the case of Q12, the Schiff base is 1,2-bis[[(dihydro-2,2,5,5-tetramethyl-3(2*H*)-furanonato)methylene]amino]ethane and the axial phosphine ligand is *tris*(3-methoxy-1-propyl)phosphine (**12**).

The mixed-ligand Q compounds were designed to allow more subtle adjustments to the lipophilic nature of the complex while increasing the reduction potential of the technetium atom to avoid forming a neutral Tc(II) complex in vivo [71]. This effort to control the reduction potential was adopted because of concern that cationic

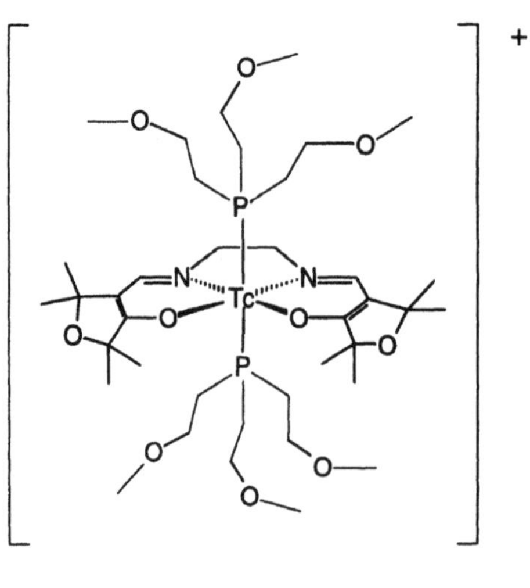

12

Tc(III)-phosphine complexes might be reduced in vivo thus limiting their retention in the heart [72]. Compared with other members of the Q series, Q12 exhibited lower binding to serum proteins and was thus more rapidly cleared from the blood and lungs [73].

2.1.4
PL-37

Another Tc(III) complex, PL-37, was synthesized by Nycomed-Amersham and shown to exhibit localization in human myocardial tissue [74]. This complex is also a bis(diphosphine) with axial chloro and N-bonded NO ligands, **13**.

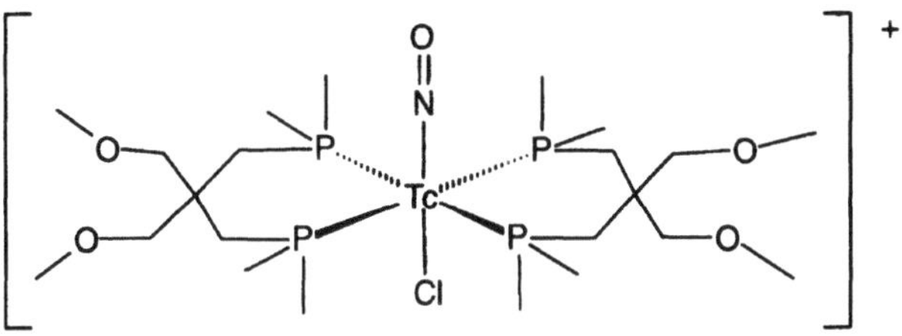

13

Although accumulation of this complex in the heart was significant, in clinical trials it produced target-to-background ratios somewhat lower than those obtained with Tc-SESTAMIBI or Tc-Tetrofosmin and was, therefore, not pursued as a commercial product [75].

Collectively, this series of ^{99m}Tc myocardial perfusion agents demonstrates that the metal center is effectively isolated from the biological environment by the ligands and actually has little influence on the biodistribution of the radiopharmaceutical. None of the lipophilic technetium cationic complexes are perfect perfusion agents because their extraction efficiency is low at the high blood flow rates that can occur during exercise. This limits their ability to accurately measure blood flow under these conditions [76]. This limitation is presumably due to the diffusion-driven nature of the cellular input function.

2.2
Neutral Myocardial Perfusion Agents

2.2.1
BATOs

Technetium-99m-teboroxime (Cardiotec, Bristol-Meyers Squibb) exemplifies another approach to imaging myocardial perfusion. This compound is a neutral, lipophilic, seven-coordinate Tc(III) complex [77]. The heart-imaging agent, $TcCl(CDO)(CDOH_2)_2BMe$ ($CDOH_2$ = cyclohexanedione dioxime) (**14**) is derived from a broad class of complexes called BATOs (boronic acid adducts of technetium dioximes) that are prepared by a template synthesis where TcO_4^{-1} is reacted with a dioxime, methyl boronic acid, and $SnCl_2$ under acidic conditions to give the neutral compound in 90% yield [78].

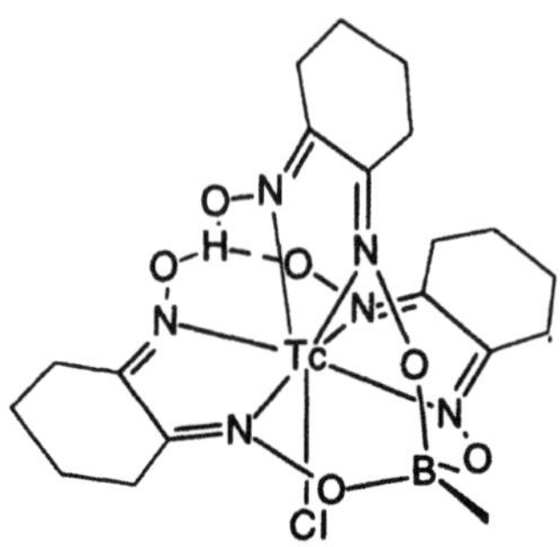

14

The BATO complexes are unique in that the three dioxime ligands are bound through six nitrogens to the Tc(III) center in a trigonal configuration with a derivatized boronic acid capping three of the nitrogens on one end. A chloride atom occupies a seventh coordination site between two of the dioximes increasing the bond angle between these dioximes to greater than the expected 120° for a symmetrical trigonal arrangement of three dioximes. This distortion makes it energetically unfavorable for a second boronic acid to cap the oximes at the other end of the molecule. The three uncapped oxime oxygens share two protons between them resulting in an overall neutral compound.

Cell culture and animal studies show this agent to have a very high first-pass extraction into myocardial tissue (90%) and more rapid washout than ^{99m}Tc-MIBI [79]. The high lipophilicity also produced very rapid blood clearance [80] with the majority of the dose accumulating in the liver [81]. In clinical studies, even though this compound accurately measures myocardial blood flow under conditions of high flow [82], the time window available for imaging is very narrow because of the rapid washout from the heart [83]. The combination of high liver uptake and potential for patient movement during the narrow imaging window limited widespread use of this agent, and it is no longer commercially available.

2.2.2
Technetium-99m-NOET

In 1994, Pasqualini et al. identified another neutral technetium complex (Tc-NOET) from a series of Tc(V) nitrido bisthiocarbamate compounds as a potential myocardial perfusion agent, **15** [84].

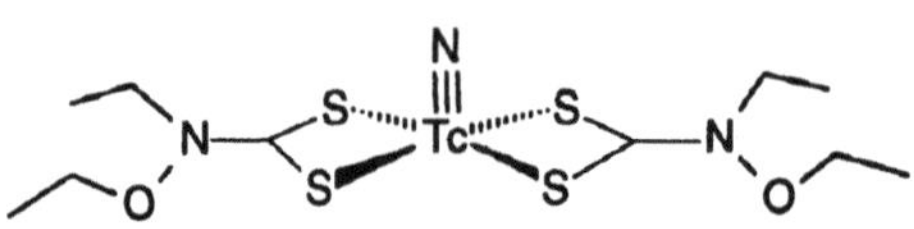

Tc-NOET **15**

Subcellular distribution studies have indicated this compound is localized predominantly in the cell membrane fraction of myocardial tissue with no evidence of chemical alteration of the ^{99m}Tc complex [85]. These results suggest that Tc-NOET heart uptake relates to coronary blood flow but, in contrast to the cationic complexes, it redistributes within the heart after its initial uptake [86–88]. Human studies confirm this [89], but the high lipophilicity also results in retention of about 25% of the injected dose in the lungs at 30 min post-injection, high uptake in the liver, and slow clearance from the blood because of binding to red blood cells [89].

2.3 Metal-Based PET Myocardial Perfusion Agents

2.3.1 *Rubidium-82 (Cardiogen)*

An agent similar in biological properties to K^+ is the positron-emitting radionuclide $^{82}Rb^+$. This isotope ($T_{1/2}$ = 75 s) is commercially available as a generator system from the long-lived (25 d) parent ^{82}Sr [90]. The ^{82}Sr is loaded as $SrCl_2$ on a chromatographic column of stannic oxide from which $^{82}Rb^+$ can be eluted every few minutes with isotonic saline [91]. The generator was approved by the FDA in 1989 (Cardiogen-82, Bracco Diagnostics). Due to the short half-life of the ^{82}Rb and to avoid any delays in handling the eluate, the generator is sold as a self-monitoring infusion system that pumps the generator eluate directly into the patient's intravenous line. Excellent PET images of myocardial perfusion can be obtained with ^{82}Rb, and sequential images can be obtained every few minutes (because of the short half-life). The short half-life of ^{82}Rb is, however, not compatible with exercise stress tests, and pharmacological stress testing is required for the detection of ischemia [92].

2.3.2 *Gallium-68 Compounds*

Gallium-68 ($T_{1/2}$ = 68 min.) is an attractive radionuclide for PET myocardial perfusion imaging because the half-life is close to optimal and, as discussed previously, the ^{68}Ge ($T_{1/2}$ = 271 d)/^{68}Ga generator [93] is commercially available (DuPont-Merck), but there are currently no FDA-approved ^{68}Ga myocardial perfusion agents. The primary problem is the chemistry of gallium itself. Gallium(III) is labile and free Ga(III) is tightly bound by transferrin in vivo (log K = 20.3) [94]. The development of Ga-based myocardial perfusion agents that resist transchelation in vivo has, therefore, proven a formidable synthetic challenge.

Despite this challenge, several neutral and cationic lipophilic gallium compounds have been prepared and tested in animals, and two conclusions can be drawn. First, as with the ^{99m}Tc compounds, neutral compounds that show acceptable uptake by the heart but are poorly retained. Cationic compounds, such as $Ga[(4,6\text{-}MeO_2sal)_3BAPEN]^+$ (**16**) are retained in the myocardium longer than the neutral compounds [95]. The cationic complex $[Ga(BAT\text{-}TECH)]^+$ (**17**) is an apparent exception, but the charge was measured in the pH range 0.9–2.3 [96]. It is probable that, at physiologic pH, this compound is neutral due to deprotonation of one of the amine nitrogens, which explains the apparently anomalous rapid washout of this cationic compound from the heart. Second, in contrast to the situation with lipophilic cationic ^{99m}Tc myocardial perfusion agents, the problem of high liver uptake with lipophilic cationic Ga complexes has yet to be resolved.

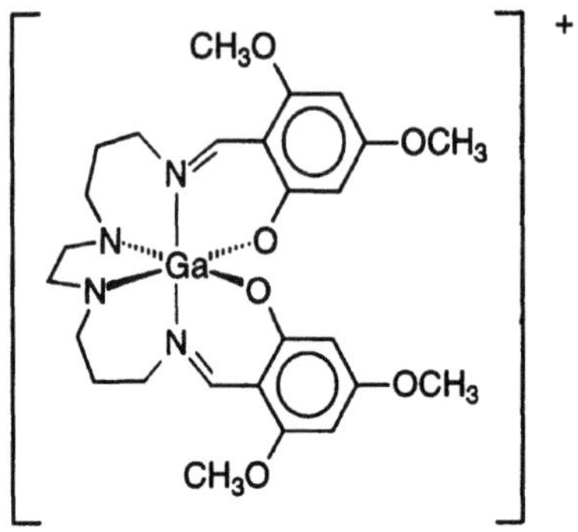

[Ga(4,6-MeO$_2$sal)$_2$BAPEN]$^+$ **16**

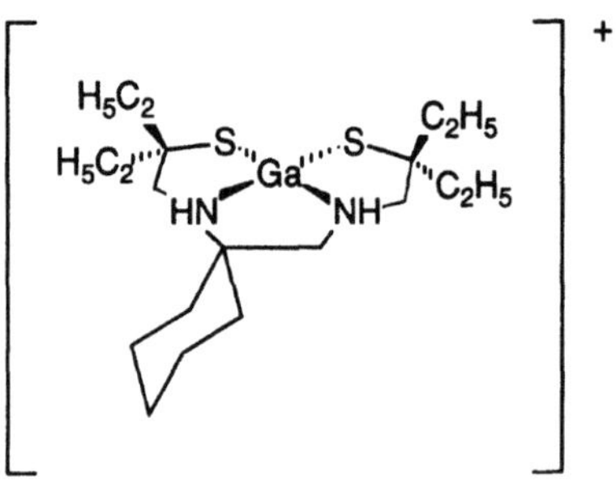

[Ga(BAT-TECH)]$^+$ **17**

2.3.3
Copper-62/64 Compounds

The ^{62}Zn ($T_{1/2}$ = 9.26 h)/^{62}Cu ($T_{1/2}$ = 9.7 min) generator [97,98] is also a promising alternative from which to develop a PET myocardial perfusion agent. The half-life of ^{62}Cu allows the diverse coordination chemistry of copper to be exploited and is compatible with a wider range of clinical studies than ^{82}Rb. This generator is now commercially available (Proportional Technologies, Houston, TX). The potential of copper-based PET radiopharmaceuticals is complemented by recent reports by Bass et al. that several other positron-emitting copper radionuclides can be efficiently produced with a medical cyclotron [6].

The focus in the development of ^{62}Cu myocardial perfusion agents has been on Cu-PTSM (**18**) [99,100]. Copper-PTSM is a neutral, lipophilic (log P = 1.97) molecule that clears relatively quickly from the blood [99,100].

The uptake by the myocardium of [^{67}Cu]Cu-PTSM in rats at 15 min post injection is 3.8 + 0.4% of the injected dose and is relatively constant at time points between 1 min and 2 h post-injection [100]. There is relatively high uptake in the liver (1.5%/g) and lungs (6.0%/g) that interferes to some degree with cardiac imaging [101].

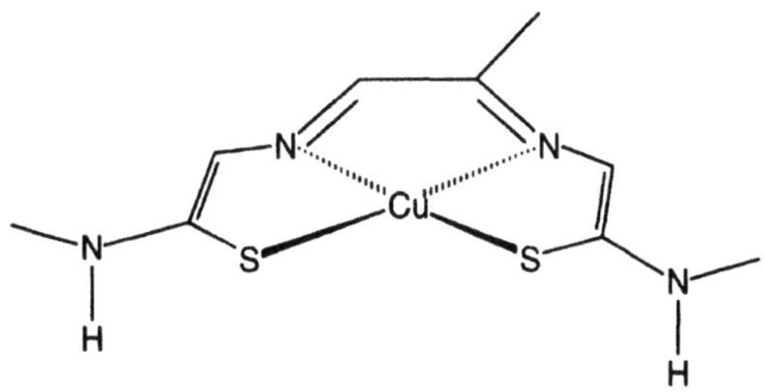

Cu-PTSM **18**

Copper-62 is retained in the heart (and brain) because the Cu(II) in the complex is reduced to Cu(I) with concomitant decomposition of the complex [102–104]. The compound also decomposes quickly in the blood so that, by one minute post-injection, none of the radioactivity in the blood is associated with [^{64}Cu]Cu-PTSM [101,105]. For quantitative PET studies, this means that measurement of blood-pool activity overestimates the true arterial input function unless the samples are analyzed to determine the fraction of the ^{62}Cu activity that is associated with ^{62}Cu-PTSM [101, 105,106]. This complicates the procedure, and a tracer that is more stable in vivo would be preferable. The extraction fraction of [^{67}Cu]Cu-PTSM by the heart is 45 ± 7% and invariant with ischemia, hyperemia, and hypoxia [107]. Green and coworkers prepared two analogs of PTSM in which the -$NHCH_3$ moieties were replaced with -NH_2 (PTS) and -$N(CH_3)_2$ ($PTSM_2$) and compared the biodistribution in rats [100]. At 15 min post-injection, the highest myocardial uptake was observed for Cu-PTSM (3.8 ± 0.4%/g) and Cu-PTS (4.3 ± 0.8%/g), although the octanol/water partition coefficient of Cu-PTS is lower than that of Cu-PTSM (0.75 and 1.95, respectively). The myocardial uptake of Cu-$PTSM_2$ is lower (0.45 ± 0.05%/g) than that of either Cu-PTSM or Cu-PTS because it is rapidly washed out of the heart. Cu-$PTSM_2$ is not reduced in vivo while both Cu-PTSM and Cu-PTS are [100]. There is, therefore, no mechanism for retention of Cu-$PTSM_2$, and it is quickly washed out of the heart.

3
Brain

Radiopharmaceuticals used in the evaluation of the brain can be considered in two separate categories, regional-cerebral-blood-flow (rCBF) agents and receptor-specific agents. Currently, the only FDA-approved ^{99m}Tc radiopharmaceuticals are rCBF agents, but the development of ^{99m}Tc-based receptor-specific agents is currently the focus of a substantial effort. The objective in the development of an rCBF agent is to have the uptake of the tracer linearly related to blood flow to the tissue at the time of tracer administration. Cerebral perfusion agents exploit the coupling of flow with metabolism allowing the evaluation of disease states that are not inherently perfusion abnormalities [108]. The most obvious application of rCBF imaging is in stroke where the structural changes only become apparent on anatomic imaging modalities (MRI and CT) after the opportunity to intervene has passed. Other diseases in which rCBF imaging is useful include dementias, epilepsy and psychiatric disorders where there are no apparent anatomic abnormalities [109,110].

The criteria for the development of an rCBF agent are seemingly quite straightforward and include low molecular weight (<500–600 Da), neutral charge, and a log *P* (octanol/water partition coefficient) between 0.5 and 2.5 [111]. These criteria address the primary problem in developing a successful rCBF agent – getting it into the brain. The brain is protected from foreign substances by unique capillary bed, the blood-brain barrier (BBB), that is different from the capillary structure in the rest of the body [112]. An rCBF agent must be able to pass through the BBB by passive diffusion. This constraint is addressed by the three criteria listed above. The lower limit on log *P* ensures that the compound is lipophilic enough to penetrate the BBB. The upper limit on log *P* arises because very lipophilic compounds bind to proteins in the blood, which inhibits diffusion into the brain. These three criteria by themselves are, however, not sufficient to ensure that the agent will be clinically useful. To be clinically useful, the compound must also: (1) be stable enough in the blood to be delivered to the brain, (2) accumulate in the brain to a sufficient degree that an image can be obtained, (3) be retained in the brain long enough for the image to be acquired, (4) not redistribute within the brain, and (5) clear quickly from the tissue surrounding the brain. Despite these rather severe constraints, several rCBF radiopharmaceuticals have been developed.

3.1 Perfusion Agents

3.1.1 *Technetium-99m-HMPAO*

The first commercially available ^{99m}Tc-based rCBF agent was based on the complex [Tc(O)(PnAO)] (**19**) [113,114].

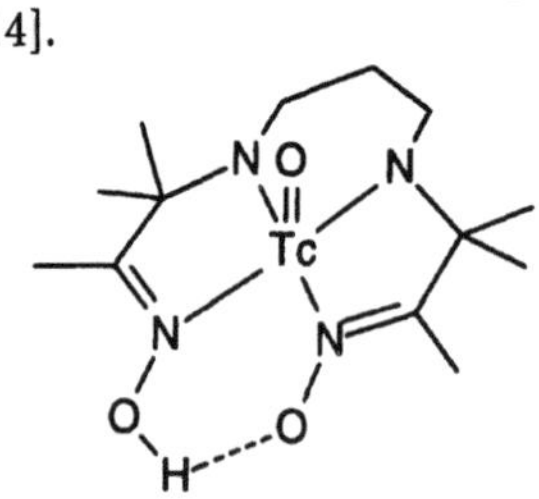

[Tc(O)(PnAO)] **19**

The molecule contains a Tc(V) = O core and a diaminedioxime ligand. When coordinated to Tc, the ligand loses a proton from one of the oxime substituents and a proton from each of the amines (three total) to form a neutral product. The prototype ^{99m}Tc-PnAO complex diffused into the brain but was not retained within the brain for a long enough period of time to allow SPECT imaging [115]. The biological properties of the ^{99m}Tc complexes of a series of derivatives of the original PnAO ligand were evaluated, and HMPAO (**20**) was chosen for further development [116,117].

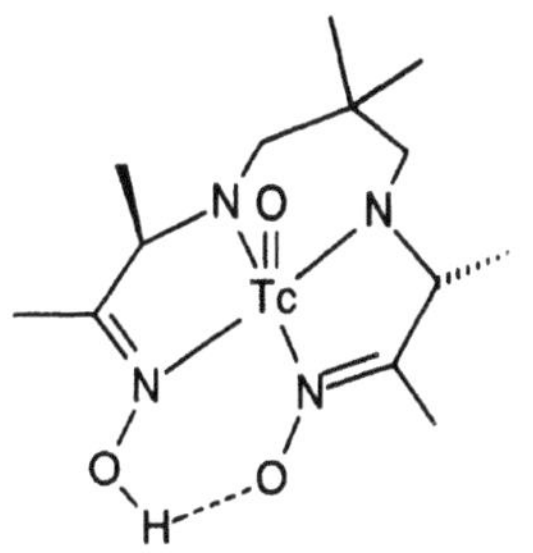

[Tc(O)(HMPAO)] **20**

It is important to note that both PnAO and HMPAO contain six methyl groups; only the arrangement around the technetium core is different. This difference, however, significantly changes the biological properties of the complex. The most important change is that the HMPAO complex is retained in the brain while the PnAO complex is not. Another consequence of the rearrangement of the methyl substituents is that the HMPAO complex exists in three stereoisomeric forms (**21**) while the PnAO ligand does not.

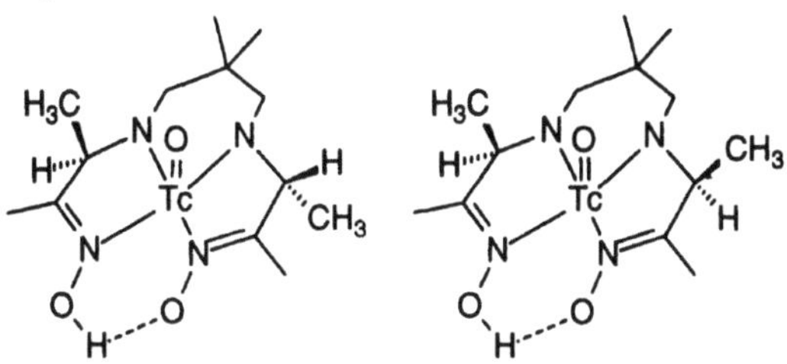

d,l-[Tc(O)(HMPAO)] *meso*-[Tc(O)(HMPAO)] **21**

Despite the fact that uptake of the complexes by the brain is a function of passive diffusion, the isomers exhibit significantly different biological properties. The *d,l* form of the complex is less stable than the *meso* form, and the (hydrophilic) decomposition product of this isomer is trapped in the brain. The *meso* form is more stable and diffuses out of the brain before it decomposes [118,119]. There is, however, no difference in the extraction of the isomers by the brain [117]. The retention of the *d,l* isomer within the brain is thought to be due to a glutathione (GSH) mediated conversion to a hydrophilic species that cannot back diffuse through the BBB [120,121]. In vitro, the reaction rate of the *meso* isomer with GSH is much slower than that of the *d,l* isomer, suggesting that it can diffuse back into the circulation before it is converted into a hydrophilic species that would be retained in the brain [120,121].

An important clinical limitation of the original ^{99m}Tc-HMPAO formulation was the instability of the complex. In contrast to most ^{99m}Tc radiopharmaceuticals which are quite stable after they are prepared, ^{99m}Tc-HMPAO could be used for only 30 min after preparation because after this time the radiochemical purity decreased to less than 85%. This was a serious limitation not only because it limited the time during which the radiopharmaceutical could be used, but also because the decomposition products accumulated in the soft tissue in the head, particularly near the base of the brain (S.T. Treves, personal communication). This decreased the image quality compared to images obtained with a freshly prepared kit. This decomposition was not observed at higher concentrations with ^{99}Tc-HMPAO, and is at least partially the result of radiolysis in the high specific-activity clinical formulations [122]. The stability of the ^{99m}Tc complex was improved by the addition of materials known to inhibit radiolytic decomposition (e.g. gentisic acid [123]), and an improved version of the kit was later introduced that included methylene blue as a free-radical scavenger [124]. The new kit can be used for up to 4 h, but by this time the radiochemical purity has decreased to 85%. This is still a shorter shelf life and lower radiochemical purity than is observed for ^{99m}Tc-bicisate (vide infra) and other ^{99m}Tc kits.

An interesting sidelight to the development of ^{99m}Tc-PnAO/HMPAO is that this is one of the few cases where the biodistribution data for the free ligand are also available [125]. These investigators demonstrated that there is no parallel between the biodistribution of PnAO and its ^{99m}Tc complex. In the absence of technetium, the free PnAO ligand does not accumulate in the brain. This demonstrates that ^{99m}Tc-PnAO is a metal-essential radiopharmaceutical. The ^{99m}Tc does not act as a label or tag for a compound that, by itself, accumulates in the target tissue but plays an essential role in determining the biological properties of the compound.

3.1.2
Technetium-99m-ECD

The Dupont-Merck radiopharmaceutical Neurolite (^{99m}Tc-ECD, ^{99m}Tc-bicisate) (22) is based on earlier efforts by Lever [126,127] and Kung [128,129] to prepare neutral Tc(V)-oxo complexes with N_2S_2 chelating ligands.

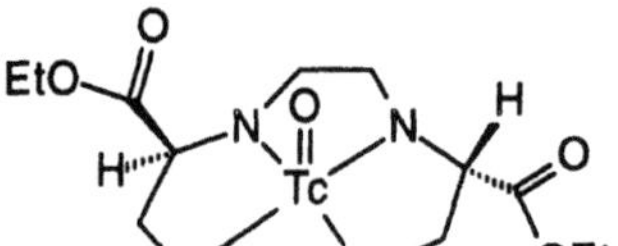

22

The prototype N_2S_2 complexes met the three criteria listed above: neutral charge, acceptable lipophilicity, and molecular weight <500 Da, but neither is useful for rCBF imaging because they are not retained within the brain. This limitation was circumvented with the development of ^{99m}Tc-bicisate [130]. The clinically important difference between this ligand and its predecessors is the addition of two ester moieties, which provide a mechanism for retention of the complex in the brain.

Technetium-99m-bicisate freely diffuses into the brain, but is trapped within the brain when the ester moieties are hydrolyzed by intracellular esterases, producing a charged complex which cannot pass out through the BBB [131]. This enzymatic reaction is stereoselective. In monkeys, the *l,l* isomer is retained in the brain, and the *d,d* isomer is not, despite the fact that both isomers diffuse into the brain equally well [131]. This observation is consistent with in vitro experiments that show the rate of hydrolysis of the *l,l* isomer by monkey-brain homogenate is much greater than that of the *d,d* isomer [131,132]. The *l,l* isomer is also hydrolyzed by esterase enzymes in the blood. The rate is slower than in the brain but fast enough to ensure rapid clearance of the radiopharmaceutical from the blood, primarily through the kidneys [133]. This explains the faster blood clearance of ^{99m}Tc-bicisate compared to ^{99m}Tc-HMPAO. At 20 min post-injection, 7–8% of the ^{99m}Tc-bicisate remains in the blood [110] compared to 17% of ^{99m}Tc-HMPAO [119]. Investigators have also demonstrated that the $^{99m}Tc\text{-}N_2S_2$ core remains intact after the hydrolysis of the ester(s) and excretion in the urine [132].

Technetium-99m-bicisate accumulates in the brain to approximately the same concentration as ^{99m}Tc-HMPAO; 4.6% at 5 min post-injection [110] versus 4.2% for ^{99m}Tc-HMPAO at 20 min post-injection [119]. Technetium-99m-bicisate washes out of the brain somewhat more quickly than ^{99m}Tc-HMPAO ($T_{1/2}$ = 17 versus 71 h), but this difference is not clinically significant compared to the half-life of ^{99m}Tc.

One important clinical advantage of ^{99m}Tc-bicisate compared to ^{99m}Tc-HMPAO is the greater stability of ^{99m}Tc-bicisate after formulation. In addition to the inconvenience, there are two clinical consequences of the lower stability of ^{99m}Tc-HMPAO. First, even with the more stable formulation, decomposition products still form and accumulate in the soft tissue surrounding the brain as described above. Second, the shorter useful life limits the utility of ^{99m}Tc-HMPAO in applications, such as identification of epileptogenic foci, where the dose must be prepared ahead of time in anticipation of injection when the patient experiences a seizure. Identification of the seizure focus can be assisted and/or confirmed by rCBF studies obtained when the tracer is injected as soon as possible after onset of the seizure, thereby providing a "snapshot" of rCBF during the seizure (ictally). In several studies, this technique has identified a probable seizure focus in >90% of the patients compared to 50% for studies obtained between seizures (interictal) and 75% for studies with [^{18}F]FDG [134–138].

3.1.3
Other ^{99m}Tc Cerebral Perfusion Agents

Chemically, one of the more interesting rCBF agents is BATO-2MP, developed by investigators at Squibb [139]. This complex, like the BATO-based myocardial perfusion agent (above), is another example of a template reaction in an "instant kit", in this case containing dimethylglyoxime, 2-methylpropyl boronic acid, stannous chloride, and several stabilizing agents. The product of this reaction is the neutral seven-coordinate complex shown in **23**.

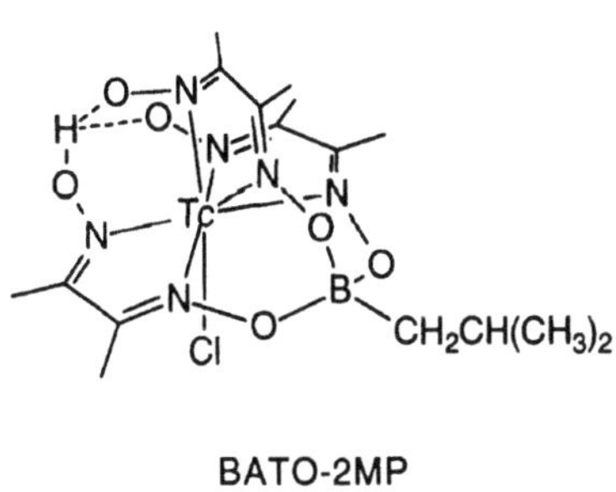

23

Despite the elegance of the chemistry, the complex clears from the brain more quickly than either ^{99m}Tc-ECD or ^{99m}Tc-HMPAO complicating image acquisition, a problem similar to that encountered with the BATO-derived myocardial perfusion agents (see Sect. 2).

The ligand MRP-20 (MR-20 = [*N*-(2-(1*H* pyrolylmethyl)-*N'*-(4-penten-3-one-2) ethane-1,2-diamine)]) was used by Morgan et al. to prepare a neutral ^{99m}Tc complex for cerebral perfusion imaging, **24** [140,141]. This compound accumulates in the brain to a somewhat lesser degree than either ^{99m}Tc-HMPAO or ^{99m}Tc-bicisate and clears from the blood more slowly than ^{99m}Tc-HMPAO.

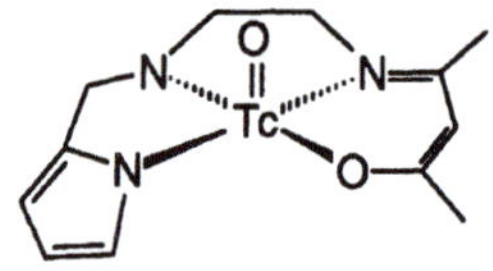

24

3.1.4
Gallium-68 Cerebral Perfusion Agents

The development of gallium-based rCBF agents is, as with the myocardial perfusion agents, motivated by the availability of the ^{68}Ge/^{68}Ga generator. The development of ^{68}Ga rCBF agents has also been hindered by the same factors that have limited the development of the myocardial perfusion agents. Martell and co-workers [142,143] and others [144] have developed ligand systems that protect gallium from in vivo transchelation but, to date, these compounds do not show sufficient uptake by the brain for use in imaging. A possible problem with these complexes is the inclusion of phenol moieties. While these groups increase the stability of the gallium complexes, they also increase the accumulation in the liver and may inhibit passage of the complex through the BBB. Phenol complexes also tend to be unstable in vivo because they are easily protonated at physiological pH.

3.1.5
Copper-62/64 Cerebral Perfusion Agents

Interest in copper radiopharmaceuticals for rCBF imaging is also, as discussed above (Sect. 2), based in part on the existence of a potentially useful generator, in this case ^{62}Zn/^{62}Cu. The copper radiopharmaceutical that has attracted the greatest interest to date is Cu-PTSM (**18**) discussed previously as a myocardial perfusion agent [145,146]. As with the ^{99m}Tc rCBF agents, Cu-PTSM is neutral allowing it to freely diffuse through the BBB. The complex is presumably retained in the brain by the same process implicated in its retention in the heart [102,147,148].

3.2
Receptor-Specific Agents

Cerebral perfusion agents are extremely useful in the diagnosis of many neurological disorders, but their utility is limited to disorders that are characterized by changes in regional cerebral perfusion. For neurological disorders not characterized by changes in perfusion or for the differential diagnosis of disorders characterized by similar changes in perfusion, receptor-specific radiopharmaceuticals are required. For PET, where a ^{11}C or ^{13}N atom can be isotopically substituted in a known ligand, or with radiohalogenated compounds, where ^{123}I or ^{18}F can be added to a known ligand, this is frequently a challenging synthetic problem. For metalloradiopharmaceuticals, where not only the metal but a chelating agent must be added, this is an even more formidable challenge because addition of the chelating agent and the metal change the size, shape, charge and lipophilicity of the molecule and thus its ability to bind to the receptor. It is not, therefore, surprising that the development of receptor-specific metalloradiopharmaceuticals lags behind the development of PET and ^{123}I-labeled agents.

3.2.1
Dopamine Transporter

Depletion of the dopamine transporter (DAT) is implicated in several neurological diseases including Parkinson's disease [149,150], and it is also the target of cocaine, **25** [151,152].

There is, therefore, considerable interest in the development of a radiopharmaceutical that can be used to assess DAT density. The ability to image this receptor has been demonstrated with several PET and ^{123}I-based SPECT tracers but, as discussed previously, neither PET nor ^{123}I-based agents are convenient for routine clinical use. Consequently, several research groups have been developing ^{99m}Tc-based DAT radiopharmaceuticals.

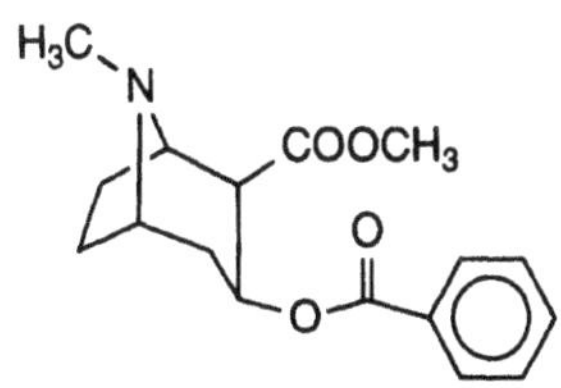

Cocaine **25**

Kung and co-workers prepared a "3 + 1" ^{99m}Tc complex for imaging the dopamine transporter [153,154]. This agent was developed by adding an *N*-ethanethiol derivative to the amine bridge of tropane. This molecule was then reacted with [^{99m}Tc]pertechnetate, tin(II) glucoheptonate, and a tridentate NS_2 co-ligand at room temperature. The yield of the desired ^{99m}Tc-NS_2-tropane complex was 40% and, after HPLC, the radiochemical purity was >95%. The rhenium complex was also prepared and was characterized by X-ray crystallography. A series of derivatives was prepared containing different alkyl and aryl substituents on the NS_2 ligand, two-carbon and three-carbon linkers (between the amine and the thiol of the tropane), and either chlorine or fluorine on the benzene ring of the tropane molecule with the molecule shown in **26** demonstrating the most favorable properties [153].

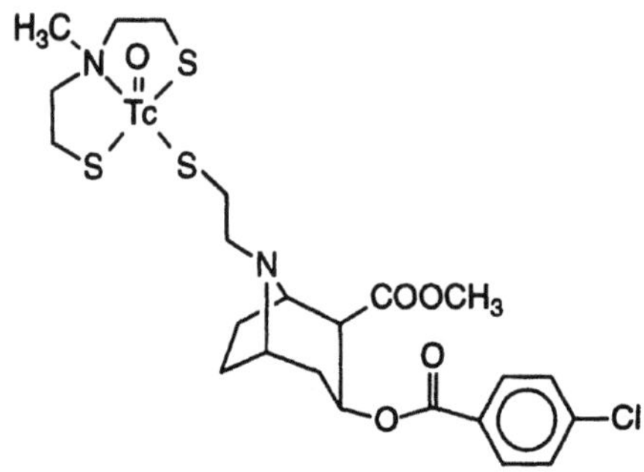

Tc-NS_2-Tropane **26**

The complex was evaluated in rats where the maximum I.D./g in the brain was found to be only 0.1% at 2 min post-injection from which it decreased to 0.05% at 1 h. The regional distribution of the complexes in the brain was also measured to determine if there was specific uptake in the striatum (ST), a region rich in dopamine transporters, versus the cerebellum (CB), which does not contain dopamine transporters. The maximum ST/CB ratio, observed at 60 min post-injection, was 3.5. Pretreatment of the rats with β-CIT, a DAT ligand, reduced the ST/CB ration to 1.0, but pretreatment with haldol, which is not a DAT ligand, did not affect the S/CB ratio. In vitro studies with the Re complex showed a binding affinity, K_i, of 0.3 nM. Unfortunately, despite the fact that the complex shows good selectivity for dopamine-rich sites in the brain and high affinity for the dopamine transporter, the overall uptake of the complex in the brain is too low for it to be useful in human studies.

These investigators have more recently adopted a second approach in which the chelating agent is attached to the tropane molecule at the 2β site [155]. In this case a quadridentate N_2S_2 chelating group was used instead of the "3 + 1" approach. Preparation of the ^{99m}Tc complexes requires heating at 100 °C or at 125 °C in an autoclave for 30 min. The yield of the reaction is 80%, and the product was purified by solvent extraction and HPLC. The molecule exists as a series of stereoisomers with two enantiomers formed by the quaternary amine nitrogen of the N_2S_2 ligand and diastereomers (*syn* and *anti*) formed by the orientation of the tropane linker relative to the oxo ligand of the technetium. Thus, it is reasonable to expect that the biological properties of these isomers are different. The isostructural Re complexes were prepared for use in the indirect characterization of the ^{99m}Tc complexes and in vitro binding studies and were characterized by mass spectrometry and infrared spectroscopy.

The biodistribution of the isomers was compared in rats, and the overall uptake in the brain was found to be significantly higher than was observed for the "3 + 1" complexes. The ^{99m}Tc complex with the highest brain uptake (0.43% I.D./g at 2 min post-injection) (TRODAT-1, 27) was chosen for further investigation.

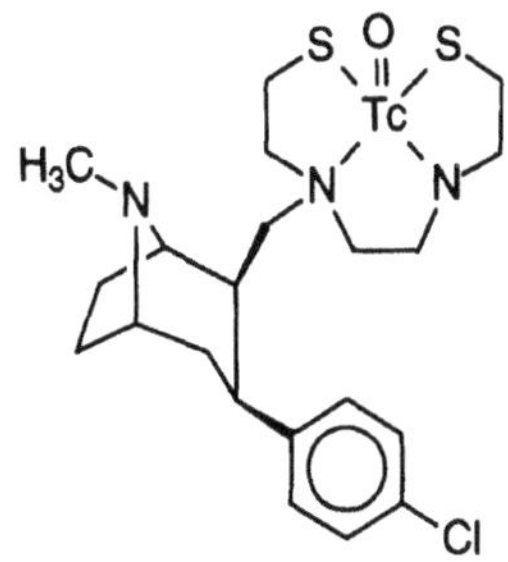

Tc-TRODAT-1 **27**

The maximum ST/CB ratio for TRODAT-1 is achieved at 4 h post injection, and, as with the "3 + 1" complex, is blocked by β-CIT but not by haldol. The K_i value for the Re complex is 14.1 nM, significantly higher than that of the "3 + 1" complex (0.3 nM) or the high-affinity DAT ligand IPT (1.29 nM) [156]. TRODAT-1 was subsequently evaluated in non-human primates where it was displaced from the striatum by CFT and in humans where it also selectively accumulated in DAT-rich regions of the brain [157].

Madras and co-workers used a trimethylene linker to attach an N_2S_2 (mono-amine mono-amide dithiol) chelating group to the amine of the tropane molecule rather than at the 2β site and prepared a group of complexes known as technepines, **28** [158,159].

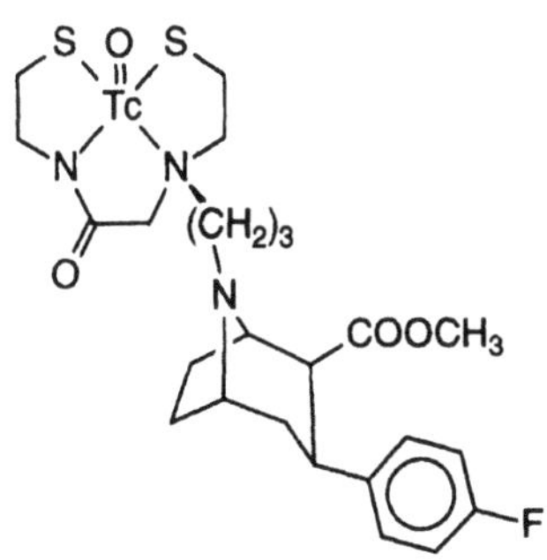

Technepine **28**

The rationale for choosing this attachment site for the chelating moiety was based on previous work indicating that binding of tropanes to the DAT were relatively insensitive to changes in this region of the molecule [160] and concern about potential lability of a ligand placed near the ester site used by Kung et al. (vide supra). The specific binding was measured using the "cold" rhenium analog which, like the ^{99m}Tc complexes, exists as a mixture of diastereomers. The diastereomers have similar DAT potency (IC_{50} = 7.38 nM, 4.04 nM) but very different serotonin (SERT) potency (66.9 and 299 nM). The DAT/SERT selectivity for the two molecules is, therefore, 9 and 74. The mixture, which could not be separated in the clinical environment, shows a DAT potency of 5.99 nM and a DAT/SERT selectivity of 21 which may be sufficient for a SPECT DAT radiopharmaceutical.

3.2.2
Muscarinic Acetylcholine Receptor

Lever and co-workers applied the DADT ligand to the problem of developing a ^{99m}Tc-based muscarinic acetylcholine receptor (mAChR) radiopharmaceutical [161]. This approach was based on the premise that the overall size of quinuclidinyl benzylate (QNB), **29**, and the technetium analog was similar if the Tc-DADT moiety replaced the benzylhydrol portion of QNB, **30**.

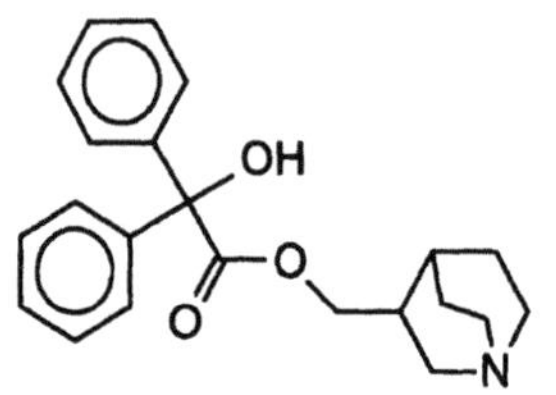

29

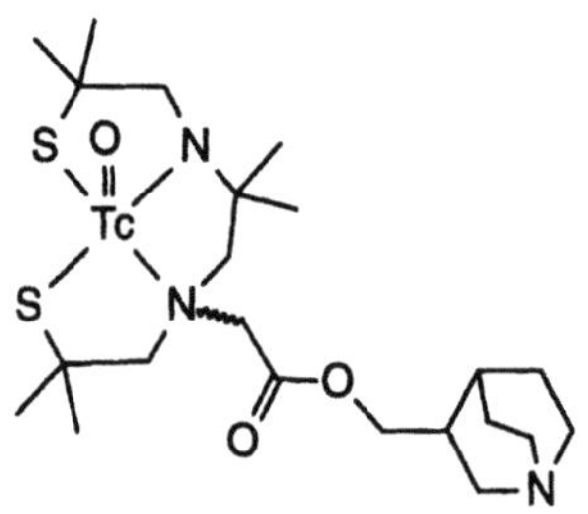

30

Examination of the space-filling models of the structure shown in this report suggests, however, that the overall shapes of QNB and the Tc-DADT-functionalized analog are quite different. The K_D values for the technetium complexes (*syn* and *anti* isomers) are 1.9 and 4.5 μM, but specific binding to the mAChR receptor is only 15% of the total, presumably due to the lipophilic nature of the technetium complexes (log P = 3.92, 3.49). In the mouse model, 0.3% of the injected dose is present in the brain at 5 min post-injection and this decreases to 0.12% at 1 h. Despite these limitations, this was the first example of a technetium neuroreceptor-binding ligand that retained affinity for the receptor.

A similar approach was adopted by Jurisson et al. who attempted to prepare an mAChR radiopharmaceutical by attaching QNB to the BATO backbone using a boronic acid derivative of QNB (**30**) in place of the isobutyl group found in the cerebral perfusion agent BATO-2mp (**23**) [162]. The ^{99m}Tc complexes of this ligand showed no measurable specific binding ($K_D > 10^{-6}$) to the mAChR receptor and high non-specific binding. The low binding affinity was attributed to the boronic acid functionalized QNB moiety, while the high non-specific binding was attributed to the hydrophobicity of the BATO moiety, estimated by the authors to increase the lipophilicity of the molecule by at least 2.5 log P units relative to that of QNB.

4 Kidney

Imaging agents are available to evaluate both the structure and function of the kidney. These agents fall into the three categories; cortical structure, glomerular filtration, and tubular secretion, although there is some overlap between the categories.

4.1 Technetium-99m-DMSA

To address the clinical questions of renal blood flow and the location and morphology of the kidneys, a cortical imaging agent is most often used because its prolonged retention in the kidneys facilitates image acquisition. The long-standing agent of choice for this application is ^{99m}Tc-DMSA (^{99m}Tc-succimer, Nycomed-Amersham). The exact chemical composition of this radiopharmaceutical is not known nor is the mechanism of retention fully understood. The radiopharmaceutical is prepared from a kit that contains *meso*-DMSA and $SnCl_2$ and is buffered at pH 4.5, a somewhat lower pH than in most kits. If the pH of the kit is raised to 7.5–8.0, the stable well-characterized Tc(V)-oxo tumor-avid compound is formed, as discussed below (Sect. 6). The low-pH agent is sometimes referred to as a Tc(III) species [163], but efforts to isolate this compound (or compounds) have been unsuccessful.

In humans, the low-pH radiopharmaceutical shows relatively slow clearance from the blood with a clearance half-time of 56 min [164] requiring a delay of four or more hours between injection and imaging. The slow blood clearance is presumably due to high protein binding (>90%) [164]. Total renal uptake is more than 40% at 4 h after injection [165].

4.2 Technetium-99m-Pentetate (DTPA)

Kits for the production of a ^{99m}Tc-DTPA complex (^{99m}Tc-Pentetate) are available from four different manufactures (CIS-US, AN-DTPA; Nycomed-Amersham, MPI DTPA; Bracco Diagnostics, Techniplex; DuPont/Syncor International, DTPA). The kits vary only slightly in their concentrations of DTPA and reducing agent ($SnCl_2$) as the complex is non-proprietary. Technetium-99m-DTPA is now used primarily for kidney imaging, including estimation of glomerular filtration rate (GFR). At one time, however, this complex was also used for brain imaging because it is excluded from the healthy brain and only accumulates where there is a breakdown in the blood brain barrier, as occurs in brain tumors or stroke. This has since been replaced, for the most part, by MRI.

The structure of the ^{99m}Tc complex formed from the DTPA kit has not been confirmed because of the tendency of technetium to form μ-oxo- and hydroxyl-bridged dimers at macroscopic concentrations. The oxidation state of the ^{99m}Tc is postulated to be either IV or V with a single DTPA per technetium under the reaction conditions present in the ^{99m}Tc-pentetate kit [166]. If it is assumed that all the oxygens are removed from pertechnetate by the stannous reducing agent, a hexadentate structure can be devised including a Tc(IV) center with binding to three nitrogens and three carboxylate oxygens leaving two free carboxylate groups, and producing a complex with an overall charge of -1, **31**.

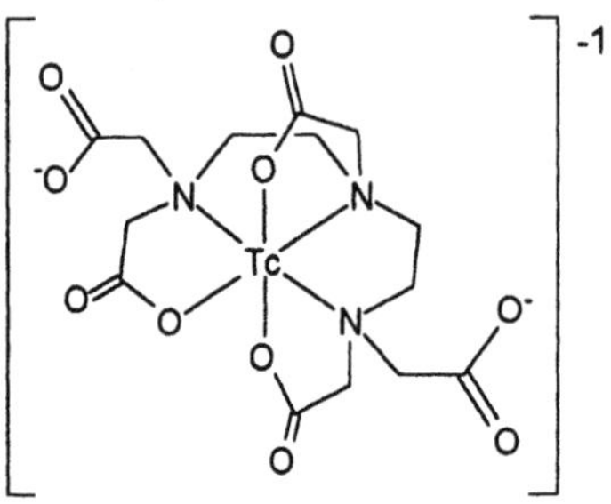

Possible Tc(IV)-DTPA Structure **31**

Alternatively, because stannous chloride more often acts as a two-electron reducing agent with technetium, a Tc(V) = O complex has been proposed, **32** [167].

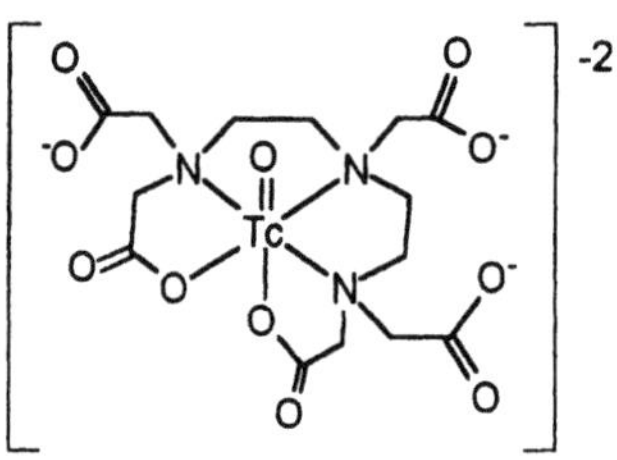

Possible Tc(V)-DTPA Structure **32**

This radiopharmaceutical is reasonably stable as it is excreted from the urine unchanged. After intravenous injection in a normal subject, this hydrophilic compound flows freely in the blood stream until it is filtered through the glomerulus in the kidneys. About 20% of the ^{99m}Tc-DTPA complex is cleared from the blood on each pass through the kidney and ultimately excreted in the urine [168].

4.3 Technetium-99m-Glucoheptonate

This radiopharmaceutical uses glucoheptonate as the ligand to form a weak hydrophilic complex with ^{99m}Tc (GlucoScan, DuPont, Inc.; Technescan, Mallinckrodt). As with ^{99m}Tc-DTPA, this agent is approved for brain imaging and renal perfusion imaging. Unlike ^{99m}Tc-DTPA, ^{99m}Tc-glucoheptonate is cleared from the blood by both glomerular filtration and tubular secretion. This agent, once removed from the reaction conditions in the kit, is not stable, and up to 15% of the injected dose is still retained in the kidneys at 4 h after injection [169]. The residual kidney activity is due to a decomposition product that binds to the proximal convoluted tubules in the renal cortex [170]. This prolonged retention limits its usefulness for the measurement of effective renal plasma flow.

The structure of the ^{99m}Tc-glucoheptonate complex is not known, but the oxidation state of Tc(V) and presence of a Tc = O core is reasonably well established suggesting the formula $[TcO(glucoheptonate)_2]^{-1}$, **33** [171].

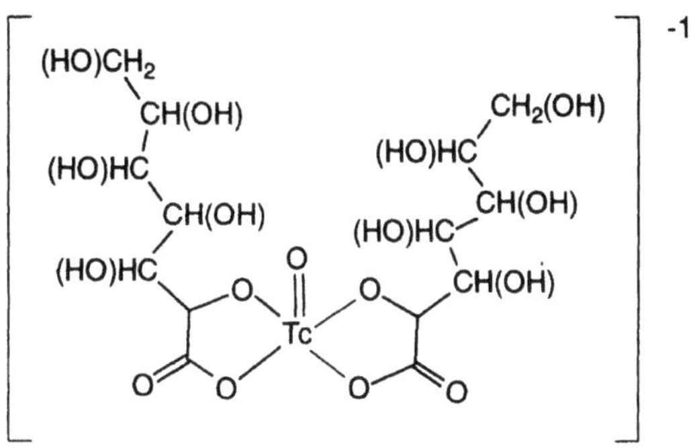

33

This radiopharmaceutical is no longer routinely used in nuclear medicine procedures but, due to its lability, it is sometimes used as a precursor to prepare ^{99m}Tc complexes of more thermodynamically stable ligands. This ligand-exchange approach to labeling molecules is often necessary in very large molecules with a dilute concentration of binding sites such as antibodies [172].

4.4 Technetium-99m-MAG$_3$

The newest commercially available renal-imaging agent is ^{99m}Tc-Mertiatide (TechneScan MAG3, Mallinckrodt), approved as a renal imaging agent and to validate renal function. The $[TcO(MAG_3)]^{-1}$ complex was developed by Fritzberg and co-workers to function as a renal secretion agent [173]. This complex shows good in vivo characteristics and, although a racemic mixture, it does not have the isomer distribution problems encountered with earlier complexes. Davison et al., using triamino acid chelating agents, had earlier demonstrated the stability of the tetradentate diamide-dimercaptide backbone with a Tc = O core [174]. These ligands, which contain a carboxylic acid group on the five-member chelate rings, produced structural isomers in which the carboxylate group is either *syn* or *anti* to the Tc = O bond [175]. The clearance of the two isomers by the renal tubules was found to be different, dependent not only on the *syn/anti* configuration but also on the position of the carboxylate group on the ligand backbone [176]. Fritzberg circumvented this problem by moving the carboxylate group to the amide nitrogen thus eliminating the chiral center [177].

To avoid intermolecular disulfide formation in the kit, the thiol function on the MAG$_3$ ligand has a benzyl protecting group which is cleaved during a 10 min heating step after addition of $^{99m}TcO_4^-$ [178]. Typically, when a $SnCl_2$ containing kit is heated with $^{99m}TcO_4^-$, the technetium is reduced to the Tc(III) state. To avoid this, the concentration of $SnCl_2$ in the ^{99m}Tc-MAG_3 kit is very low (less than 0.2 mg) and air is added to the kit before heating to oxidize excess stannous ion.

The $TcO(MAG_3)^{-1}$ complex exhibits square-pyramidal geometry with the oxo group in the apical position and the Tc(V) atom situated above the plane containing the sulfur and the three amide nitrogens [179], **34**.

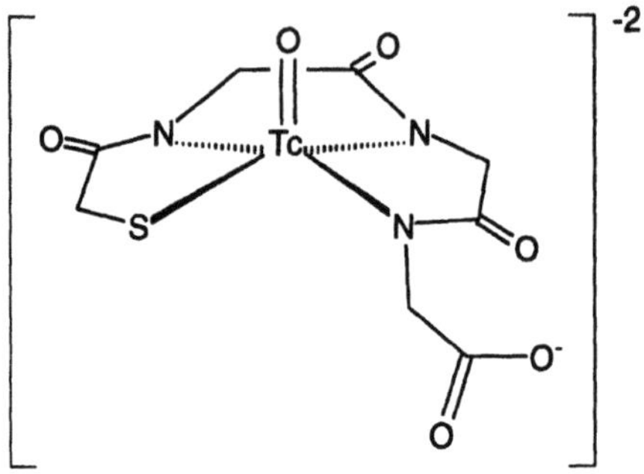

34

The carboxylate moiety is deprotonated at physiological pH, but it is not coordinated to the metal [180]. A free carboxylic acid group next to a carbonyl oxygen appears to be required for a compound to act as a substrate for the anionic receptors in the renal tubules and be actively excreted, as was observed for the hippuran analogs [181]. After intravenous injection, the ^{99m}Tc-MAG_3 complex is rapidly bound (90%) to plasma proteins. Despite being protein-bound, 50% of the activity in the blood is actively excreted by the renal tubules into the urine on each pass through the kidneys. This property makes it the current agent of choice for measuring effective renal plasma flow, replacing [^{131}I]orthoiodohippuran (OIH), despite the fact that the plasma clearance is only 60% that of OIH, the "gold standard" for renal plasma flow measurement [182–184].

4.5 Other ^{99m}Tc Renal Agents

Another potential renal function agent that is now being evaluated in humans is ^{99m}Tc-L,L-EC, **35** [185]. This compound is the fully hydrolyzed product of the neutral lipophilic technetium rCBF agent ^{99m}Tc-*l*,*l*-ethylenedicysteine ethylester (^{99m}Tc-bicisate), discussed previously (Sect. 3).

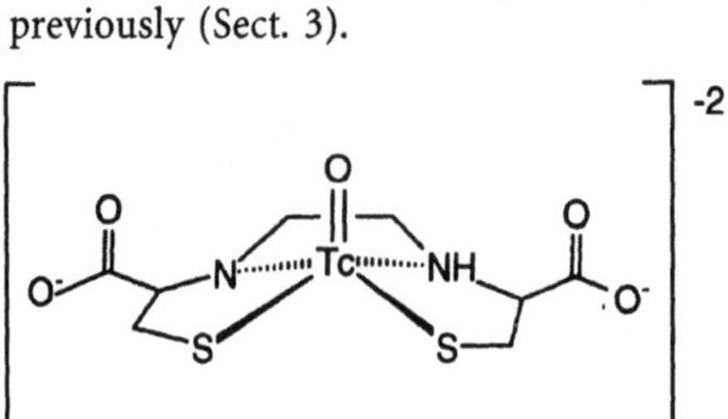

35

The dihydrolyzed product is hydrophilic with an overall charge of –2 and was initially detected as a metabolite of the ethyl ester compound [186]. It was later shown that this compound was actively secreted through the renal tubules [187]. The plasma clearance of Tc-L,L-EC is 39% higher than Tc-MAG_3 in humans and better correlates with OIH [188]. This complex is also reported to have lower liver accumulation than Tc-MAG_3.

Another renal agent that is also a polycarboxylic acid is ^{99m}Tc-CAMI (**36**), the fully hydrolyzed derivative of the ester-isonitrile complex $[Tc(CNCH_2COOCH_3)_6]^+$ [189]. Hydrolysis of the six pendant methyl ester groups produces a hydrophilic species with an overall charge of −5 that has shown promise as an agent for measuring effective renal plasma flow [189].

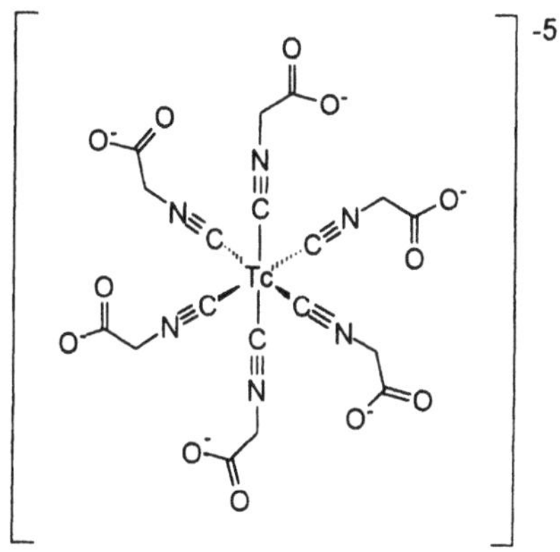

36

5 Liver

The liver contains two major types of cells; parenchymal (polygonal) cells which make up approximately 85% of the cell population, and reticuloendothelial (Kupffer) cells which make up the remaining 15% [190]. The polygonal cells are the primary metabolic cells of the liver and Kupffer cells, as part of the reticuloendothelial system (RES), phagocytize foreign particles. Ultrasound and CT have generally replaced nuclear medicine as the modalities of choice for the morphological evaluation of the RES [191], but nuclear medicine is more sensitive than ultrasound or CT for the evaluation of non-RES diseases such as acute cholesystitis and obstruction of the common bile duct [192]. Obstructions of the common bile duct can be diagnosed using a hepatobiliary scan immediately after impaction, but 72 h are required before ductal dilatation is visible on ultrasound [192]. The hepatobiliary scan is also the only diagnostic procedure that provides a real-time assessment of the progress of bile from the liver, through the gall bladder, and into the small intestine [193]. The hepatobiliary scan is performed with ^{99m}Tc iminodiacetic acid derivatives including mebrofenin and disofenin, the development of which has been discussed previously [167,191,194–196].

A more recent development in hepatobiliary imaging is the preparation of the receptor-binding agent ^{99m}Tc-NGA [197]. Technetium-99m-NGA binds to the asialoglycoprotein receptor which is present only on the liver hepatocytes. Changes in receptor concentration occur in acute and chronic liver diseases including cirrhosis and hepatitis [198]. The compound is prepared by electrolytic reduction of [^{99m}Tc]pertechnetate in the presence of human serum albumin to which the galactosyl moiety has been covalently attached [197]. Investigators in Japan have developed a similar product: ^{99m}Tc-GSA in which galactosyl and DTPA moieties are attached to HSA [199]. In contrast to ^{99m}Tc-NGA, this material has been formulated as a kit in which the ^{99m}Tc is reduced by Sn(II) rather than an electrolytic procedure. Neither agent is commercially available in the US at this time.

6 Tumors

The development of diagnostic and therapeutic oncologic radiopharmaceuticals holds perhaps the greatest potential for nuclear medicine and also the greatest challenge. The potential is great because therapeutic radiopharmaceuticals hold the promise of being able to deliver the mythical "magic bullet" directly to malignant cells. The challenge is great for several reasons. First, cancer is not a single disease, and the problems associated with the development of an effective diagnostic or therapeutic radiopharmaceutical can vary greatly between different types of malignancies. Second, the risks associated with inadequate sensitivity are greater than with other radiopharmaceuticals. One may accept a 10% chance that a diagnostic procedure will miss early stages of heart disease, but that level of risk may be considered too high for breast or prostate cancer where the consequence of a missed diagnosis may be progression from treatable to non-treatable (and possibly fatal) disease. Similarly, an antibiotic that kills 90% of the target bacteria might be acceptable because the body's own defenses can then deal with the remaining infection but, in oncology, anything less than complete sterilization of the lesion is generally considered unacceptable.

6.1 Small Molecules

Probably the most commonly used radiopharmaceutical in oncology is [^{67}Ga]gallium citrate, which is used primarily in the assessment of tumor viability after treatment of Hodgkin's and non-Hodgkin's lymphoma [200–202]. The citrate functions as a chelating agent to prevent hydrolysis of the gallium in the vial. After injection, gallium is rapidly transchelated to transferrin for which it has an extremely high affinity (log $K = 20.3$) [94]. As the transferrin complex, ^{67}Ga is an in vivo detector of transferrin receptors and an indicator of tumor cell proliferation [203]. The physical and biological properties of ^{67}Ga are less than optimal for a radiopharmaceutical. The half-life is relatively long ($T_{1/2} = 3.3$ d), the photon yield is low, and the photon energies (93 keV, 36%; 185 keV, 20%; 300 keV, 16%; 394 keV, 5%) not well matched to most current imaging systems. These limitations have led to considerable interest in the development of other oncologic radiopharmaceuticals.

6.1.1
Thallium-201

Thallium-201 was observed to accumulate in tumors shortly after its first use as a myocardial perfusion agent [204,205]. The physical limitations of ^{201}Tl, described in Sect. 3, also apply to its use as a tumor-imaging agent. Despite these limitations, there remains considerable interest in the use of ^{201}Tl for this purpose [206,207], most recently in the evaluation of breast cancer [208,209].

6.1.2
Technetium-99m-MIBI

Because ^{99m}Tc-MIBI was developed as a replacement for ^{201}Tl, it is not surprising that it also accumulates in tumors, although the mechanism of uptake of the two agents is different, as described previously. As with ^{201}Tl, there is interest in the use of ^{99m}Tc-MIBI for the evaluation of breast cancer [210–212] and DuPont-Merck has recently received FDA approval to market ^{99m}Tc-MIBI (Cardiolite) under a new name (Miraluma) as a supplement to mammography when the mammographic results are not definitive [213].

6.1.3
Technetium(V)-DMSA

An unusual example of a tumor-avid radiopharmaceutical is $[^{99m}Tc(V)O(DMSA)_2]^-$ [214], an oxidized form of the Tc(III) renal agent that can be prepared by modification of the standard Tc-DMSA kit [215,216]. There is also a report of a non-Sn(II) formulation with significantly lower renal uptake [217]. The Tc(V) version of the radiopharmaceutical is not approved in the US but in other countries, primarily Japan, it is used to evaluate medullary thyroid carcinoma [218], and there are reports of its use in the detection of brain metastases and breast cancer [216,219]. There are also reports suggesting that this compound may be useful in imaging intra-abdominal deposits of amyloid protein [220] and, when prepared with $^{186/188}$Re, as a radiotherapeutic agent [221,222].

The technetium complex exists as three isomeric monomers, *syn-endo*, *syn-exo*, and *anti*, based on studies of the ^{99}Tc complex [223] and the rhenium analog [221,222], **37**.

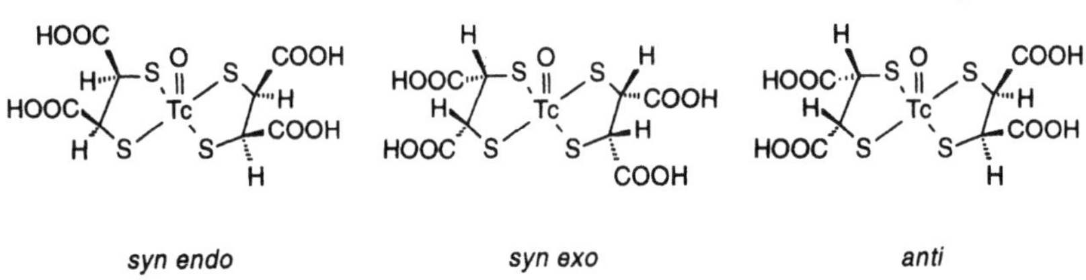

37

This differs from previous reports that suggested that the complex exists as a polymer [224], although this seems unlikely at no-carrier-added concentrations of ^{99m}Tc. The crystal structure of the *syn-endo* $(Et_4N)[ReO(DMSA)_2] \cdot 1.5\ H_2O$ has been determined and shows the expected square-pyramidal $ReOS_4$ core [222]. There are apparently no reports of the biodistribution of the separated isomers.

6.1.4
Somatostatin Derivatives

OctreoScan (Mallinckrodt) is an ^{111}In-labeled tumor-imaging agent developed from somatostatin, **38**, an endogenous 14 amino acid peptide that regulates hormone secretion [225,226].

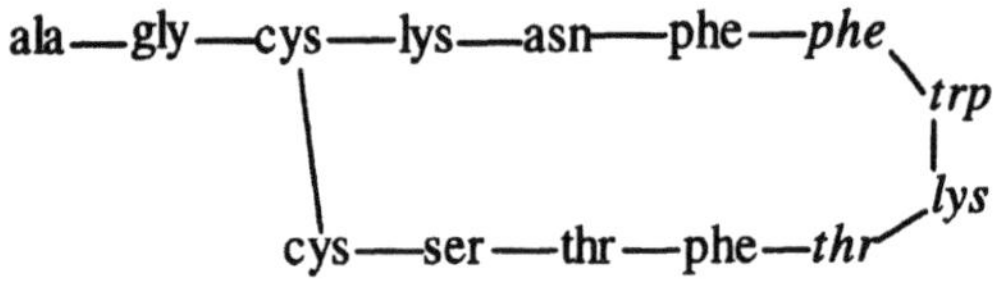

Somatostatin **38**

In many neuroendocrine tumors, hypersecretion of hormones causes a variety of medical problems in addition to those caused by the tumor itself. The biological half-life of somatostatin itself is, however, too short (<5 min) for therapeutic use, so a synthetic form of the peptide that degrades more slowly in vivo was developed. This effort led to Octreotide (**39**) a cyclic eight amino acid somatostatin analog that includes a D-Trp in place of the original L-Trp, a D-Phe N-terminus, and a Thr(ol) C-terminus, all of which slow the rate at which peptidases degrade the drug in vivo [227].

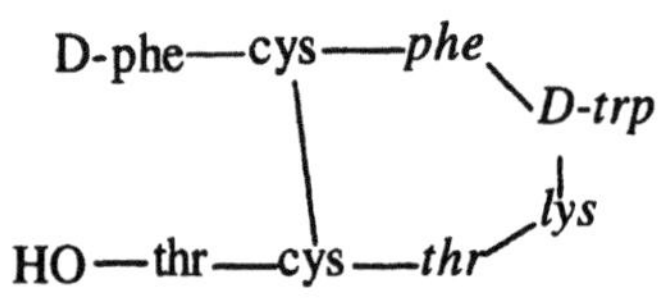

Octreotide **39**

The imaging agent OctreoScan (**40**) is the Octreotide molecule to which a DTPA moiety has been added to chelate the ^{111}In label [228]. The DTPA anhydride technique was used to couple the ligand to the peptide [228]. This approach sacrifices one of the arms of the DTPA ligand to act as the linker to the peptide leaving DTTA rather than DTPA as the chelating agent. This does not, apparently, adversely affect the biological properties of OctreoScan, but proved a limitation in subsequent studies of the ^{90}Y analog where the absence of the fifth carboxylate donor led to significant loss of ^{90}Y from the complex [229].

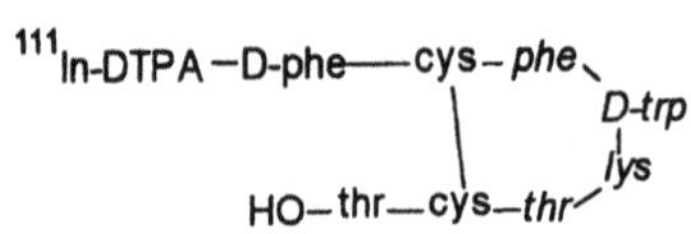

[^{111}In-DTPA]-Octreotide (OctreoScan) **40**

OctreoScan can be used to image a relatively large number of different types of malignancies that express somatostatin receptors including neuroblastoma, carcinoids, small-cell lung cancer, gastrinomas, paragangliomas, and possibly breast cancer and malignant lymphomas [230–235].

As discussed previously, the imaging characteristics of ^{111}In are less than ideal but, in this instance, the options are limited because stannous ion, which is used in most "instant kits" to reduce pertechnetate, also reduces disulfide bonds. In Octreotide, the disulfide bond not only holds the cyclic peptide in the required conformation but, if reduced, would provide competing binding sites for ^{99m}Tc. A second concern is that OctreoScan images are frequently acquired at 24 h or more post-injection. This is not a problem with ^{111}In but, with ^{99m}Tc, 90% of the injected activity is lost to radioactive decay after 24 h. The first limitation has been circumvented by investigators at Diatide who have developed new somatostatin-receptor-avid cyclic peptides in which the ring is closed by an amide rather than a disulfide bond. A second unique feature of these molecules is that ^{99m}Tc is complexed not by DTPA but by a tripeptide external to the receptor binding site. Two different chelating groups were employed, as shown in **41** and **42**.

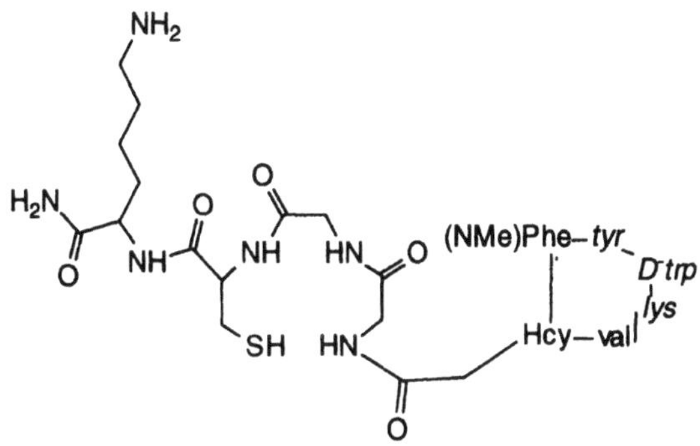

P587 **41**

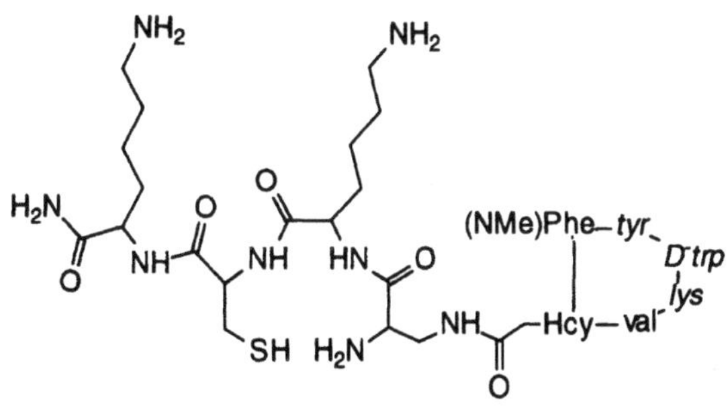

P829 **42**

These groups share an N_3S core but convey different charges to the molecule which result in some differences between the biological properties of the two ^{99m}Tc complexes [236,237]. The receptor-binding segment of the peptide is also slightly different from that in Octreotide in that it contains a Tyr-D-Trp-Lys-Val sequence rather than Phe-D-Trp-Lys-Thr at the binding site. The binding of the Re-labeled peptide to the somatostatin receptor is similar to that of OctreoScan, 0.3 nM (P829) versus 1.2 nM (OctreoScan) [237]. The high affinity of the Re-labeled peptide suggests that the $^{186/188}$Re analogs might be useful as radiotherapeutic agents [237].

Mäcke and co-workers have prepared an Octreotide derivative that contains a PnAO chelating moiety in place of DTPA [238,239], but both the tumor uptake and tumor/normal tissue ratios are lower than those observed for OctreoScan or the Diatide compounds [238]. This difference might reasonably be attributed to the much higher lipophilicity of the PnAO versus DTPA or peptidyl chelating moieties.

Anderson and co-workers at Washington University have prepared a $^{64/67}$Cu-labeled octreotide derivative in which the DTPA ligand is replaced by a TETA or CPTA ligand [240]. The CPTA-derived product had somewhat higher affinity for the somatostatin receptor than the TETA derivative, but the liver clearance of the CPTA is unsatisfactory compared to the TETA derivative. The affinity of the ^{64}Cu-TETA derivative for the receptor is five-times higher than that of the original ^{111}In-DTPA complex. The uptake of all three complexes in tumors is similar, but the tumor/normal tissue ratios are significantly higher for OctreoScan® than for either of the ^{64}Cu-labeled compounds.

6.2 Labeled Antibodies and Fragments

Several radiolabeled antibodies and fragments are now commercially available for oncologic imaging. These include CEA-Scan (Immunomedics), an anti-carcinoembryonic antigen (CEA) Fab′ fragment of the IMMU-4 antibody which is directly labeled with ^{99m}Tc (i.e. without the addition of an exogenous chelating agent) for colorectal cancer [241]; Verluma (Dupont-Merck), the ^{99m}Tc-labeled diamidedithiol-functionalized Fab fragment of the murine antibody NR-LU-10 for staging small-cell lung cancer [242]; ProstaScint (Cytogen), an ^{111}In-labeled DTPA-functionalized CYT-356 murine antibody for preoperative staging of prostate cancer [243,244]; and OncoScint (Cytogen), an ^{111}In-labeled DTPA-functionalized B72.3 murine antibody for imaging colorectal and ovarian carcinoma [244,245]. It is important to note that the more recently developed agents (CEA-Scan and Verluma) are based on antibody fragments labeled with ^{99m}Tc while ProstaScint and OncoScint are based on intact murine antibodies labeled with ^{111}In. The primary reason for this (and the primary limitation of intact antibodies) is that intact antibodies clear very slowly from non-target tissue and, as a result, the patient typically cannot be imaged until at least 24 h post-injection. If the intact antibody were labeled with ^{99m}Tc, at 24 h post-injection, the amount remaining would be less than 10% of that injected. This is not a limitation with ^{111}In ($T_{1/2}$ = 2.8 d), but it is inconvenient both for the patient who must return to the clinic for a second time and the referring physician who must wait 24 h to obtain the results of the study.

A complete description of all of the labeled-antibody literature would include a discussion of the relative merits of different labeling methods; radionuclides, both diagnostic and therapeutic; and the relative merits of the different antibodies themselves. A comprehensive discussion of this topic is beyond the scope of this work, and the reader is referred to recent review articles on this topic [246–252]. In the context of metal-based radiopharmaceuticals, we include a brief discussion of some of the issues relevant to the attachment of metallic radionuclides to monoclonal antibodies.

6.2.1
Protein-Targeted Radiometal Complexes

The advantages of a targeting moiety in association with a metallic radionuclide complex are obvious. The selective delivery of γ and β^+ emitters for diagnosis or β^- and α emitters for therapeutic applications might greatly enhance sensitivity, decrease toxicity to surrounding tissues, and increase efficacy. Development and subsequent investigation of monoclonal antibodies, their fragments, associated engineered immunoproteins, as well as a host of receptor targeting proteins for this purpose, have been proposed to fulfill this ideal [253–255]. However, their use places stringent requirements upon the radioactive metal ion complexes directed by these vectors. While myriad challenges exist concerning the delivery portion of such constructs, these challenges are moot without first adequately addressing the chemistry required for use with the chosen radiometal ion. In general, two problems are involved. The first concerns the method of linking the metal complex to the protein. The second concerns the complexation or coordination chemistry of the metal ion and the minimum criteria that must be met for success (vide infra).

Linking the Metal Complex to the Protein (or not...). Two paradigms govern the linking of radiometal complexes to proteins. The first is to irreversibly link the complex to the protein via a covalent bond that is exceedingly stable in vivo and prevents loss of the complex from the protein. Although there is a rich variety of chemistry suitable for linking molecules to reactive functional groups available in proteins, the vast majority of efforts have centered on either directly targeting the ε-amines of lysine residues or the N-terminal amine [256–258], coupling to a thiol chemically introduced to the protein [259], or oxidative generation of aldehyde groups from carbohydrate structures that are then used in reductive amination reactions to introduce chelating agents [260,261]. The amines may be functionalized by either acylation with a variety of active esters forming an amide [256,257,262], alkylation with reactive halides that partially retain the amine character [259], or with isothiocyanates to form a thiourea [258]. Thiols, available for derivatization with either a reactive halide [259], a maleimide derivative, or reactive vinyl group [263–265], generate a thioether linkage product. A subset of this methodology is to insert a metabolically cleavable functional group within the linker moiety. This option has been proposed to improve background clearance kinetics of antibody conjugates and to reduce deleterious toxicity from circulating reagent that fails for a variety of reasons to reach its target [263,266–268].

Another approach to linking radiometal complexes to proteins does not actually attach the metal complex to the targeting vector, but delivers it in vivo via employment of the concept of "pre-targeting" [269]. For example, the targeting vector is cross-linked to either avidin or streptavidin and allowed to proceed to its target. After an appropriate time has elapsed for maximum uptake and penetration, excess reagent in circulation is cleared through the liver, and the radiometal complex conjugated to biotin is added. The biotin conjugate then penetrates deeply and rapidly into tissues where it binds to the (strept)avidin. This two-step approach is the advantage of this method; the activity either arrives at the target or is cleared rapidly minimizing the dose to non-target tissue. This advantage is counter-balanced by the rapid whole-body clearance and subsequent excretion of the majority of the radiometal-biotin complex, which means that large amounts of activity must be administered to attain the high targeting percentages required for therapeutic applications.

Complexing the Metal Ion. The second aspect of the development of a metal-labeled biomolecule is the sequestration of the metal ion. This requires a chelating agent that complements the coordination chemistry of the radiometal ion chosen for use. A list, by no means exhaustive, of radionuclides that have been proposed for use in either radioimmunoimaging or therapy includes: ^{111}In, ^{90}Y, ^{64,67}Cu, ^{177}Lu, ^{153}Sm, ^{212,213}Bi, ^{212}Pb, ^{66}Ga, ^{225}Ac, ^{223}Ra, ^{99m}Tc, ^{186,188}Re. Regardless of the paradigm followed for delivery of the metal complex to the target tissue, the absolute requirement remains: *the complex must be of adequate kinetic stability to prevent loss of the radiometal from the complex in vivo.* If the metal is lost from a diagnostic agent, the resulting images may be unacceptable. If the metal is lost from a therapeutic agent, the result may be delivery of an unacceptable radiation dose to non-target tissue.

Development of chelating agents for these radionuclides essentially began by abstraction of well-established chemical reagents from the inorganic chemistry literature. What was introduced into these reagents was a methodology or a reactive functional group by which they could then be linked to protein. The acyclic EDTA and DTPA chelators, **43**, were recognized very early to be effective chelating agents for a variety of metal ions such as In(III), Y(III), lanthanides, and several other radionuclides with which they have high thermodynamic stability constants [270].

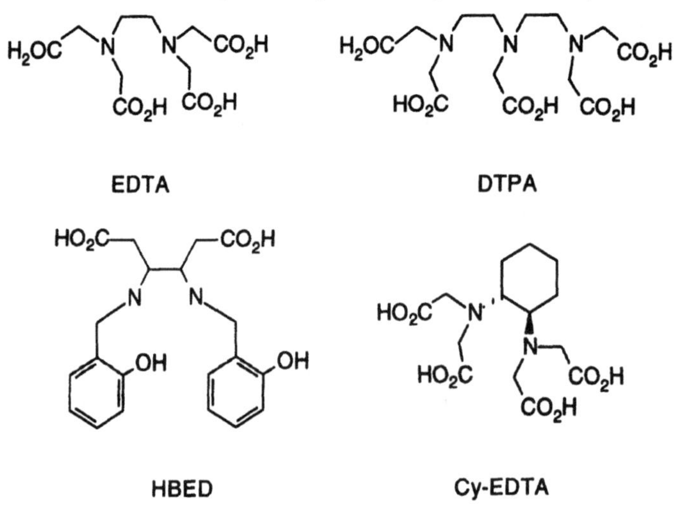

43

The metal complexes of these ligands possess a hexa- or octadentate coordination sphere that might meet the requirements for maintaining kinetically inert complexes in vivo while retaining reasonable complex formation times, an important consideration for short half-life radionuclides. Also, being acyclic chelating reagents, they possessed some limited flexibility to assume a suitable geometry for their metal-binding sites to favorably align with the orbital arrangements of the M(+2,+3) ion, allowing for the various ionic radii of the radionuclides (80, 90, and 103 pm for In(III), Y(III), and Bi(III), respectively) [271]. Thus, one finds the stability constants of the DTPA complexes of In(III), Y(III) and Bi(III) to be $10^{29.0}$, $10^{22.13}$ and $10^{35.6}$, respectively, providing the impetus for the development of two derivatives for linking DTPA to protein [270]. These two simple approaches, the dianhydride of DTPA [257] and the carbonic anhydride of DTPA [256], both utilize one of the carboxylate arms to link the ligand to protein via an amide group (Scheme 1).

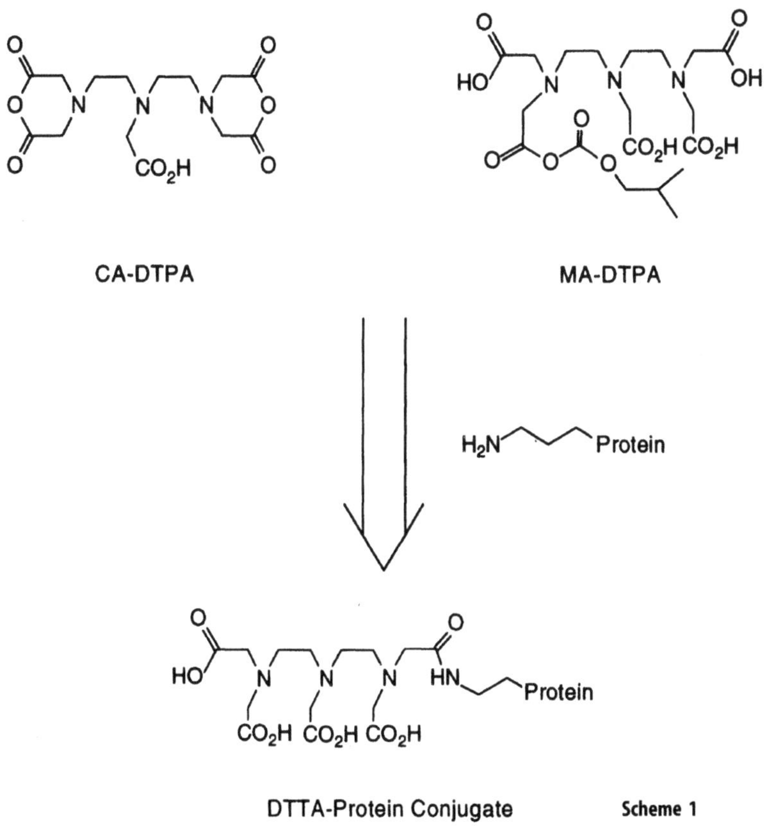

Scheme 1

This approach, however, decreases the denticity of the ligand to heptadentate, forming a DTTA. This compromise in denticity was initially overlooked as contributing to complex instability because of the extreme ease of use associated with these reagents. With ^{111}In, which forms complexes of coordination number 6, 7, or 8, the DTTA forms what might be termed an adequately stable complex for in vivo experiments with these reagents. However, attempts to transfer this methodology to lanthanides and ^{90}Y quickly demonstrated the limitations of chelating agents of inadequate denticity and prompted the development of bifunctional reagents that preserved the full octadentate character of DTPA. Use of DTPA reagents with an aryl isothiocyanate group inserted into the ligand structure for protein linkage substantially increased the in vivo stability of ^{88}Y complexes compared to bifunctional EDTA and DTTA forming reagents [272]. Subsequently, numerous C- and N-functionalized DTPA derivatives, **44**, were reported in the literature [273–275].

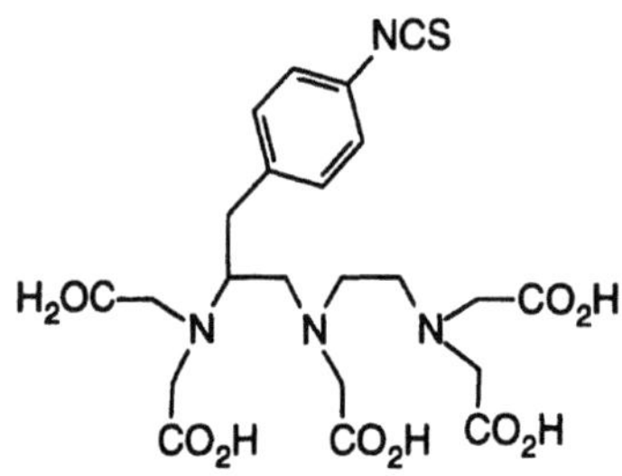

1B-DTPA

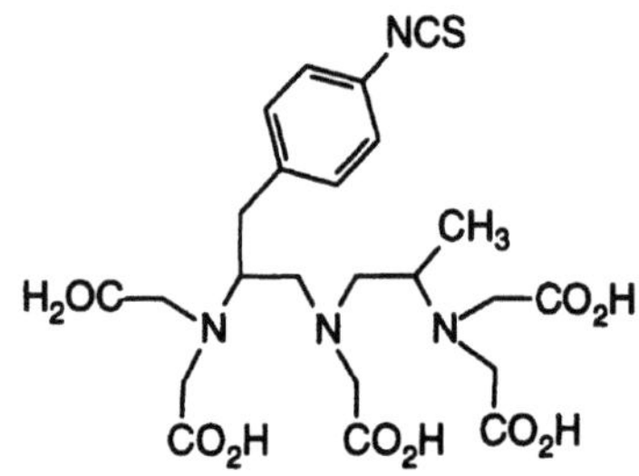

1B4M-DTPA

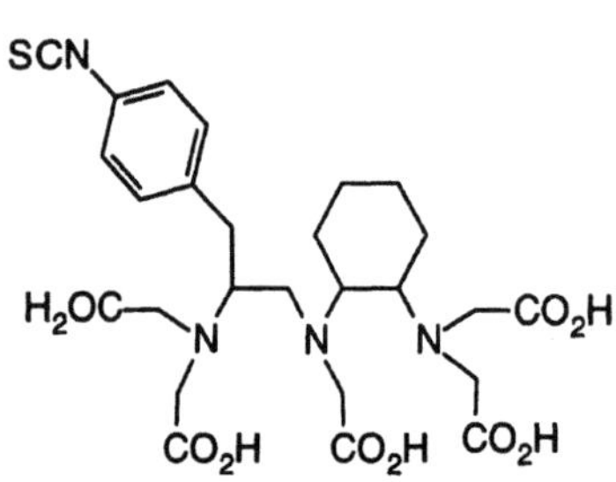

CHX-DTPA

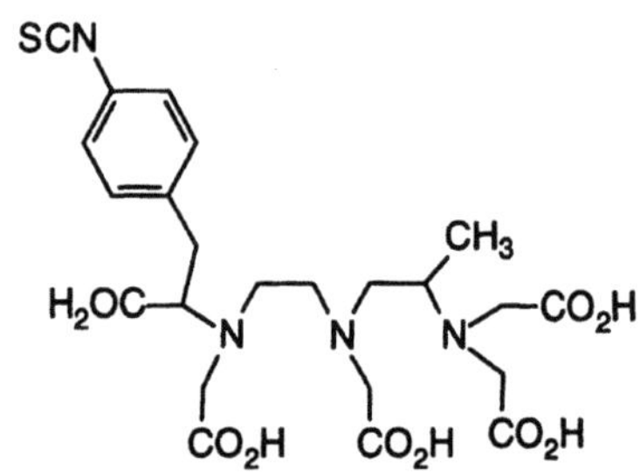

N-Functionalized DTPA **44**

Additional C-functionalization of the DTPA ligand with methyl groups was shown to increase the in vivo stability of the ^{88}Y and ^{153}Sm complexes [276], a result that was clearly indicated from the chemical literature from a combination of both inductive and steric effects [277]. Despite a small loss of metal from the ligand, and the fact that these complexes are not completely kinetically inert in vivo, they have been employed with success in the clinic [278,279]. Additionally, although the stability of the Bi(III) complexes of C-functionalized DTPA derivatives is very high, none of them have demonstrated acceptable in vivo stability.

This result demonstrates that thermodynamic stability constants in and of themselves are not a particularly good indicator of the ability of a chelating agent to sequester a metal ion in a biological environment and that they may best be employed as a relative guide within a metabolically similar family of elements. Another exceedingly important point is that it is unreasonable to expect that a single chelating agent can used with all radionuclides of interest.

Review of the chemical literature revealed that the stability constants of Cy-EDTA, **43**, complexes with In(III), Y(III) and Bi(III) were very close to those of the corresponding DTPA complexes [270]. Whole-body clearance studies of ^{90}Y complexes have demonstrated this stability within the limitations of such an experiment, i.e. free complexes as opposed to protein conjugates [280]. Additionally, reports of the use of heptadentate and pentadentate reagents based on *trans*-cyclohexyl-DTPA and *trans*-cyclohexyl-EDTA demonstrated their potential for use with ^{111}In, but not with Pb(II) isotopes [281,282]. Recently, a report of a fully hexadentate bifunctional *trans*-cyclohexyl-EDTA has appeared, although with insufficient information to evaluate the in vivo stability with any radionuclide of medical interest [283].

Functionalization to merge the *trans*-cyclohexyl substructure with the previously reported bifunctional DTPA provided the CHX-DTPA ligands, **44**, now extensively studied with ^{111}In, ^{90}Y and Bi(III) radionuclides [284].These ligands are substantial improvements over prior DTPA reagents because the pre-organizational geometry conferred by the *trans*-cyclohexyl unit provides not only adequate stability for use with $^{212,213}Bi$ but also retains the rapid complex formation kinetics required for use with these isotopes. It should be noted that these ligands exist in several stereochemical configurations and that very significant differences have been reported for the in vivo stability with ^{88}Y, not only between the diastereomers, but between the enantiomers as well [285]. The criterion of "correct" stereochemistry as applied to the design of bifunctional chelating agents has been regarded as inconsequential biologically, but may turn out to be of profound importance in the design of future reagents. Another very important point to be emphasized is that serum stability and transchelation experiments, standards in the field to predict in vivo stability, failed to detect the differences between the enantiomers. The effect was only observed in animal experiments. This points out the fact that these in vitro assays provide, at best, only relative measures of in vivo stability.

The use of acyclic hexadentate ligands had been investigated principally through bifunctional EDTA derivatives. However, due to the lack of stability of their complexes with therapeutic radionuclides in vivo [272], hexadentate ligands of greater stability were sought. This led to the development of bifunctional derivatives of HBED, **43**, which have exceptionally high stability constants for Fe(III), In(III) and Ga(III). Based on this template, several bifunctional derivatives have been synthesized that allow conjugation of the ligand to the protein via active ester, reactive halide, and aryl isothiocyanate functional groups [286–289]. Also, efforts have been made to adjust the lipophilicity of the complex formed by manipulation of alkyl substituents on the phenol ring [287].

Concurrent with the development of the acyclic bifunctional chelating agents, those searching for other reagents that might provide a truly kinetically inert complex for radionuclides, ^{90}Y in particular, recognized the inherent potential of the macrocyclic polyaminocarboxylates ligands such as NOTA (**45**), DOTA (**46**), and TETA (**47**).

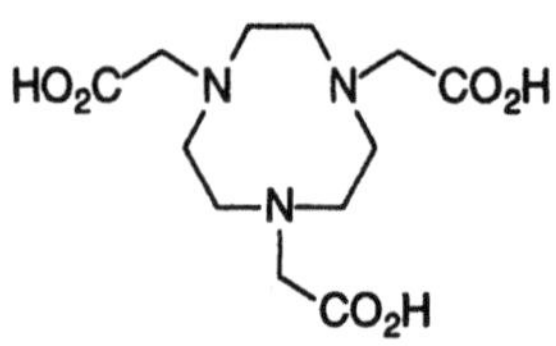

NOTA **45**

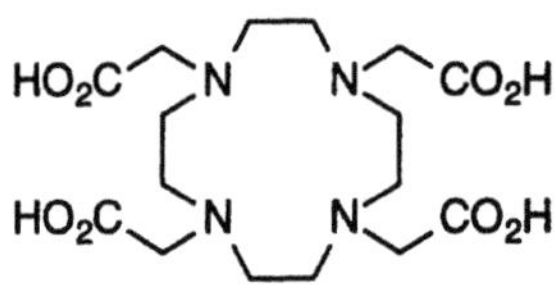

DOTA **46**

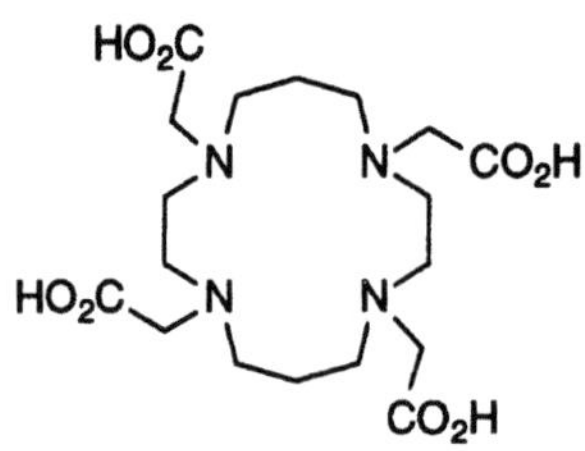

TETA **47**

These three ligands possess several inherent advantages over acyclic ligands. In general, their macrocyclic nature confers a high degree of pre-organization and limits the conformational disorder thus lowering the energy cost of forming metal complexes. Also, with this trio of ligands and their homologs, a broad range of both denticity and cavity size is available to accommodate different metal ions and thus provide the basis for additional bifunctional chelating agents suitable for in vivo applications.

NOTA, for example, being hexadentate and possessing an appropriate cavity size for Ga(III), has been shown to be not only the optimal ligand for Ga(III), but also to form complexes that are stable even under acidic conditions exceeding those encountered during metabolic degradation of proteins [290,291]. This property insures that the radionuclide remains chelated and promotes eventual clearance of the isotope from the body. Despite preferring coordination numbers of 7, 8, or even 9, In(III) also forms very stable complexes with NOTA, due to the optimal cavity size [292]. An additional advantage gained through the use of NOTA is that complexes with M(+3) ions are neutral in charge, a condition proposed to promote more rapid clearance from non-target tissue [290,291]. To this end, several bifunctional C- and N-functionalized NOTA derivatives, **48**, have been reported in the literature that effectively exploit these properties [290,293,294].

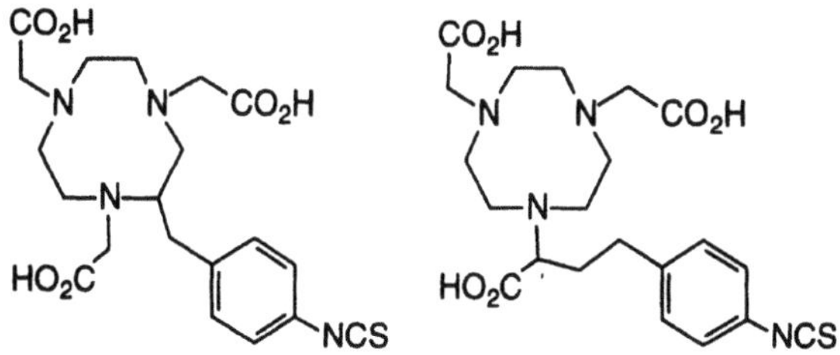

C-Functionalized NOTA N-Functionalized NOTA **48**

The larger octadentate macrocycles DOTA and TETA and, where potentially useful, their polyamine precursors, have seen extensive investigation for use with ^{111}In, ^{90}Y, ^{177}Lu, ^{153}Sm and Cu radionuclides. The use of DOTA as a complexing agent for lanthanides has a substantial history in the coordination chemistry literature and that it has been shown to be an excellent ligand for these metals is quite unsurprising [295]. Thus, numerous syntheses of both C- and N-functionalized DOTA derivatives, **49**, and closely related analogs have been reported in the literature [296–301], which are all claimed to form kinetically inert complexes with not only Y(III) and lanthanides but also with Bi(III) [302].

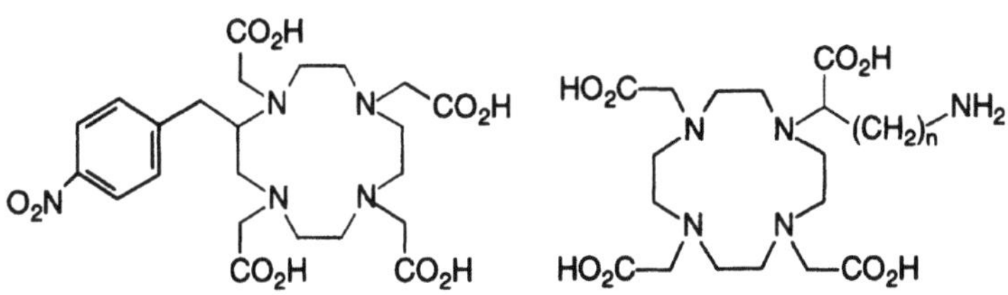

C-Functionalized-DOTA N-Functionalized-DOTA

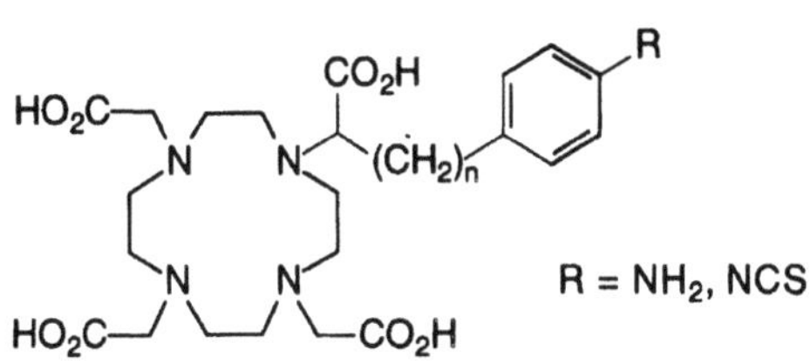

PA-DOTA **49**

This high kinetic stability has also been a limitation to the widespread use of DOTA, because the complex formation rates with M(+3) ions are significantly slower than with the acyclic ligands. Bifunctional DOTA ligands have also been reported to be highly sensitive to the presence of M(II) ion contamination of the radionuclide, principally Ca(II) [303]. Thus, DOTA, despite forming a kinetically inert complex with Bi(III), was found difficult to use because of the combination of slow formation rates and the half-life of ^{212}Bi ($T_{1/2}$ = 60 min) [304]. An attempt was made to use DOTA as a complexing agent for Pb(II), which would allow the use of ^{203}Pb and ^{212}Pb for imaging and therapy, based on a report that the Pb-DOTA complex was inert at physiological pH [305]. Unfortunately, these attempts were unsuccessful because: (1) the DOTA[Pb(II)] complex is labile under the acidic conditions present during metabolic processing and (2) during the β^- decay event of ^{212}Pb, ca. 33% of the ^{212}Bi (produced by decay of ^{212}Pb) is lost from the complex [306]. Both properties contribute to unacceptable toxicity despite finding positive therapeutic results [307,308]. Preliminary results using "pre-targeting" methodology, however, appear to minimize these drawbacks [309].

Attempts to circumvent the slow formation kinetics have been driven by the goal of developing a clinically useful DOTA reagent. The mechanism of complex formation has been proposed to involve a two-step process wherein the first step is a rapid reversible capture of the metal ion via electrostatic attraction followed by a second, slow, irreversible step that involves reorganization of the ring conformation and deprotonation to allow the ion to become encapsulated within the coordination sphere of the chelator [310]. Attempts have been made to increase the rate of the reaction by introducing additional pendent donor groups, **50**, to assist with capture of the metal and to hinder the reversible first step of the reaction and promote the second, complex-forming, step [265,311].

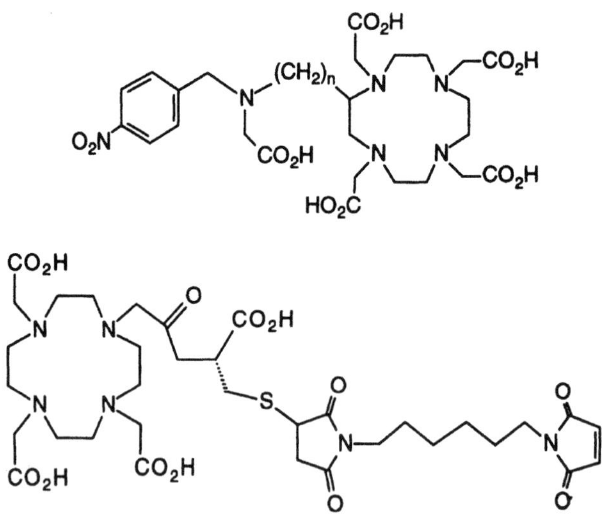

50

Some success has been reported with this strategy and, in one case, this has included introduction of a cleavable linker into the chelating agent, the actual advantage of which has yet to be fully explored. Another approach has been to replace one of the carboxylates with an amide as the protein-linking group, **51**. This still maintains the requisite denticity, produces a neutral complex with M(+3), and may improve formation rates by eliminating a deprotonation step during complex formation [312,313].

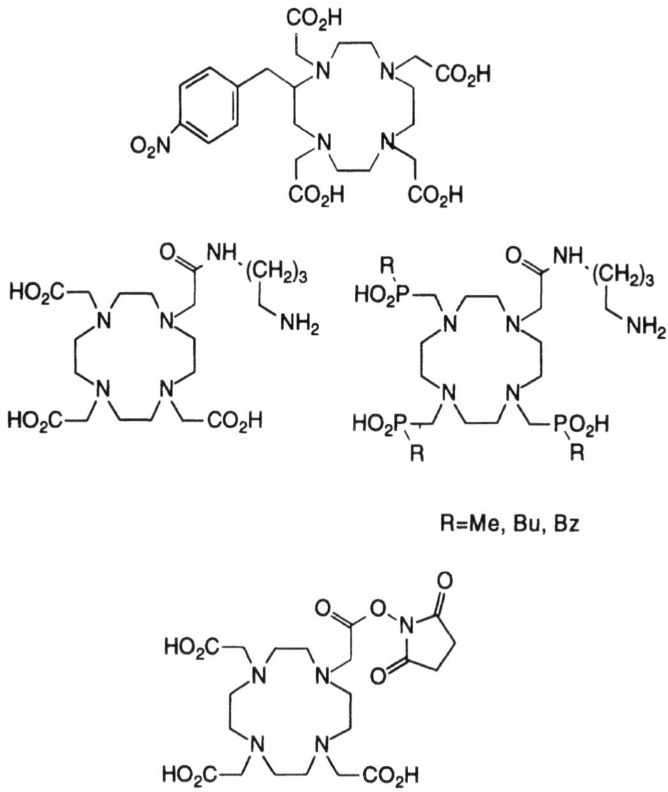

51

The isotopes of Cu(II) (^{61,62,64,67}Cu) have received extensive investigation due to their attractive range of both emission qualities and half-lives. The vast majority of efforts towards the use of Cu(II) radiopharmaceuticals have focused on use of derivatives of NOTA, DOTA, TETA, and the polyamine precursor of TETA, cyclam. There are numerous challenges associated with the use of Cu(II) radionuclides in vivo, namely the inherent acid lability of many Cu(II) complexes, the redox chemistry of Cu(I/II), and the added complication that copper is also an element recognized and utilized in mammalian systems. The use of bifunctional chelating agents with isotopes of Cu has recently been extensively reviewed [314]. Despite numerous claims to the effect that the macrocyclic ligands provide kinetically inert complexes suitable for sequestering Cu(II) in - vivo, the acid lability of these complexes has been corroborated by recent reports of intracellular transchelation to metal-binding proteins clearly indicating that further investigation into the stability of the complexes is warranted [315,316].

It should be noted at this point that the discussion of macrocyclic chelating agents has been limited to the polyaminocarboxylate derivatives. This clearly need not be the case and the polyamine platforms upon which these ligands are based allow for substantial variation in donor groups. Considerable efforts have been directed towards the introduction of phosphonates and phosphinates, either as uniform donor groups or mixed with carboxylic acid group derivative for the sequestration of lanthanide ions [249].

An alternate chelating technology that has been employed for sequestration of Cu(II) radionuclides in vivo, and to some extent In(III) and Ga(III), has been the use of bifunctional porphyrin derivatives [317–321]. N-Benzylated derivatives that deform the planar geometry of the ligand obviate the slow kinetics of complex formation with simultaneous debenzylation [317]. A variety of methods for conjugation of protein have also been employed. Limitations to the use of these reagents centers on the inherently specialized nature of the chemistry as well as the difficulties associated with the aggregation properties of porphyrins [321].

Recently, attempts to address chelation and radiolabeling of proteins have been directed towards use of the elements Ac and Ra. Actinium-225 decays through three α emissions to ^{213}Bi which, in turn, decays through a split α,β decay, potentially bringing an enormous cytotoxic force to bear upon single cells. While it is obvious that no chelate will withstand the destructive force of this decay process, ^{225}Ac might still find use in a rapid localization and internalization scenario or as an adjuvant, or intracavitary administration which would minimize ^{213}Bi delivery to the kidney. Radium-223 has a similar emission potential with the fourth α decay coming from ^{209}Pb. Unfortunately, Pb(II) is very difficult to deal with in vivo and localizes in the blood and marrow at the concentrations associated with these applications. Nonetheless, the therapeutic potential is of such interest that reports of calix[4]arene derivatives suitable for forming stable complexes with either ^{225}Ac or ^{223}Ra have recently appeared [322–324]. While no data have been made available concerning actual in vivo stability, these reports of suitability are based on estimated thermodynamic stability constant studies, an important physical characteristic, but one that, as noted above, is not without its limitations. Further chemical and in vivo studies are warranted not only from a purely chemical standpoint but also due to the potential utility of these isotopes. Also, the acid-catalyzed lability properties of M(II) ions could make Ra(II) extremely challenging to adequately sequester for in vivo applications. The calix[4]arene derivatives possess adequate cavity size for these ions; however, other structures to accommodate these large ions should be investigated. Despite its popularity, the cavity of DOTA is probably too small for either element. If one extrapolates from lanthanide data, DTPA derivatives ($10^{14.2}$ for Ac[EDTA], $10^{15.47}$ for Ac[CyDTA]) [325] are unlikely to be adequate for complexing Ac for in vivo applications and preliminary experiments tend to confirm this prediction [326]. Pursuit of these two radionuclides thus remains speculative until the required chemistry is unambiguously defined.

For the reasons discussed in the Introduction, the use of ^{99m}Tc and ^{186,188}Re in this context is of considerable interest, both attached to biological materials and as free complexes. A frequently used arrangement of donor elements is the acyclic N_XS_Y type wherein X + Y = 4 and Y is equal to at least unity, and there is an additional functional group ("R") for attachment of the complex to a protein, **52**.

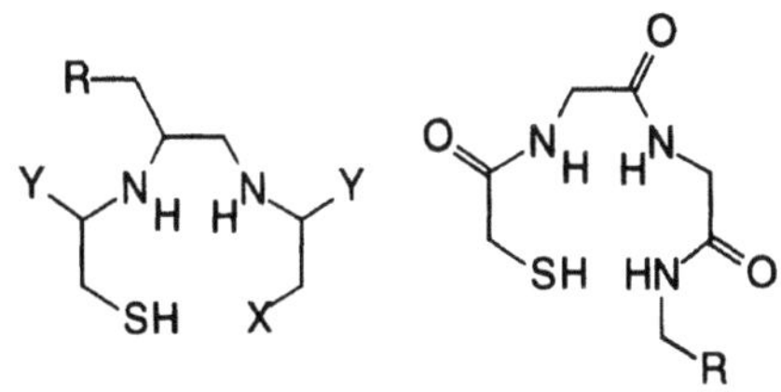

Y = H_2 or O
X = SH or NH_2
R = Linking Group **52**

The nitrogen donors may be amines, amides, or a combination of both, with the sulfur donor nearly always being a thiol. A large amount of literature and numerous reviews on this aspect of Tc (and analogously Re) coordination chemistry are available [167,327–335]. Of particular recent interest in this context is the use of created, or natural binding sites or domains, either in proteins or peptides, for the sequestration of technetium (or rhenium) via the polyamide structure and associated amino acid functional groups [237,336]. This subject was previously discussed in the context of the Diatide Octreotide derivatives. Efforts employing combinatorial chemistry technology to achieve these goals have also been reported [336]. If successful, this approach might not only eliminate the need to use bifunctional chelating agents with Tc and Re, but might also ultimately provide methods for creating binding sites for other metal ions of interest. This area of investigation might then complement or be incorporated into on-going efforts to optimize the targeting and clearance of the biological delivery vectors.

6.3 Multidrug Resistance

Multidrug resistance (mdr) is a phenomenon in which a malignant lesion is either resistant to or becomes resistant to a specific group of chemotherapeutic agents. Of the several types of drug resistance, the most frequently encountered, mdr1, is characterized by overexpression of the *MDR1* gene and increased concentration of its gene product, P-glycoprotein (Pgp), a 170 kDa transmembrane glycoprotein that acts as an ATP-dependent efflux pump to reduce the intracellular concentration of these drugs to non-toxic levels [337–339]. The chemotherapeutics in the mdr group include doxorubicin, daunorubicin, and paclitaxel, some of the most effective chemotherapeutic agents currently available. The presence or development of mdr in a malignancy means that these agents are no longer effective. Unfortunately, this is not clinically apparent until the patient fails to respond at which time the disease may have progressed beyond the stage where another agent might be effective. There is, consequently, considerable interest in the development of a sensitive diagnostic test for mdr and also of reversal agents that block the function of Pgp and increase intercellular concentration of the chemotherapeutics.

In 1993, Piwnica-Worms and co-workers observed that the myocardial perfusion agent ^{99m}Tc-MIBI is a substrate for Pgp and thus might be useful for the evaluation of mdr [340]. Subsequently, it has been reported that several cationic ^{99m}Tc myocardial perfusion agents are also substrates for Pgp [341,342], but the neutral myocardial perfusion agents (e.g. BATOs) are not [343]. Radiolabeled complexes that are substrates for Pgp may show differences in tumor uptake and clearance thereby providing a tool with which to measure the functional status of mdr in vivo. A study with this objective was undertaken by Del Vecchio and co-workers who observed that the clearance rate of ^{99m}Tc-MIBI from breast cancer lesions correlated with the Pgp concentration in the lesions [344–346] and that the rate of ^{99m}Tc-MIBI clearance from the lesion was inversely correlated with the response of the lesions to chemotherapy with mdr chemotherapeutics [347]. Piwnica-Worms has reported that the biodistribution of ^{99m}Tc-MIBI in patients is significantly altered by the introduction of PSC 833, an mdr reversal agent developed by Sandoz [348]. This suggests that, in addition to monitoring mdr, these radiopharmaceuticals might also be used to monitor the effectiveness of reversal agents. If this is confirmed in a larger clinical study, it will provide a real-time method with which to evaluate mdr reversal agents and chemotherapeutics, something that is not currently available.

As described previously, DuPont-Merck recently received FDA approval to market ^{99m}Tc-MIBI as a second-line diagnostic agent to assist in the evaluation of patients with abnormal mammograms or palpable breast lesions [213]. The use of ^{99m}Tc-MIBI for this purpose is complicated by the fact that ^{99m}Tc-MIBI is a substrate for Pgp. A significant fraction (the percentage varies with the type of assay used) of all breast tumors are drug resistant at initial presentation [349,350], and these lesions may not be detected by ^{99m}Tc-MIBI because it will be excreted from tumor cells that express Pgp. Because these are also the lesions that are least likely to respond to chemotherapy, imaging in combination with reversal agents may be required in these cases.

In addition to the ^{99m}Tc-based compounds, there is also an effort to develop $^{62/64}$Cu and ^{68}Ga [351] mdr imaging agents that can be used with PET for more accurate measurement of efflux rates. The (cationic) copper-based compounds include Cu(I) phosphines [352] and Cu(II) complexes of lipophilic derivatives of the ligand PreH, **53** [353,354], discussed previously (see Sect. 2). These radionuclides have an advantage over ^{11}C-PET radiopharmaceuticals for this application because the half-life of ^{11}C (20.5 min) is significantly shorter than the efflux rates measured by Del Vecchio et al. (204 min) [347].

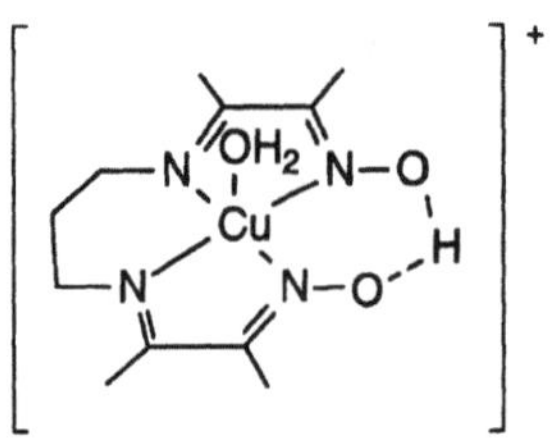

53

6.4 Bone-Pain Palliation

More than one million new cases of cancer are diagnosed each year in the US, including more than 175,000 cases of breast cancer and more than 180,000 cases of prostate cancer [355]. More than 80% of the patients who eventually die from these cancers have osseous (bone) metastases [356]. These metastatic lesions are the most common cause of cancer pain, and this pain is often difficult to manage [357]. The first step in pain management for these patients is non-steroidal anti-inflammatories (NSAIDs) followed by increasingly more potent opiates including morphine. When these options no longer provide adequate relief, external beam radiotherapy can be used on solitary metastatic sites and is effective in 60–90% of patients [358–360]. When there are multiple metastatic sites, however, external beam therapy is no longer a viable option. In these cases, targeted radiotherapy using unsealed sources can be very useful.

The history of the development of these radiotherapeutics extends back to 1950 when strontium and phosphorus radionuclides were first used for this purpose [361,362]. Phosphorus-32 (primarily as orthophosphate) has a long research history in this application but has been abandoned because of concern about hematological toxicity due to the radiation dose delivered to the bone marrow by the 1.71 MeV β^- particle, approximately 20% more energetic than that of the ^{89}Sr (Table 2). Hematological toxicity is a serious concern because chemotherapy patients frequently have depleted marrow reserves and a correspondingly reduced ability to produce white blood cells (leukocytes) and platelets (thrombocytes). Conserving the remaining marrow is, therefore, extremely important in these patients.

Table 2. Physical properties of radiopharmaceuticals for bone-pain palliation

Agent	$T_{1/2}$ (d)	Particle energy (β_{max}, MeV)	Photon energy (keV)
^{32}P	14.3	1.71	–
^{89}SrCl	50.5	1.46	–
^{186}Re-HEDP	3.8	1.07	137 (9%)
^{153}Sm-EDTMP	2.0	0.8	103 (29%)
^{117m}Sn-DTPA	13.6	0.13, 0.16[a]	159 (86%)

[a] Conversion electrons

6.4.1 *Strontium-89*

Despite its long research history, ^{89}Sr (Metastron, Amersham) was only recently approved by the FDA for use in palliation of bone pain. Strontium (as SrCl) accumulates in the bone as an analog of calcium [363] and has a high affinity for growing bone such as is found at metastatic bone lesions. For the treatment of painful metastatic sites in the bone from prostate and breast cancer, the overall response rate is approximately 80% with 10 to 15% of patients experiencing complete relief from pain [364]. There is a delay of 2 to 4 weeks between injection and response, but the duration of response is 3 to 6 months [364]. The primary side effect is transient myelosuppression from which the patient typically recovers in 6 weeks. An interesting clinical observation is that, after injection of ^{89}Sr, the concentrations of prostate specific antigen and alkaline phosphatase in the blood of prostate cancer patients frequently decrease. This would suggest that there is a therapeutic effect (other than analgesia) associated with injection of ^{89}Sr, but no increases in patient survival were observed [365,366]. There is, however, a decrease in the rate of appearance of new metastatic sites in the bone suggesting that the ^{89}Sr has some therapeutic effect on the metastatic sites but not on the primary tumor. The lack of therapeutic effect on the primary tumor is not surprising when one considers that there is no reason to presume that ^{89}Sr, a bone-avid radionuclide, would accumulate in the primary tumor (breast or prostate).

6.4.2
Rhenium-186-HEDP

An HEDP (54) complex with ^{186}Re (Etidronate, Mallinckrodt) was first evaluated for use in the treatment of bone pain by Mathieu [367] and Eisenhut [368], but the efforts by these investigators were complicated by the chemical instability of their preparations [369]. The development of the radiopharmaceutical was based on the chemical similarity of technetium and rhenium, and the biodistribution of ^{186}Re-HEDP is very similar to that of the ^{99m}Tc bone-imaging MDP (55) complex [358,360].

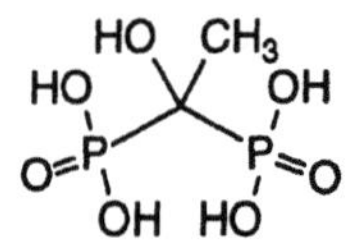

HEDP **54**

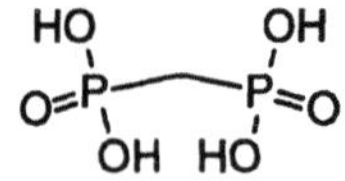

MDP **55**

The ^{186}Re radiopharmaceutical, like ^{99m}Tc-MDP, is not a single discreet complex but rather a mixture of oligomers [369]. Because of this, it was necessary to chromatograph the product prior to use in order to isolate the component with the highest affinity for bone and most rapid clearance from non-target tissue [370]. In more recent reports, a kit formulation is described with no reference to chromatographic purification [371]. This formulation requires heating (10 min at 98–100 °C) and includes 3 mg of gentisic acid (an antioxidant) which apparently eliminates at least some of the contaminants present in the earlier formulation.

The ^{186}Re-HEDP complex is incorporated into the bone because of the interaction of the phosphonate ligand with the hydroxyapatite matrix of growing bone. The therapeutic effect is similar to that of ^{89}Sr with up to 75% of patients responding and 15% experiencing complete pain relief [371,372]. Pain relief begins within 2 weeks, often within days, and lasts for 5 to 8 weeks. Experiments performed comparing ^{186}Re-HEDP and ^{99m}Tc-MDP showed that the response was not a placebo effect [373]. The primary toxicity is transient myelosuppression with a decrease in platelet counts of about 50% at 4 weeks after treatment and recovery by 8 weeks [371]. As with ^{89}Sr, there is a decrease in alkaline phosphatase concentration suggesting some transient tumoricidal effect but no corresponding increase in patient survival [359,371,372].

Two important differences between ^{186}Re-HEDP and ^{89}Sr are the time to response and duration of response. With ^{186}Re-HEDP, many patients respond within 24 to 48 h, and all who respond do so within two weeks. In contrast, the time to response for ^{89}Sr is between two and four weeks. The other side of the coin is that the duration of response for ^{186}Re-HEDP is shorter than that of ^{89}Sr, 7 to 8 weeks versus three to six months. Not surprisingly, these differences parallel the differences in the physical half-lives of the two radionuclides, 3.8 d (^{186}Re) versus 50.5 d (^{89}Sr). The radiation dose is delivered more quickly with ^{186}Re which produces a faster but shorter duration response.

It is interesting to note that the radiation dose delivered to the bone metastases is less than 10 Gy, but a dose of 40–80 Gy is typically required for cell killing [374]. This suggests that a mechanism other than direct destruction of the cancerous cells is responsible for the observed clinical response. Recently, there was also a report of the possible use of ^{188}Re-HEDP for treatment of painful osseous metastases [375]. The higher energy of the ^{188}Re β^- particle (2.12 MeV, $T_{1/2}$ = 17 h) suggests, however, that myelosuppression might be more of a problem with ^{188}Re than with ^{186}Re and other radionuclides that emit lower energy electrons. The advantage of ^{188}Re versus ^{186}Re is that it can be obtained from the $^{188}W/^{188}Re$ generator [376].

6.4.3 *Samarium-153-EDTMP*

The second coordination compound that is available for the treatment of bone pain is a ^{153}Sm complex with EDTMP (**56**) (Quadramet, DuPont Pharma). The crystal structure of the complex has not been obtained, but it has been characterized by elemental analysis, NMR, and fast atom bombardment mass spectrometry (FAB-MS) as $Rb_5[Sm(EDTMP)] \cdot 3H_2O$ and is a monomer in solution [167].

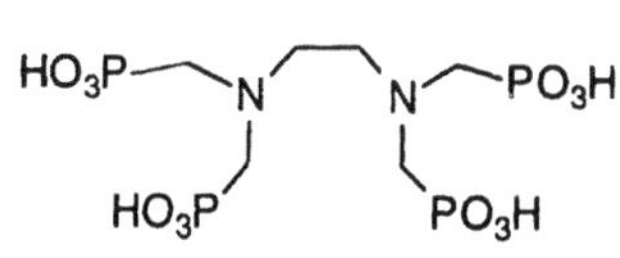

EDTMP **56**

As with ^{186}Re-HEDP, the time to therapeutic response is shorter than that observed for ^{89}Sr (less than 2 weeks) and the duration of response is shorter (2 to 3 months) [377–379], consistent with the shorter half-lives of these radionuclides compared to ^{89}Sr. As with the other palliative agents, the response rate is 60–90%, and there is transient myelotoxicity from which the patient typically recovers six to eight weeks after treatment [377–379]. There is also some evidence of a short-term tumoricidal effect at higher injected doses [379].

6.4.4 *Tin(IV)-117m-DTPA*

The most recent palliative agent to enter clinical trials is ^{117m}Sn(IV)-DTPA [380], although Durbin reported as early as 1960 that Sn(citrate) delivered more than 30% of the injected dose to the bone [363]. Tin-117m-DTPA is somewhat different from other radiopharmaceuticals in this group in that the radiation dose is delivered by conversion electrons rather than by β^- particles. The conversion electrons emitted by ^{117m}Sn have considerably lower energy than the β^- particles emitted by the other agents (Table 2).

The hypothesis of these investigators is that the lower energy (and shorter range) electrons deliver more of the radiation dose to the lesions and less to the marrow. This complex also differs from the other radiopharmaceuticals in this group in that it contains neither phosphate-like ligands (MDP, HEDP or EDTMP) nor is it a calcium analog (^{89}Sr). The reason for its accumulation in bone remains undetermined but may be due to transchelation of the tin from the DTPA ligand to phosphate in the hydroxyapatite bone matrix and the insolubility of tin(IV) phosphates [380]. In a recent clinical trial that included 39 patients [381], 30% of the patients experienced complete pain relief, 45% experienced a greater than 50% decrease in pain, and 25% experienced a less than 50% decrease or no decrease in pain. The mean duration of response was 98 d (range: 20–402 d). These results are similar to those obtained with the β^- emitters described above. Phase II and III clinical trials are now being conducted (Diatide, Londonderry, NH).

Although the clinical response is similar to that obtained with the other therapeutic radiopharmaceuticals, there is a significant difference in myelotoxicity. For ^{117m}Sn-DTPA, however, early evidence suggests that myelotoxicity is lower than that incurred with the β^--emitters. For WBCs, the decrease was significantly lower than for all three β^- emitters [381]. For platelets, the decrease for ^{117m}Sn was less than that for ^{89}Sr but not significantly different than that for either ^{186}Re-HEDP or ^{153}Sm-EDTP [381]. A lower percentage of patients experienced moderate to severe myelotoxicity with ^{117m}Sn than for the other radiopharmaceuticals, and no patients experienced more than mild thrombocytopenia [381]. In comparison, 15–100% of the patients treated with ^{89}Sr, ^{186}Re or ^{153}Sm experienced moderate or greater WBC toxicity, and 25–60% experienced moderate or greater platelet toxicity. If these results are borne out in larger clinical studies, it will validate the hypothesis that conversion electrons can be used to deliver a more localized therapeutic radiation dose than β^--emitters.

6.4.5 *Bismuth-212-DOTMP*

The need to reduce radiation exposure to the bone marrow can also be addressed by using α-emitting radionuclides. Using a methylene phosphonic acid derivative of

DOTA, **57**, investigators prepared a ^{212}Bi complex whose biodistribution in mice is very similar to that of ^{153}Sm-EDTMP [382]. A limitation of ^{212}Bi is that the half-life (1.0 h) may be too short for this application. These investigators suggest that this limitation may be circumvented by using ^{212}Pb-DOTMP (^{212}Pb, $T_{1/2}$ = 10.6 h) as an in vivo generator of ^{212}Bi.

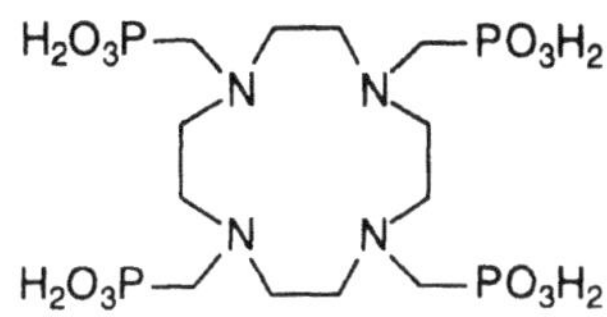

DOTMP **57**

7 Hypoxia

Hypoxia is a condition in which tissue demand for oxygen exceeds the available supply, usually because of inadequate tissue perfusion. It is an intermediate state between normal oxygenation (normoxia) and the complete absence of oxygen (anoxia). An imaging agent that localizes in hypoxic tissue could be used to identify regions of myocardial ischemia, brain tissue at risk from strokes, and malignant tissue where the tumor's need for oxygen outgrows its supply. When combined with therapeutic radionuclides, these agents might also prove useful in radionuclide therapy of hypoxic tumors that do not respond to conventional radiation therapy.

The challenge in developing an imaging agent for hypoxic tissue is that, by definition, the blood supply and, therefore, the ability to deliver the tracer to the hypoxic tissue is diminished. Because of this, accumulation of the tracer in the tissue must occur over a somewhat longer period of time than is required for normoxic tissue. For this to occur, the tracer must remain in circulation for a prolonged period of time compared to the more common objective of eliminating the radiopharmaceutical from the blood as quickly as possible after injection. This makes evaluation of potential hypoxia agents somewhat more complicated than other radiopharmaceuticals. Essentially any radiopharmaceutical that remains in the circulation for a prolonged period of time will appear to accumulate to some degree in hypoxic tissue simply due to the fact that it is cleared more slowly from this tissue than from tissue that is adequately perfused.

The imaging agents that have been developed for hypoxia are, generally, based on neutral nitroimidazoles which undergo a stepwise reduction to amines after passive diffusion into hypoxic cells [383]. As has usually been the case in nuclear medicine, the first radiopharmaceuticals based on these compounds were the radiohalogenated derivatives; ^{82}Br [384,385], ^{18}F [386,387], and ^{131}I [388,389]. Recently, several ^{99m}Tc derivatives have been developed that show promise in this application, some of which contain nitroimidazoles substituents and some that do not.

7.1 BATO-Nitroimidazole Derivatives

Using the same BATO framework from which the myocardial and cerebral perfusion agents were developed, investigators at Bristol-Myers Squibb prepared a BATO-imidazole adduct in which the nitroimidazole moiety replaced the 2-methylpropyl substituent on the boronic acid (**23**) [390]. The nitroimidazole substituent of the BATO molecule was reduced by xanthine oxidase (XOD) under anaerobic conditions in vitro, albeit at a rate considered less than optimal for imaging purposes. The slower rate of reduction was attributed to the possibility that the high lipophilicity and bulk of the BATO molecules compared to nitroimidazole might interfere with access of the nitro group to the active site of the enzyme [383,390].

7.2 Technetium-99m-PnAO Derivatives

Additional efforts by the Bristol-Myers Squibb group led to the development of TcO(PnAO-1-(2-nitroimidazole) (PnAO-1-(2-nitroimidazole) = 3,3,9,9-tetramethyl-1-(2-nitro-1*H*-imidazol-1-yl)-4,8-diazaundecane-2,10-dione dioxime) (BMS-181321, **58**) [391,392]. The complex is a 2-nitroimidazole derivative of TcO(PnAO) (**19**) [113] (see Sect. 3) and exists as two enantiomers.

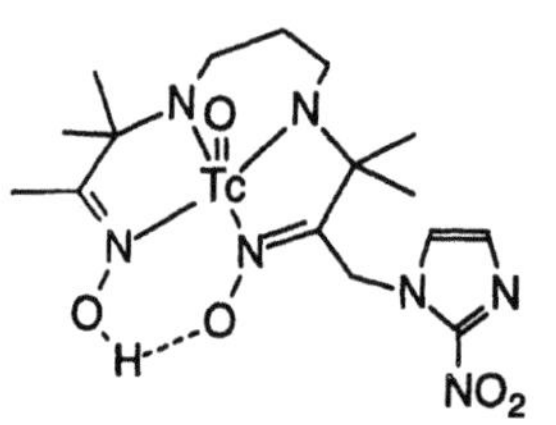

Tc(O)(PnAO-1-(2-nitroimidazole)) (BMS-181321) **58**

Because the biological properties of enantiomers frequently differ, Linder et al. resolved the isomers to evaluate possible differences in the imaging properties [392]. They observed, however, that the resolved isomers rapidly racemized in the presence of water precluding biological studies.

In vitro studies demonstrate that BMS-181321 is reduced to a polar species that is retained within the cell [393]. In vivo studies have been carried out in animals that suggest that BMS-181321 can be used to delineate ischemic areas in myocardium [394,395], brain [396] and tumors [397]. In two studies, investigators compared the retention of BMS-181321 and a non-nitroimidazole-containing PnAO derivative (6-methylpropyleneamine oxime) in rat [398] and rabbit [394] models of ischemic myocardium and observed that BMS-181321 quickly washed out of normoxic myocardium but was selectively retained in hypoxic myocardium. The 6-methyl derivative was not retained in either tissue. The accumulation in hypoxic tissue is, however, relatively small compared to the uptake of BMS-181321 by the liver (which can obscure the apex of the heart) [399]. At this time, no human studies have been reported for this compound.

7.3 Technetium-99m-HL91

A second amine oxime derivative that is being evaluated as a hypoxia-imaging agent is ^{99m}Tc-HL91 (**59**).

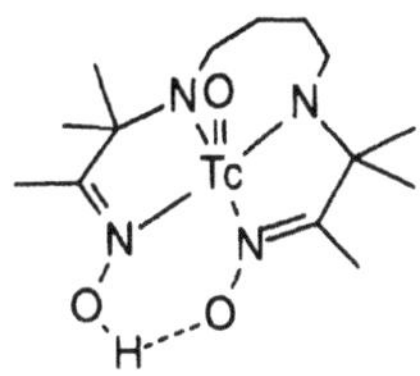

59

This compound is simply the 1,4-butanediamine oxime derivative of PnAO (**19**) containing a fourth methylene group in the "bridge" between the amines [400]. The addition of this single methylene group changes the reduction potential enough to allow the compound to be reduced in hypoxic environments. The details of its retention mechanism have not been reported, so it is unclear what chemical change causes this behavior.

In open-chest studies in dogs, ^{99m}Tc-HL91 was successfully used to image regional myocardial ischemia and appears to clear from the liver more quickly than BMS-181321 [401]. In human studies of hypoxic tumors that included ten patients, the compound accumulated in the lesions in seven patients in whom the lesion was also identified by [^{18}F]FDG PET, two patients where PET studies were not available, and did not identify one lesion that was weakly positive by FDG/PET [402]. The SPECT images included in this study also suggest that liver uptake of ^{99m}Tc-HL91 is low compared to that of BMS-181321.

7.4 Copper-62-ATSM

Another example of a potential hypoxia-imaging agent derived from a rCBF agent is ^{62}Cu-ATSM. This complex is an analog of Cu-PTSM (**18**) with an additional methyl substituent, **60** [403].

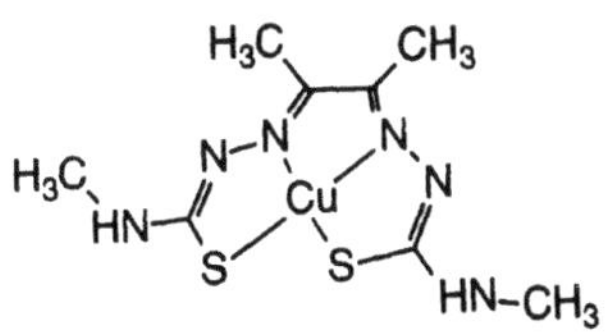

60

As with ^{99m}Tc-HL91, the compound does not contain a nitroimidazole substituent, but is itself reduced under hypoxic conditions. This system exploits the tendency of the copper atom in Cu(II)-bisthiosemicarbazone complexes to be reduced to Cu(I) in vivo, as described above (see Sects. 2 and 3). This complicates the use of Cu-PTSM for myocardial and cerebral perfusion imaging, but the sensitivity of the reduction potential to ligand substitution [404] allows the "tuning" of the redox potential for use in hypoxia imaging.

In isolated perfused rat hearts, retention was increased 3–4-fold in hypoxic versus normoxic conditions [403]. Although ^{62}Cu has the advantage of being generator produced, the half-life of ^{62}Cu may be too short for use in this application; ^{64}Cu, because of its longer half-life, might prove more compatible with this biological process.

8 Other

8.1 Pulmonary

An interesting development in pulmonary imaging has been the introduction of Technegas for ventilation imaging [405–407]. Technegas is produced by heating a solution of $^{99m}TcO_4^-$ in a specially designed graphite crucible (Tetley Medical, Ltd), first at low temperature to evaporate the water and then at 2550 °C in an argon atmosphere to produce an aerosol [408]. The aerosol is inhaled by the patient to produce images of the lungs that reflect ventilation, the distribution of air in the lungs during inhalation.

The structure of the aerosol was, at one time, reported to include Tc-fullerenes, as might be expected of a carbon aerosol produced under these conditions [409], but a recent report by Senden and co-workers concludes that the aerosol is instead composed of platelets of technetium metal contained within a thin layer of graphitic carbon [410].

8.2 Inflammation, Infection and Deep-Vein Thrombosis

The development of new imaging agents for infection/inflammation and deep-vein thrombosis has been the focus of a major research effort recently, primarily by industry. Most of this effort has been focused on peptide-based agents that use intrinsic chelators; that is, a portion of the peptide is constructed so that it acts as a chelating agent for ^{99m}Tc as described for Diatide's ^{99m}Tc-Octreotide cancer agent. The remainder of the peptide is designed to target a specific receptor. For example, the deep-vein thrombosis agents typically include the sequence Arg-Gly-Asp (RGD) or a mimetic and bind to the GP IIb/IIIa receptor, which is involved in the formation of blood clots. A large number of reviews covering this topic have recently been published, and the reader is referred to these for a comprehensive discussion of this topic [411–416].

References

1. Perrier C, Segre' E (1937) J Chem Phys 5:712
2. Perrier C, Segre' E (1947) Nature 159:24
3. Cotton FA, Wilkinson G (1972) Advanced inorganic chemistry. Interscience, New York, p 974
4. Eckelman WC, Richards P (1971) Nucl Med (Stuttg) 10:245

5. Lederer CM, Shirley V, Browne E, Dairiki JM, Doebler RE, Shihab-Eldin AA, Jardine LJ, Tuli JK, Buyrn AB (1978) Table of isotopes, 7th edn. Wiley, New York
6. Bass LA, McCarthy DW, Jones LA, Cutler PD, Shefer RE, Klinkowstein RE, Schwarz SW, Cutler CS, Lewis JS, Anderson CJ, Welch MJ (1997) J Labelled Compd Radiopharm 40:325
7. McCarthy DW, Shefer RE, Klinkowstein RE, Bass LA, Margeneau WH, Cutler CS, Anderson CJ, Welch MJ (1997) Nucl Med Biol 24:35
8. Nickles RJ (1991) Acta Radiol Suppl 376:69
9. Lagunas-Solar MC (1983) Int J Nucl Med Biol 10:75
10. Grant PM, O'Brien HA, Kahn M (1975) J Inorg Nucl Chem 37:413
11. Richards P (1960) Congresso Nucleare. Rome, p 223
12. Cotton FA, Wilkinson G (1988) Advanced inorganic chemistry, 5th edn. Wiley, New York
13. Sorenson JA, Phelps ME (1987) Physics in nuclear medicine, 2nd edn. Grune & Stratton, Orlando
14. Lacy JL, LeBlanc AD, Babich JW, Bungo MW, Latson LA, Lewis RM, Poliner LR, Jones RH, Johnson PC (1984) J Nucl Med 25:1003
15. Schwaiger M, Ziegler S (1997) Nuklearmedizin 36:3
16. Strauss HW, Harrison K, Langan JK, Lebowitz E, Pitt B (1975) Circulation 51:641
17. Pitt B, Strauss HW (1976) Am J Cardiol 37:797
18. Atkins HL, Budinger TF, Lebowitz E, Ansari AN, Greene MW, Fairchild RG, Ellis KJ (1977) J Nucl Med 18:133
19. Deutsch E, Bushong W, Glavan KA, Elder RC, Sodd VJ, Scholz KL, Fortman DL, Lukes SJ (1981) Science 214:85
20. Fergusson JE, Nyholm RS (1959) Nature 183:1039
21. Lewis J, Nyholm RS, Rodney GA (1965) J Chem Soc:1483
22. Fergusson JE, Hickford JH (1970) Aust J Chem 23:453
23. Deutsch E, Glavan KA, Sodd VJ, Nishiyama H, Ferguson DL, Lukes SJ (1981) J Nucl Med 22:897
24. Nishiyama H, Deutsch E, Adolph RJ, Sodd VJ, Libson K, Saenger EL, Gerson MC, Gabel M, Lukes SJ, Vanderheyden JL, Fortman DL, Scholz KL, Grossman LW, Williams CC (1982) J Nucl Med 23:1093
25. Gerson MC, Deutsch EA, Nishiyama H, Libson KF, Adolph RJ, Grossman LW, Sodd VJ, Fortman DL, Vanderheyden JL, Williams CC et al. (1983) Eur J Nucl Med 8:371
26. Linder KE, Dewan JC, Davison A (1989) Inorg Chem 28:3820
27. Saha GB (1992) Fundamentals of nuclear pharmacy. Springer, Berlin Heidelberg New York
28. Eckelman WC, Steigman J (1991) Nucl Med Biol 18:3
29. Glavan KA, Kronauge JF (1983) E.R. Squibb & Sons Inc, US Patent 4,374,821
30. Neirinckx RD, Glavan KA, Eakins MN, Kronauge JF (1983) In: Raynaud C (ed) Nuclear medicine and biology advances, vol 1. Pergamon Press, Oxford, p 698
31. Cohen AI, Glavan KA, Kronauge JF (1983) Biomed Mass Spect 10:287
32. Kronauge JF, Glavan KA (1984) E.R. Squibb & Sons Inc, US Patent 4,481,184
33. Linder KE (1985) EI DuPont de Nemours and Co, US Patent 4,512,967
34. Deutsch E, Ketring AR, Libson K, Vanderheyden JL, Hirth WW (1989) Nucl Med Biol 16:191
35. Gerson MC, Deutsch EA, Libson KF, Adolph RJ, Ketring AR, Vanderheyden JL, Williams CC, Saenger EL (1984) Eur J Nucl Med 9:403
36. Wester DW, Coveney JR, Nosco DL, Robbins MS, Dean RT (1991) J Med Chem 34:3284
37. Dean RT, Miller FW, Wester DW (1986) Mallinckrodt Inc, US Patent 4,582,700
38. Gerundini P, Savi A, Gilardi MC, Margonato A, Vicedomini G, Zecca L, Hirth W, Libson K, Bhatia JC, Fazio F et al (1986) J Nucl Med 27:409
39. Deutsch E, Hirth W (1987) J Nucl Med 28:1491
40. Baldas J, Bonnyman J, Poier PM, William GA, Mackay MF (1981) J Chem Soc Dalton Trans 1798
41. Archer CM, Dilworth JR, Jobanputra P, Parrot SJ, Thompson RM, Kelly JD (1990) In: Mazzi U (ed) Technetium and rhenium in chemistry and nuclear medicine, vol 3. Raven Press, New York, p 405
42. Kelly JD, Higley B, Archer CM, Canning LR, Chiu KW, Edwards B, Forster AM, Gill HK, Latham IA, Pickett RD, Webbon P, Edwards PG, Imran A, Griffiths DV, York DC, Mahoney PM, Tonkinson DJ, Dilworth JR, Lahiri A (1990) In: Mazzi U (ed) Technetium and rhenium in chemistry and nuclear medicine, vol 3. Raven Press, New York, p 405
43. Zanelli GD, Cook N, Lahiri A, Ellison D, Webbon P, Woolley G (1988) J Nucl Med 29:62

44. Zanelli GD, Lahiri A, Patel N, Smith T, Brady F, Radfar G, Cook ND, Crawley JC, Silvester DJ (1987) Eur J Nucl Med 13:12
45. Abrams MJ, Davison A, Jones AG, Costello CE, Pang H (1983) Inorg Chem 22:2798
46. Tulip TH, Calabrese J, Kronauge JF, Davison A, Jones AG (1986) In: Mazzi U (ed) Technetium and rhenium in chemistry and nuclear medicine, vol 2. Raven Press, New York, p 119
47. Carpenter APJ, Maheu LJ, Patz MA, Tulip TH, Linder KE, Subramanyam V, Thompson JS (1960) EI Du Pont de Nemours and Co, US Patent 4,894,445
48. Holman BL, Jones AG, Lister-James J, Davison A, Abrams MJ, Kirshenbaum JM, Tumeh SS, English RJ (1984) J Nucl Med 25:1350
49. Jones AG, Davison A, Kronauge J, Abrams MJ (1988) Massachusetts Institute of Technology and President and Fellows of Harvard College, US Patent 4,735,793
50. Kronauge JF, Noska MA, Davison A, Holman BL, Jones AG (1992) J Nucl Med 33:1357
51. Kronauge JF, Davison A, Roseberry AM, Costello CE, Maleknia S, Jones AG (1991) Inorg Chem 30:4265
52. Kronauge JF, Leon AS, Verdera ES, Balter HS, Leon ET, Mut F, Oliveira MC, Garcia FA, Holman BL, Davison A et al (1992) J Nucl Med 33:1949
53. Holman BL, Sporn V, Jones AG, Sia ST, Perez-Balino N, Davison A, Lister-James J, Kronauge JF, Mitta AE, Camin LL et al (1987) J Nucl Med 28:13
54. Piwnica-Worms D, Kronauge JF, Holman BL, Lister JJ, Davison A, Jones AG (1988) J Nucl Med 29:55
55. Carroll TR, Maheu LJ, Patz MA, Tulip TH, Linder KE (1988) Eur J Nucl Med 14:231
56. Mousa SA, Cooney JM, Williams SJ (1987) J Am Coll Cardiol 9:137A
57. Kronauge JF, Kawamura M, Holman BL, Davison A, Jones AG, Costello CE, Zeng C (1990) In: Mazzi U (ed) Technetium and rhenium in chemistry and nuclear medicine, vol 3. Raven Press, New York, p 677
58. Wackers FJ, Berman DS, Maddahi J, Watson DD, Beller GA, Strauss HW, Boucher CA, Picard M, Holman BL, Fridrich R (1989) J Nucl Med 30:301
59. Piwnica-Worms D, Kronauge JF, Chiu ML (1990) Circulation 82:1826
60. Piwnica-Worms D, Chiu ML, Kronauge JF (1992) Circulation 85:1531
61. Backus M, Piwnica-Worms D, Hockett D, Kronauge J, Lieberman M, Ingram P, LeFurgey A (1993) Am J Physiol 265:C178
62. Piwnica-Worms DP, Kronauge JF, LeFurgey A, Backus M, Hockett D, Ingram P, Lieberman M, Holman BL, Jones AG, Davison A (1994) Magn Reson Imaging 12:641
63. Piwnica-Worms D, Kronauge JF, Holman BL, Davison A, Jones A (1989) Invest Radiol 24:25
64. Carvalho PA, Chiu ML, Kronauge JF, Kawamura M, Jones AG, Holman BL, Piwnica-Worms D (1992) J Nucl Med 33:1516
65. Kelly JD, Chiu KW, Latham IA (1991) Amersham International plc, US Patent 5,045,302
66. Kelly JD, Forster AM, Higley B, Archer CM, Booker FS, Canning LR, Chiu KW, Edwards B, Gill HK, McPartlin M, Nagle K, Latham I, Pickett R, Storey A, Webbon P (1993) J Nucl Med 34:222
67. Platts EA, North TL, Pickett RD, Kelly JD (1995) J Nucl Cardiol 2:317
68. Higley B, Smith FW, Smith T, Gemmell HG, Das GP, Gvozdanovic DV, Graham D, Hinge D, Davidson J, Lahiri A (1993) J Nucl Med 34:30
69. Jain D, Wackers FJ, Mattera J, McMahon M, Sinusas AJ, Zaret BL (1993) J Nucl Med 34:1254
70. Deutsch E, Vanderheyden JL, Gerundini P, Libson K, Hirth W, Colombo F, Zecca L, Savi A, Fazio F (1987) J Nucl Med 28:1870
71. Rossetti C, Vanoli G, Paganelli G, Kwiatkowski M, Zito F, Colombo F, Bonino C, Carpinelli A, Casati R, Deutsch K, et al (1994) J Nucl Med 35:1571
72. Kirchhoff JR, Heineman WR, Deutsch E (1987) Inorg Chem 26:3108
73. Gerson MC, Lukes J, Deutsch E, Biniakiewicz D, Washburn LC, Walsh RA (1995) J Nucl Cardiol 2:224
74. Chiu KW, Kelly JD, Latham IA, Griffiths DV, Edwards PG (1990) Amersham International plc, US Patent 4,916,214
75. Lahiri A, Higley B, Crawley JCW, Chiu KW, Edwards B, Smith T, Griffiths DV, Archer CM, Latham IA, Kelly JD (1989) J Nucl Med 30:818
76. Meleca MJ, McGoron AJ, Gerson MC, Millard RW, Gabel M, Biniakiewicz D, Roszell NJ, Walsh RA (1997) J Nucl Med 38:1847
77. Treher EN, Francesconi LC, Gougoutas JZ, Malley MF, Nunn AD (1989) Inorg Chem 28:3411

78. Stewart RE, Schwaiger M, Hutchins GD, Chiao PC, Gallagher KP, Nguyen N, Petry NA, Rogers WL (1990) J Nucl Med 31:1183
79. Kronauge JF, Chiu ML, Cone JS, Davison A, Holman BL, Jones AG, Piwnica WD (1992) Nucl Med Biol 19:141
80. Rumsey WL, Rosenspire KC, Nunn AD (1992) J Nucl Med 33:94
81. Narra RK, Feld T, Nunn AD (1992) J Nucl Med 33:88
82. Leppo JA, DePuey EG, Johnson LL (1991) J Nucl Med 32:2012
83. Li QS, Solot G, Frank TL, Wagner HN, Jr., Becker LC (1991) J Nucl Med 32:1968
84. Pasqualini R, Duatti A, Bellande E, Comazzi V, Brucato V, Hoffschir D, Fagret D, Comet M (1994) J Nucl Med 35:334
85. Uccelli L, Giganti M, Duatti A, Bolzati C, Pasqualini R, Cittanti C, Colamussi P, Piffanelli A (1995) J Nucl Med 36:2075
86. Fagret D, Marie PY, Brunotte F, Giganti M, Le Guludec D, Bertrand A, Wolf JE, Piffanelli A, Chossat F, Bekhechi D et al (1995) J Nucl Med 36:936
87. Ghezzi C, Fagret D, Arvieux CC, Mathieu JP, Bontron R, Pasqualini R, de Leiris J, Comet M (1995) J Nucl Med 36:1069
88. Johnson G, 3rd, Nguyen KN, Pasqualini R, Okada RD (1997) J Nucl Med 38:138
89. Johnson G, 3rd, Nguyen KN, Liu Z, Gao P, Pasqualini R, Okada RD (1997) J Nucl Cardiol 4:217
90. Neirinckx RD, Kronauge JF, Loberg MD (1983) Int J Appl Radiat Isot 34:721
91. Mullani NA (1984) J Nucl Med 25:1190
92. Williams BR, Mullani NA, Jansen DE, Anderson BA (1994) J Nucl Med 35:1586
93. Loc'h C, Maziere B, Comar D (1980) J Nucl Med 21:171
94. Harris WR, Recoraro VL (1983) Biochem 22:292
95. Tsang BW, Mathias CJ, Green MA (1993) J Nucl Med 34:1127
96. Kung HF, Liu BL, Mankoff D, Kung MP, Billings JJ, Francesconi L, Alavi A (1990) J Nucl Med 31:1635
97. Zweit J, Goodall R, Cox M, Babich JW, Potter GA, Sharma HL, Ott RJ (1992) Eur J Nucl Med 19:418
98. Fujibayashi Y, Matsumoto K, Yonekura Y, Konishi J, Yokoyama A (1989) J Nucl Med 30:1838
99. Green MA (1987) Nucl Med Biol 14:59
100. Green MA, Klippenstein DL, Tennison JR (1988) J Nucl Med 29:1549
101. Beanlands RS, Muzik O, Mintun M, Mangner T, Lee K, Petry N, Hutchins GD, Schwaiger M (1992) J Nucl Med 33:684
102. Barnhart BA, Green MA (1991) Nucl Med Biol 18:865
103. Fujibayashi Y, Wada K, Taniuchi H, Yonekura Y, Konishi J, Yokoyama A (1993) Biol Pharm Bull 16:146
104. Fujibayashi Y, Taniuchi H, Wada K, Yonekura Y, Konishi J, Yokoyama A (1995) Ann Nucl Med 9:1
105. Mathias CJ, Welch MJ, Raichle ME, Mintun MA, Lich LL, McGuire AH, Zinn KR, John EK, Green MA (1990) J Nucl Med 31:351
106. Mathias CJ, Bergmann SR, Green MA (1993) Nucl Med Biol 20:343
107. Shelton ME, Green MA, Mathias CJ, Welch MJ, Bergmann SR (1989) J Nucl Med 30:1843
108. Lagreze HL, Levine RL (1987) Am J Physiol Imaging 2:208
109. Ell PJ, Jarritt PH, Costa DC, Cullum ID, Lui D (1987) Semin Nucl Med 17:214
110. Leveille J, Demonceau G, De RM, Rigo P, Taillefer R, Morgan RA, Kupranick D, Walovitch RC (1989) J Nucl Med 30:1902
111. Dishino DD, Welch MJ, Kilbourn MR, Raichle ME (1983) J Nucl Med 24:1030
112. Rapoport SI (1976) The blood-brain barrier in physiology and medicine. Raven Press, New York
113. Volkert WA, Hoffman TJ, Seger RM, Troutner DE, Holmes RA (1984) Eur J Nucl Med 9:511
114. Troutner DE, Volkert WA, Hoffman TJ, Holmes RA (1984) Int J Appl Radiat Isot 35:467
115. Holm S, Andersen AR, Vorstrup S, Lassen NA, Paulson OB, Holmes RA (1985) J Nucl Med 26:1129
116. Leonard JP, Nowotnik DP, Neirinckx RD (1986) J Nucl Med 27:1819
117. Neirinckx RD, Canning LR, Piper IM, Nowotnik DP, Pickett RD, Holmes RA, Volkert WA, Forster AM, Weisner PS, Marriott JA et al (1987) J Nucl Med 28:191

118. Nowotnik DP, Canning LR, Cumming SA, Harrison RC, Higley B, Nechvatal G, Pickett RD, Piper IM, Bayne VJ, Forster AM et al (1985) Nucl Med Commun 6:499
119. Sharp PF, Smith FW, Gemmell HG, Lyall D, Evans NT, Gvozdanovic D, Davidson J, Tyrrell DA, Pickett RD, Neirinckx RD (1986) J Nucl Med 27:171
120. Ballinger JR, Reid RH, Gulenchyn KY (1988) J Nucl Med 29:1998
121. Neirinckx RD, Burke JF, Harrison RC, Forster AM, Andersen AR, Lassen NA (1988) J Cereb Blood Flow Metab 8:S4
122. Tubergen K, Corlija M, Volkert WA, Holmes RA (1991) J Nucl Med 32:111
123. Hung JC, Volkert WA, Holmes RA (1989) Nucl Med Biol 16:675
124. Medi-Physics, Inc (1996) Ceretec Product Insert. Arlington Heights, IL
125. McKenzie EH, Volkert WA, Holmes RA (1985) Int J Nucl Med Biol 12:133
126. Lever SZ, Burns HD, Kervitsky TM, Goldfarb HW, Woo DV, Wong DF, Epps LA, Kramer AV, Wagner HN Jr (1985) J Nucl Med 26:1287
127. Epps LA, Burns HD, Lever SZ, Goldfarb HW, Wagner HN Jr (1987) Nucl Inst Meth 38:661
128. Kung HF, Guo YZ, Yu CC, Billings J, Subramanyam V, Calabrese JC (1989) J Med Chem 32:433
129. Kung HF, Ohmomo Y, Kung MP (1990) Semin Nucl Med 20:290
130. Vallabhajosula S, Zimmerman RE, Picard M, Stritzke P, Mena I, Hellman RS, Tikofsky RS, Stabin MG, Morgan RA, Goldsmith SJ (1989) J Nucl Med 30:599
131. Walovitch RC, Hill TC, Garrity ST, Cheesman EH, Burgess BA, O'Leary DH, Watson AD, Ganey MV, Morgan RA, Williams SJ (1989) J Nucl Med 30:1892
132. Walovitch RC, Cheesman EH, Maheu LJ, Hall KM (1994) J Cereb Blood Flow Metab 14:S4
133. Holman BL, Hellman RS, Goldsmith SJ, Mena IG, Leveille J, Gherardi PG, Moretti JL, Bischof-Delaloye A, Hill TC, Rigo PM et al (1989) J Nucl Med 30:1018
134. Devous MDS (1994) J Nucl Med 35:1094
135. Spencer SS (1994) Epilepsia 35:S72
136. Grunwald F, Menzel C, Pavics L, Bauer J, Hufnagel A, Reichman K, Sakowski R, Elger CE, Biersack HJ (1994) J Nucl Med 35:1896
137. Menzel C, Steidele S, Grunwald F, Hufnagel A, Pavics L, Elger CE, Biersack HJ (1996) J Nucl Med 37:1106
138. Packard AB, Roach PJ, Davis RT, Carmant L, Davis R, Riviello J, Holmes G, Barnes PD, O'Tuama LA, Bjornson B, Treves ST (1996) J Nucl Med 37:1101
139. Narra RK, Nunn AD, Kuczynski BL, DiRocco RJ, Feld T, Silva DA, Eckelman WC (1990) J Nucl Med 31:1370
140. Morgan GF, Deblaton M, Clemens P, Van Den Broeck P, Bossuyt A, Thornback JR (1991) J Nucl Med 32:500
141. Bossuyt A, Morgan GF, Deblaton M, Pirotte R, Chirico A, Clemens P, Vandenbroeck P, Thornback JR (1991) J Nucl Med 32:399
142. Sun Y, Anderson CJ, Pajeau TS, Reichert DE, Hancock RD, Motekaitis RJ, Martell AE, Welch MJ (1996) J Med Chem 39:458
143. Madsen SL, Welch MJ, Motekaitis RJ, Martell AE (1992) Nucl Med Biol 19:431
144. Moore DA, Fanwick PE, Welch MJ (1990) Inorg Chem 29:672
145. Herrero P, Markham J, Weinheimer CJ, Anderson CJ, Welch MJ, Green MA, Bergmann SR (1993) Circulation 87:173
146. Green MA, Mathias CJ, Welch MJ, McGuire AH, Perry D, Fernandez RF, Perlmutter JS, Raichle ME, Bergmann SR (1990) J Nucl Med 31:1989
147. Mathias CJ, Welch MJ, Perry DJ, McGuire AH, Zhu X, Connett JM, Green MA (1991) Nucl Med Biol 18:807
148. Taniuchi H, Fujibayashi Y, Okazawa H, Yonekura Y, Konishi J, Yokoyama A (1995) Biol Pharm Bull 18:1126
149. Kaufman MJ, Madras BK (1991) Synapse 9:43
150. Hantraye P, Brownell AL, Elmaleh D, Spealman RD, Wullner U, Brownell GL, Madras BK, Isacson O (1992) Neuroreport 3:265
151. Reith ME, Meisler BE, Sershen H, Lajtha A (1986) Biochem Pharmacol 35:1123
152. Madras BK, Kaufman MJ (1994) Synapse 18:261
153. Meegalla S, Plossl K, Kung MP, Chumpradit S, Stevenson DA, Frederick D, Kung HF (1996) Bioconjug Chem 7:421

154. Meegalla S, Plossl K, Kung MP, Svenson DA, Liable-Sands LM, Rheingold AL, Kung HF (1995) J Am Chem Soc 117:11037
155. Meegalla SK, Plossl K, Kung MP, Chumpradit S, Stevenson DA, Kushner SA, McElgin WT, Mozley PD, Kung HF (1997) J Med Chem 40:9
156. Kung MP, Stevenson DA, Plossl K, Meegalla SK, Beckwith A, Essman WD, Mu M, Lucki I, Kung HF (1997) Eur J Nucl Med 24:372
157. Kung HF, Kim HJ, Kung MP, Meegalla SK, Plossl K, Lee HK (1996) Eur J Nucl Med 23:1527
158. Madras BK, Jones AG, Mahmood A, Zimmerman RE, Garada B, Holman BL, Davison A, Blundell P, Meltzer PC (1996) Synapse 22:239
159. Meltzer PC, Blundell P, Jones AG, Mahmood A, Garada B, Zimmerman RE, Davison A, Holman BL, Madras BK (1997) J Med Chem 40:1835
160. Meltzer PC, Liang AY, Madras BK (1996) J Med Chem 39:371
161. Lever SZ, Baidoo KE, Mahmood A, Matsumura K, Scheffel U, Wagner HN, Jr (1994) Nucl Med Biol 21:157
162. Jurisson SS, Pirro J, DiRocco RJ, Rosenspire KC, Jagoda E, Nanjappan P, Eckelman WC, Nowotnik DP, Nunn AD (1995) Nucl Med Biol 22:269
163. Ikeda I, Inoue O, Kurata K (1976) Int J Appl Radiat Isot 27:681
164. Enlander D, Weber PM, dosRemedios LV (1974) J Nucl Med 15:743
165. Medi-Physics, Inc (1993) Technetium Tc99m Succimer Product Insert. Arlington Heights, IL
166. Eckelman WC, Meinken G, Richards P (1972) J Nucl Med 13:577
167. Jurisson S, Berning D, Jia W, Ma DS (1993) Chem Rev 93:1137
168. Harbert JC, Andrich MP, Peller PJ (1996) In: Harbert JC, Eckelman WC, Neumann RD (eds) Nuclear medicine: diagnosis and therapy. Thieme Medical, New York, p 713
169. Mallinckrodt Medical, Inc (1991) Technescan Gluceptate Product Insert. St. Louis, MO
170. Mallinckrodt Medical, Inc (1982) Technescan Product Insert. St. Louis, MO
171. de Kieviet W (1981) J Nucl Med 22:703
172. Kasina S, Rao TN, Srinivasan A, Sanderson JA, Fitzner JN, Reno JM, Beaumier PL, Fritzberg AR (1991) J Nucl Med 32:1445
173. Eshima D, Taylor A, Jr., Fritzberg AR, Kasina S, Hansen L, Sorenson JF (1987) J Nucl Med 28:1180
174. Davison A, Jones AG, Orvig C, Sohn M (1981) Inorg Chem 20:1629
175. Jones AG, Davison A, LaTegola MR, Brodack JW, Orvig C, Sohn M, Toothaker AK, Lock CJ, Franklin KJ, Costello CE, Carr SA, Biemann K, Kaplan ML (1982) J Nucl Med 23:801
176. Kasina S, Fritzberg AR, Eshima D, Johnson DL, Jones AG, Lister-James J, Davison A, Brodack JW (1984) J Nucl Med 23:16
177. Fritzberg AR, Kuni CC, Klingensmith WCd, Stevens J, Whitney WP (1982) J Nucl Med 23:592
178. Rao TN, Adhikesavalu D, Camerman A, Fritzberg AR (1990) J Amer Chem Soc 112:5798
179. Kasina S, Fritzberg AR, Johnson DL, Eshima D (1986) J Med Chem 29:1933
180. Fritzberg AR, Kasina S, Eshima D, Johnson DL (1986) J Nucl Med 27:111
181. Despopoulos A (1965) J Theor Biol 8:163
182. Jafri RA, Britton KE, Nimmon CC, Solanki K, Al-Nahhas A, Bomanji J, Fettich J, Hawkins LA (1988) J Nucl Med 29:147
183. Russell CD, Thorstad B, Yester MV, Stutzman M, Baker T, Dubovsky EV (1988) J Nucl Med 29:1189
184. Russell CD, Thorstad BL, Yester MV, Stutzman M, Dubovsky EV (1988) J Nucl Med 29:1931
185. Verbruggen AM, Nosco DL, Van Nerom CG, Bormans GM, Adriaens PJ, De Roo MJ (1992) J Nucl Med 33:551
186. Stoffel M, Jamar F, Van Nerom C, Verbruggen A, Mourad M, Leners N, Squifflet JP, Beckers C (1994) J Nucl Med 35:1951
187. Stoffel M, Jamar F, Van Nerom C, Verbruggen A, Besse T, Squifflet JP, Beckers C (1996) Eur J Nucl Med 23:365
188. Gupta NK, Bomanji JB, Waddington W, Lui D, Costa DC, Verbruggen AM, Ell PJ (1995) Eur J Nucl Med 22:617
189. Kronauge JF, Barbarics E, Davison A, Jones AG (1993) J Labelled Compd Radiopharm 32:62
190. Oppenheim BE (1996) In: Sandler MP, Coleman RE, Wackers FJT, Patton JA, Gottschalk A, Hoffer PB (eds) Diagnostic nuclear medicine, vol 1. Williams & Wilkins, Baltimore, p 749

191. Treves ST, Jones AG, Markisz J (1995) In: Treves ST (ed) Pediatric nuclear medicine, Springer, Berlin Heidelberg New York, p 466
192. Fink-Bennett D (1996) In: Sandler MP, Coleman RE, Wackers FJT, Patton JA, Gottschalk A, Hoffer PB (eds) Diagnostic nuclear medicine, vol 1. Williams & Wilkins, Baltimore, p 759
193. Zerbib E (1995) Annales de Chirurgie 49:637
194. Chervu LR, Nunn AD, Loberg MD (1982) Semin Nuc Med 12:5
195. Clarke MJ, Podbielski L (1987) Coord Chem Rev 78:253
196. Dewanjee MK (1990) Semin Nuc Med 20:5
197. Vera DR, Stadalnik RC, Krohn KA (1985) J Nucl Med 26:1157
198. Stadalnik RC, Kudo M, Eckelman WC, Vera DR (1993) Invest Radiol 28:64
199. Torizuka K, Ha-Kawa SK, Ikekubo K, Suga Y, Tanaka Y, Hino M, Ito H, Yamamoto K, Yonekura Y (1991) Kaku Igaku 28:1321
200. Front D, Bar-Shalom R, Israel O (1997) Semin Nuc Med 27:68
201. van Amsterdam JA, Kluin-Nelemans JC, van Eck-Smit BL, Pauwels EK (1996) Ann Hematol 72:202
202. Front D, Israel O (1995) Semin Nuc Med 25:60
203. Weiner RE (1996) Nucl Med Biol 23:745
204. Salvatore M, Carratu L, Porta E (1976) Radiology 121:487
205. Cox P, Belfer A, von der Pompe W (1976) Br J Radiol 49:767
206. Biersack HJ, Briele B, Hotze AL, Oehr P, Qian L, Mekkawy MA, Shih WJ (1992) Ann Nucl Med 6:131
207. Tonami N (1993) Kaku Igaku 30:449
208. Waxman AD, Ramanna L, Memsic LD, Foster CE, Silberman AW, Gleischman SH, Brenner RJ, Brachman MB, Kuhar CJ, Yadegar J (1993) J Nucl Med 34:18
209. Khalkhali I, Mena I, Diggles L (1994) Eur J Nucl Med 21:357
210. Khalkhali I, Cutrone J, Mena I, Diggles L, Venegas R, Vargas H, Jackson B, Klein S (1995) J Nucl Med 36:1784
211. Khalkhali I, Iraniha S, Diggles LE, Cutrone JA, Mishkin FS (1997) Q J Nucl Med 41:231
212. Waxman AD (1997) Semin Nuc Med 27:40
213. DuPont-Merck, Inc (1997) Miraluma Product Insert. Billerica, MA
214. Ohta H, Yamamoto K, Endo K, Mori T, Hamanaka D, Shimazu A, Ikekubo K, Makimoto K, Iida Y, Konishi J et al (1984) J Nucl Med 25:323
215. Babbar A, Kashyap R, Chauhan UP (1991) J Nucl Biol Med 35:100
216. Chauhan UP, Babbar A, Kashyap R, Prakash R (1992) Nucl Med Biol 19:825
217. Dadachova E, Chapman J (1998) Nucl Med Commun 19:173
218. Endo K (1995) Kaku Igaku 32:1125
219. Ambrus E, Rajtar M, Ormandi K, Sera T, Toszegi A, Lang J, Pavics L, Csernay L (1997) Anticancer Res 17:1599
220. Kobayashi H, Sakahara H, Itoh T, Kudoh T, Takagi T, Shibuya K, Nomura S, Hosono M, Endo K, Konishi J (1993) J Nucl Med 34:815
221. Bisunadan MM, Blower PJ, Clarke SE, Singh J, Went MJ (1991) Int J Radiat Appl Inst 42:167
222. Singh J, Reghebi K, Lazarus CR, Clarke SE, Callahan AP, Knapp FF Jr, Blower PJ (1993) Nucl Med Commun 14:197
223. Blower PJ, Singh J, Clarke SE (1991) J Nucl Med 32:845
224. Yokoyama A, Hata N, Horiuchi K, Masuda H, Saji H, Ohta H, Yamamoto K, Endo K, Torizuka K (1985) Int J Nucl Med Biol 12:273
225. Brazeau P, Vale W, Burgus R, Ling N, Butcher M, Rivier J, Guillemin R (1973) Science 179:77
226. Hellman B, Lernmark A (1969) Endocrinology 84:1484
227. Bauer W, Briner U, Doepfner W, Haller R, Huguenin R, Marbach P, Petcher TJ, Pless J (1982) Life Sci 31:1133
228. Bakker WH, Krenning EP, Reubi JC, Breeman WA, Setyono HB, de Jong M, Kooij PP, Bruns C, van Hagen PM, Marbach P et al (1991) Life Sci 49:1593
229. Smith-Jones PM, Stolz B, Albert R, Ruser G, Briner U, Mäcke HR, Bruns C (1998) Nucl Med Biol 25:181
230. Kwekkeboom DJ, Krenning EP (1997) Eur Radiol 7:1103
231. Valkema R, Steens J, Cleton FJ, Pauwels EK (1996) J Cancer Res Clin Oncol 122:513

232. Krenning EP, Kwekkeboom DJ, Bakker WH, Breeman WA, Kooij PP, Oei HY, van Hagen M, Postema PT, de Jong M, Reubi JC et al (1993) Eur J Nucl Med 20:716
233. Lamberts SW, Reubi JC, Krenning EP (1993) Acta Oncol 32:167
234. Krenning EP, Bakker WH, Kooij PP, Breeman WA, Oei HY, deJong M, Reubi JC, Visser TJ, Bruns C, Kwekkeboom DJ et al (1992) J Nucl Med 33:652
235. Krenning EP, Kwekkeboom DJ, Reubi JC, Van Hagen PM, van Eijck CH, Oei HY, Lamberts SW (1992) Metabolism 41:83
236. Vallabhajosula S, Moyer BR, Lister-James J, McBride BJ, Lipszyc H, Lee H, Bastidas D, Dean RT (1996) J Nucl Med 37:1016
237. Pearson DA, Lister-James J, McBride WJ, Wilson DM, Martel LJ, Civitello ER, Taylor JE, Moyer BR, Dean RT (1996) J Med Chem 39:1361
238. Mäcke HR, Smith-Jones P, Maina T, Stolz B, Albert R, Bruns C, Reist H (1993) Horm Metab Res Suppl 27:12
239. Maina T, Stolz B, Albert R, Bruns C, Koch P, Mäcke H (1994) Eur J Nucl Med 21:437
240. Anderson CJ, Pajeau TS, Edwards WB, Sherman EL, Rogers BE, Welch MJ (1995) J Nucl Med 36:2315
241. Moffat FL Jr, Pinsky CM, Hammershaimb L, Petrelli NJ, Patt YZ, Whaley FS, Goldenberg DM (1996) J Clin Oncol 14:2295
242. Breitz HB, Tyler A, Bjorn MJ, Lesley T, Weiden PL (1997) Clin Nucl Med 22:615
243. Mardirossian G, Brill AB, Dwyer KM, Kahn D, Nelp W (1996) J Nucl Med 37:1583
244. Neal CE, Meis LC (1994) Semin Nuc Med 24:272
245. Bohdiewicz PJ, Scott GC, Juni JE, Fink-Bennett D, Wilner F, Nagle C, Dworkin HJ (1995) Clin Nucl Med 20:230
246. Meares C, Wensel T (1984) Accts Chem Res 17:202
247. Goldenberg DM (1990) In: McGuire WL (ed) Cancer treatment and research, vol 51. Kluwer, Boston
248. Magerstadt M (1991) Antibody conjugates and malignant disease. CRC Press, Boca Raton
249. Parker D (1991) Chem Rev 91:1441
250. Liu Y, Wu C (1991) Pure Appl Chem 63:427
251. Goldenberg DM (1994) Cancer therapy with radiolabeled antibodies. CRC Press, Boca Raton
252. Sautter-Bihl M-L, Bihl H, Wannenmacher M (1996) Systemic radiotherapy with monoclonal antibodies. Springer, Berlin Heidelberg New York
253. Kohler G, Milstein C (1975) Nature 256:495
254. Hu S, Shively L, Raubitschek A, Sherman M, Williams LE, Wong JY, Shively JE, Wu AM (1996) Cancer Res 56:3055
255. Adams GP (1998) In vivo 12:11
256. Krejcarek GE, Tucker KL (1977) Biochem Biophys Res Commun 77:581
257. Hnatowich DJ, Layne WW, Childs RL (1982) Int J Appl Radiat Isot 33:327
258. Brechbiel MW, Gansow OA, Atcher RW, Schlom J, Esteban J, Simpson DE, Colcher D (1986) Inorg Chem 25:2772
259. McCall MJ, Diril H, Meares CF (1990) Bioconjug Chem 1:222
260. Rodwell JD, Alvarez VL, Lee C, Lopes AD, Goers JW, King HD, Powsner HJ, McKearn TJ (1986) Proc Natl Acad Sci USA 83:2632
261. Govindan SV, Goldenberg DM, Griffiths GL, Leung SO, Losman MJ, Hansen HJ (1995) Cancer Res 55:5721s
262. Fritzberg AR (1987) Nuklearmedizin 26:7
263. Arano Y, Wakisaka K, Mukai T, Uezono T, Motonari H, Akizawa H, Kairiyama C, Ohmomo Y, Tanaka C, Ishiyama M, Sakahara H, Konishi J, Yokoyama A (1996) Nucl Med Biol 23:129
264. Cox JPL, Craig AS, Helps IM, Jankowski KJ, Parker D, Eaton MAW, Millican AT, Millar K, Beeley NRA, Boyce BA (1990) J Chem Soc Perkins Trans 1 2567
265. Lewis MR, Shively JE (1998) Bioconjug Chem 9:72
266. Paik CH, Yokoyama K, Reynolds JC, Quadri SM, Min CY, Shin SY, Maloney PJ, Larson SM, Reba RC (1989) J Nucl Med 30:1693
267. Deshpande SV, Denardo SJ, Meares CF, McCall MJ, Adams GP, Denardo GL (1989) Nucl Med Biol 16:587
268. Faivre-Chauvet A, Gestin JF, Mease RC, Sai-Maurel C, Thedrez P, Slinkin M, Meinken GE, Srivastava SC, Chatal JF (1993) Nucl Med Biol 20:763

269. Goodwin DA (1991) Antib Immunoconjug Radiopharm 4:427
270. Martell AE, Smith RM (1974) Critical stability constants. Plenum Press, New York
271. Shannon RD (1976) Acta Crystallogr A 32:751
272. Roselli M, Schlom J, Gansow OA, Raubitschek A, Mirzadeh S, Brechbiel MW, Colcher D (1989) J Nucl Med 30:672
273. Brechbiel MW, Gansow OA (1991) Bioconjug Chem 2:187
274. Westerberg DA, Carney PL, Rogers PE, Kline SJ, Johnson DK (1989) J Med Chem 32:236
275. Keana JFW, Mann JS (1990) J Org Chem 55:2868
276. Izard ME, Boniface GR, Hardiman KL, Brechbiel MW, Gansow OA, Walkers KZ (1992) Bioconjug Chem 3:346
277. Margerum DW, Cayley GR, Weatherburn DC, Pagenkopf GK (1978) In: Martell AE (ed) Coodination Chemistry, ACS Monograph 174, vol 2. American Chemical Society, Washington DC
278. Waldmann TA, White JD, Carrasquillo JA, Reynolds JC, Paik CH, Gansow OA, Brechbiel MW, Jaffe ES, Fleisher TA, Goldman CK et al (1995) Blood 86:4063
279. Knox SJ, Goris ML, Trisler K, Negrin R, Davis T, Liles TM, GrilloLopez A, Chinn P, Varns C, Ning SC, Fowler S, Deb N, Becker M, Marquez C, Levy R (1996) Clin Cancer Res 2:457
280. Kroll H, Korman S, Siegel E, Hart HE, Rosoff B, Spencer H, Laszlo D (1957) Nature 180:919
281. Srivastava SC, Mease RC, Meinken GE, Mausner LF, Steplewski Z (1990) Acta Radiol Suppl 374:135
282. Srivastava SC, Mease RC (1991) Nucl Med Biol 18:589
283. Loussouarn A, Duflos M, Benoist E, Chatal JF, Lebaut G, Gestin JF (1998) J Chem Soc Perkins Trans 1 237
284. Brechbiel MW, Pippin CG, McMurry TJ, Milenic D, Roselli M, Colcher D, Gansow OA (1991) J Chem Soc Chem Commun 1169
285. Wu C, Kobayashi H, Sun B, Yoo TM, Paik CH, Gansow OA, Carrasquillo JA, Pastan I, Brechbiel MW (1997) Bioorg Med Chem 5:1925
286. Mathias CJ, Sun YZ, Connett JM, Philpott GW, Welch MJ, Martell AE (1990) Inorg Chem 29:1475
287. Schwarz SW, Mathias CJ, Sun JY, Dilley WG, Wells SA Jr, Martell AE, Welch MJ (1991) Nucl Med Biol 18:477
288. Schuhmacher J, Klivenyi G, Hull WE, Matys R, Hauser H, Kalthoff H, Schmiegel WH, Maier-Borst W, Matzku S (1992) Nucl Med Biol 19:809
289. Schuhmacher J, Klivenyi G, Matys R, Stadler M, Regiert T, Hauser H, Doll J, Maier-Borst W, Zoller M (1995) Cancer Res 55:115
290. Craig AS, Parker D, Adams H, Bailey NA (1989) J Chem Soc Chem Commun 1793
291. Wu C, Jagoda E, Brechbiel M, Webber KO, Pastan I, Gansow O, Eckelman WC (1997) Bioconjug Chem 8:365
292. Lee J, Garmestani K, Wu C, Brechbiel MW, Chang HK, Choi CW, Gansow OA, Carrasquillo JA, Paik CH (1997) Nucl Med Biol 24:225
293. McMurry TJ, Brechbiel M, Wu C, Gansow OA (1993) Bioconjug Chem 4:236
294. Brechbiel MW, McMurry TJ, Gansow OA (1993) Tetrahedron Lett 34:3691
295. Desreux JF (1980) Inorg Chem 19:1319
296. Moi MK, Meares CF, Denardo SJ (1988) J Am Chem Soc 110:6266
297. Cox JPL, Jankowski KJ, Kataky R, Parker D, Beeley NRA, Boyce BA, Eaton MAW, Millar K, Millican AT, Harrison A, Walker C (1989) J Chem Soc Chem Commun 797
298. McMurry TJ, Brechbiel M, Kumar K, Gansow OA (1992) Bioconjug Chem 3:108
299. Kline SJ, Betebenner DA, Johnson DK (1991) Bioconjug Chem 2:26
300. Schlom J, Siler K, Milenic DE, Eggensperger D, Colcher D, Miller LS, Houchens D, Cheng R, Kaplan D, Goeckeler W (1991) Cancer Res 51:2889
301. Aime S, Botta M, Ermondi G, Terreno E, Anelli PL, Fedeli F, Uggeri F (1996) Inorg Chem 35:2726
302. Ruegg CL, Anderson-Berg WT, Brechbiel MW, Mirzadeh S, Gansow OA, Strand M (1990) Cancer Res 50:4221
303. Delgado R, Fausto da Silva JJR (1982) Talanta 29:815
304. Huneke RB, Pippin CG, Squire RA, Brechbiel MW, Gansow OA, Strand M (1992) Cancer Res 52:5818
305. Kumar K, Magerstadt M, Gansow OA (1989) J Chem Soc Chem Commun 145

306. Mirzadeh S, Kumar K, Gansow OA (1993) Radiochim Acta 60:1
307. Ruble G, Wu C, Squire RA, Ganswo OA, Strand M (1996) Int J Radiat Oncol Biol Phys 34:609
308. Horak E, Hartmann F, Garmestani K, Wu C, Brechbiel M, Gansow OA, Landolfi NF, Waldmann TA (1997) J Nucl Med 38:1944
309. Fritzberg AR (1998) Neo-Rx Corporation. personal comunication
310. Wu SL, Horrocks WD (1995) Inorg Chem 34:3724
311. Takenouchi K, Tabe M, Watanabe K, Hazato A, Kato Y, Shionoya M, Koike T, Kimura E (1993) J Org Chem 58:6895
312. Srivastava SC, Mausner LF, Mease RC, Meinken GE, Joshi V, Kolsky K, Sweet M, Steplewski Z (1995) Int J Pharmocogn 33:92
313. Pulukkody KP, Norman TJ, Parker D, Royle L, Broan CJ (1993) J Chem Soc Perkins Trans 2:605
314. Blower PJ, Lewis JS, Zweit J (1996) Nucl Med Biol 23:957
315. Pippin CG, Kumar K, Mirzadeh S, Gansow OA (1991) J Labeled Cmpd Radiopharm 30:221
316. Rogers BE, Anderson CJ, Connett JM, Guo LW, Edwards WB, Sherman EL, Zinn KR, Welch MJ (1996) Bioconjug Chem 7:511
317. Mercer-Smith JA, Cole DA, Roberts JC, Lewis D, Behr MJ, Lavallee DK (1989) Adv Exp Med Biol 258:103
318. Bedel-Cloutour CH, Maneta-Peyret L, Pereyre M, Bezian JH (1991) J Immunol Methods 144:35
319. Bedel-Cloutour CH, Mauclaire L, Saux A, Pereyre M (1996) Bioconjug Chem 7:617
320. Sakata I, Takata H, Ikeuchi M, Hirata N, Inui H, Nakajima S, Yamauchi H, Koshimizu K (1990) Yakugaku Zasshi 110:916
321. Bhalgat MK, Roberts JC, MercerSmith JA, Vessella RL, Lavallee DK (1997) Drug Deliv 4:13
322. Chen X, Ji M, Fisher DR, Wai CM (1998) 215th Meeting of the American Chemical Society. Dallas, TX, Abst NUCL032
323. Fisher DR (1998) 215th Meeting of the American Chemical Society. Dallas, TX, Abst NUCL017
324. Chen XY, Ji M, Fisher DR, Wai CM (1998) J Chem Soc Chem Commun 377
325. (1981) Gmelin Handbook of Inorganic Chemistry, vol 1. Springer, Berlin Heidelberg New York
326. Kennel SJ (1998) Oak Ridge National Laboratory. personal comunication
327. Fritzberg AR, Berninger RW, Hadley SW, Wester DW (1988) Pharmaceutical Research 5:325
328. Franz J, Volkert WA, Barefield EK, Holmes RA (1987) Nucl Med Biol 14:569
329. Stahl W, Kuhlmann L, Wiesner M, Walch A (1994) Bioorg Med Chem Lett 4:2597
330. John CS, Schlemper EO, Hosain P, Paik CH, Reba RC (1992) Nucl Med Biol 19:269
331. Pillai MRA, John CS, Lo JM, Troutner DE, Corlija M, Volkert WA, Holmes RA (1993) Nucl Med Biol 20:211
332. van Gog FB, Visser GW, Klok R, van der Schors R, Snow GB, van Dongen GA (1996) J Nucl Med 37:352
333. Goldrosen MH, Biddle WC, Pancook J, Bakshi S, Vanderheyden JL, Fritzberg AR, Morgan AC, Jr., Foon KA (1990) Cancer Res 50:7973
334. Fritzberg AR (1997) Rhenium, Rhenium Alloys, Proc Int Symp 479
335. Eckelman WC (1995) Eur J Nucl Med 22:249
336. Malin R, Steinbrecher R, Jannsen J, Semmler W, Noll B, Johannsen B, Frommel C, Hohne W, Schneidermergener J (1995) J Am Chem Soc 117:11821
337. Bradley G, Juranka PF, Ling V (1988) Biochim Biophys Acta 948:87
338. Pastan I, Gottesman MM (1991) Annu Rev Med 42:277
339. Gottesman MM, Pastan I (1993) Annu Rev Biochem 62:385
340. Piwnica-Worms D, Chiu ML, Budding M, Kronauge JF, Kramer RA, Croop JM (1993) Cancer Res 53:977
341. Crankshaw CL, Marmion M, Burleigh BD, Deutsch E, Piwnica-Worms D (1995) Society of Nuclear Medicine, 42nd Annual Meeting. Minneapolis, MN, p 130
342. Ballinger JR, Bannerman J, Boxen I, Firby P, Hartman NG, Moore MJ (1996) J Nucl Med 37:1578
343. Bae KT, Piwnica-Worms D (1997) Q J Nucl Med 41:101
344. Del Vecchio S, Ciarmiello A, Salvatore M (1998) Eur J Nucl Med 25:192
345. Del Vecchio S, Ciarmiello A, Potena MI, Carriero MV, Mainolfi C, Botti G, Thomas R, Cerra M, G DA, Tsuruo T, Salvatore M (1997) Eur J Nucl Med 24:150
346. Del Vecchio S, Ciarmiello A, Pace L, Potena MI, Carriero MV, Mainolfi C, Thomas R, G DA, Tsuruo T, Salvatore M (1997) J Nucl Med 38:1348

347. Ciarmiello A, Del Vecchio S, Silvestro P, Potena MI, Carriero MV, Thomas R, Botti G, D'Aiuto G, Salvatore M (1998) J Clin Oncol 16:1677
348. Luker GD, Fracasso PM, Dobkin J, Piwnica-Worms D (1997) J Nucl Med 38:369
349. Goldstein LJ (1995) Curr Probl Cancer 19:65
350. Gianni L (1997) Semin Oncol 24:S10
351. Green MA (1990) J Nucl Med 31:1641
352. Zweit J, Lewis J, Dearling J, Rooney B, Blower P, Coley H, Kelland L (1997) J Nucl Med 38:133P
353. Packard AB, Day PJ, Kronauge JF, Treves ST (1995) J Labelled Compd Radiopharm 37:776
354. Packard AB, Day PJ, Kronauge JF, Wen P, Treves ST (1997) J Labelled Compd Radiopharm 40:484
355. (1998) American Cancer Society, http://www.cancer.org/bottomcancinfo.html
356. Stoll BA (1983) In: Stoll BA (ed) Bone metastases: monitoring and treatment. Raven Press, New York, p 1
357. Twycross RG, Fairfield S (1982) Pain 14:303
358. Serafini AN (1994) Int J Radiat Oncol Biol Phys 30:1187
359. Lewington VJ (1996) Phys Med Biol 41:2027
360. McEwan AJ (1997) Semin Nuc Med 27:165
361. Pecher C (1942) Univ Calif Publ Pharmacol 2:117
362. Lawrence JA, Wasserman LR (1950) Ann Intern Med 33:41
363. Durbin PW (1960) Health Phys 2:225
364. Robinson RG, Preston DF, Schiefelbein M, Baxter KG (1995) JAMA 274:420
365. Porter AT, McEwan AJ (1993) Semin Oncol 20(2):38
366. Porter AT (1994) Eur Urol 26:20
367. Mathieu L, Chevalier P, Galy G, Berger M (1979) Int J Appl Radiat Isot 30:725
368. Eisenhut M (1982) Int J Appl Radiat Isot 33:99
369. Deutsch E, Libson K, Vanderheyden JL, Ketring AR, Maxon HR (1986) Nucl Med Biol 13:465
370. Maxon HR, 3rd, Schroder LE, Thomas SR, Hertzberg VS, Deutsch EA, Scher HI, Samaratunga RC, Libson KF, Williams CC, Moulton JS, et al (1990) Radiology 176:155
371. De Klerk JM, Zonnenberg BA, Blijham GH, Van Het Schip AD, Hoekstra A, Han SH, Quirijnen JM, Van Dijk A, Van Rijk PP (1997) Anticancer Res 17:1773
372. Maxon HR, 3rd, Thomas SR, Hertzberg VS, Schroder LE, Englaro EE, Samaratunga R, Scher HI, Moulton JS, Deutsch EA, Deutsch KF et al (1992) Semin Nuc Med 22:33
373. Maxon HR 3rd, Schroder LE, Hertzberg VS, Thomas SR, Englaro EE, Samaratunga R, Smith H, Moulton JS, Williams CC, Ehrhardt GJ et al (1991) J Nucl Med 32:1877
374. Zonnenberg BA (1996) Mallinckrodt, http://www.mallinckrodt.nl/nucmed/nuclear.htm
375. Maxon HR, III, Schroder LE, Washburn LC, Thomas SR, Samaratunga RC, Biniakiewicz D, Moulton JS, Cummings D, Ehrhardt GJ, Morris V (1998) J Nucl Med 39:659
376. Knapp FF, Jr, Beets AL, Guhlke S, Zamora PO, Bender H, Palmedo H, Biersack HJ (1997) Anticancer Res 17:1783
377. Turner JH, Claringbold PG (1991) Eur J Cancer 27:1084
378. Holmes RA (1992) Semin Nuc Med 22:41
379. Collins C, Eary JF, Donaldson G, Vernon C, Bush NE, Petersdorf S, Livingston RB, Gordon EE, Chapman CR, Appelbaum FR (1993) J Nucl Med 34:1839
380. Srivastava SC, Meinken GE, Richards P, Som P, Oster ZH, Atkins HL, Brill AB, Knapp FF, Jr., Butler TA (1985) Nucl Med Biol 12:167
381. Srivastava SC, Atkins HL, Krishnamurthy GT, Zanzi I, Silberstein EB, Meinken G, Mausner LF, Swailem F, T DA, Cabahug CJ, Lau Y, Park T, Madajewicz S (1998) Clin Cancer Res 4:61
382. Hassfjell SP, Bruland OS, Hoff P (1997) Nucl Med Biol 24:231
383. Nunn A, Linder K, Strauss HW (1995) Eur J Nucl Med 22:265
384. Rasey JS, Krohn KA, Freauff S (1982) Radiat Res 91:542
385. Jette DC, Wiebe LI, Chapman JD (1983) Int J Nucl Med Biol 10:205
386. Jerabek PA, Patrick TB, Kilbourn MR, Dischino DD, Welch MJ (1986) Nucl Inst Meth 37:599
387. Grierson JR, Link JM, Mathis CA, Rasey JS, Krohn KA (1989) J Nucl Med 30:343
388. Biskupiak JE, Grierson JR, Rasey JS, Martin GV, Krohn KA (1991) J Med Chem 34:2165
389. Martin GV, Biskupiak JE, Caldwell JH, Rasey JS, Krohn KA (1993) J Nucl Med 34:918
390. Linder KE, Chan YW, Cyr JE, Nowotnik DP, Eckelman WC, Nunn AD (1993) Bioconjug Chem 4:326

391. Rumsey WL, Cyr JE, Raju N, Narra RK (1993) Biochem Biophys Res Commun 193:1239
392. Linder KE, Chan YW, Cyr JE, Malley MF, Nowotnik DP, Nunn AD (1994) J Med Chem 37:9
393. Jayatilak PG, Chan Y-W, Patel BP, Wen M-L, Rumsey WL (1994) J Nucl Med 35:152P
394. Di Rocco RJ, Bauer AA, Pirro JP, Kuczynski BL, Belnavis L, Chan YW, Linder KE, Narra RK, Nowotnik DP, Nunn AD (1997) Nucl Med Biol 24:201
395. Weinstein H, Reinhardt CP, Leppo JA (1998) J Nucl Med 39:598
396. Di Rocco RJ, Kuczynski BL, Pirro JP, Bauer A, Linder KE, Ramalingam K, Cyr JE, Chan YW, Raju N, Narra RK, et al. (1993) J Cereb Blood Flow Metab 13:755
397. Ballinger JR, Kee JW, Rauth AM (1996) J Nucl Med 37:1023
398. Kusuoka H, Hashimoto K, Fukuchi K, Nishimura T (1994) J Nucl Med 35:1371
399. Rumsey WL, Cyr JE, Raju N, Patel B, Chan E-Y, Narra RK (1993) J Nucl Med 34:146P
400. Archer CM, Edwards B, Kelly JD, King AC, Burke JF, Riley ALM (1995) In: Nicolini M, Bandoli G, Mazzi U (eds) Technetium and rhenium in chemistry and nuclear medicine, vol 4. Raven Press, New York, p 535
401. Okada RD, Johnson G, III, Nguyen KN, Liu Z, Edwards B, Archer CM, North TL, King AC, Kelly JD (1998) Circulation 97:2557
402. Cook GJ, Houston S, Barrington SF, Fogelman I (1998) J Nucl Med 39:99
403. Fujibayashi Y, Taniuchi H, Yonekura Y, Ohtani H, Konishi J, Yokoyama A (1997) J Nucl Med 38:1155
404. John EK, Green MA (1990) J Med Chem 33:1764
405. Salvatori M (1997) Rays 22:51
406. James JM, Testa HJ (1995) Nucl Med Commun 16:802
407. Burch WM (1993) Nucl Med Commun 14:559
408. Burch WM, Sullivan PJ, McLaren CJ (1986) Nucl Med Commun 7:865
409. Mackey DW, Burch WM, Dance IG, Fisher KJ, Willett GD (1994) Nucl Med Commun 15:430
410. Senden TJ, Moock KH, Gerald JF, Burch WM, Browitt RJ, Ling CD, Heath GA (1997) J Nucl Med 38:1327
411. Liu S, Edwards DS, Barrett JA (1997) Bioconjug Chem 8:621
412. Lister-James J, Moyer BR, Dean RT (1997) Q J Nucl Med 41:111
413. Hom RK, Katzenellenbogen JA (1997) Nucl Med Biol 24:485
414. Wong E, Fauconnier T, Bennett S, Valliant J, Nguyen T, Lau F, Lu LFL, Pollak A, Bell RA, Thornback JR (1997) Inorg Chem 36:5799
415. Lister-James J, Moyer BR, Dean T (1996) Q J Nucl Med 40:221
416. Thakur ML (1995) Nucl Med Commun 16:724

Polyoxometalates and Fullerenes as Anti-HIV Agents

Jeffrey T. Rhule[1], Craig L. Hill[2], Zhanmiao Zheng[3], Raymond F. Schinazi[4]*

Department of Chemistry, Emory University, Atlanta, GA 30322, USA
E-mail: [1]*jrhule@emory.edu,* [2]*chill@emory.edu,* [3]*zzheng@emory.edu*

*Veterans Affairs Medical Center, and Laboratory of Biochemical Pharmacology, Department of Pediatrics, Emory University School of Medicine, Decatur, GA 30033 USA
E-mail: [4]*rschina@emory.edu*

The lack of proofreading ability of HIV reverse transcriptase leads to errors and the development of drug-resistant HIV variants. This has prompted a concerted research effort to find new efficacious chemotherapies that are not cross-resistant with currently approved antiviral agents. Two potential classes of compounds that have shown promise as anti-HIV agents are polyoxometalates and fullerenes. Both classes of compounds have shown substantial activity against HIV and low cytotoxicity. Several complementary spectroscopic and other experiments collectively provide strong evidence that the polyoxometalates enter cells. They appear to exhibit multiple modes of action: interference with surface protein function (e.g. HIV gp120) and inhibition of viral enzymes (HIV reverse transcriptase and protease). While some fullerenes, rendered water soluble by covalent modification with organic groups, are also active against HIV in cell culture, their biochemistry and molecular biology is less well understood than that of polyoxometalates. The few fullerene derivatives evaluated to date appear to exhibit virucidal and protease inhibition activity.

Keywords. Polyoxometalates (POMs), Fullerenes, Human immunodeficiency virus (HIV), Antiviral

List of Abbreviations

AIDS	acquired immune deficiency syndrome
AZT	3′-azido-3′-deoxythymidine (Zidovudine)
CD4	T-helper membrane molecule that recognizes class II MHC molecule
CEM	human T-cell lymphoma
DNA Poly α	deoxyribonucleic acid α polymerase
EC_{50}	effective concentration, concentration which inhibits virus by 50%
gp120	glycoprotein 120 (a 120-kDa HIV protein)
gp160	glycoprotein 160 (a 160-kDa HIV protein)
H9	human T-cell lymphoma
HIV	human immunodeficiency virus
HIVP	human immunodeficiency virus protease
HIV RT	human immunodeficiency virus reverse transcriptase
HPA-23	$(NH_4)_{18}[NaSb_9W_{21}O_{86}]$
IC_{50}	inhibitory concentration, concentration where cellular toxicity is 50%
PBMC	peripheral blood mononuclear cells
POM(s)	polyoxometalate(s)
TF228.1.16	gp160 transfected B-cell lymphoma
Vero	normal African green monkey kidney cells

1
Polyoxometalates as Anti-HIV Agents

1.1
Introduction to Polyoxometalates: Basic Chemistry and Structures

Polyoxometalates (POMs) are large inorganic structures akin to clusters. There are two general classes of POMs based on their structural composition: isopolyanions which are composed of d^0 metal cations and oxygen atoms; and heteropolyanions which contain p-, d-, or f-block heteroatoms in addition to the metal cations and oxygen [1–3]. Because most of elements in the periodic table can serve as heteroatoms in POM structures, a large number of POMs with varying properties have already been reported. Furthermore, those reported constitute a modest percentage of those that are potentially accessible via syntheses that are increasingly defensible. The use of POMs as antiviral agents was first reported in 1972 by Raybaud et al. [4]. Since then, more than 200 POMs have been evaluated for antiviral activity and other pharmacological indicators. Several structural classes of POMs are illustrated in Fig. 1.

The majority of antiviral investigations have involved POMs of the common Keggin structure (A in Fig. 1), while the most recent investigations have focused on the double-Keggin POMs (C in Fig. 1) [5,6].

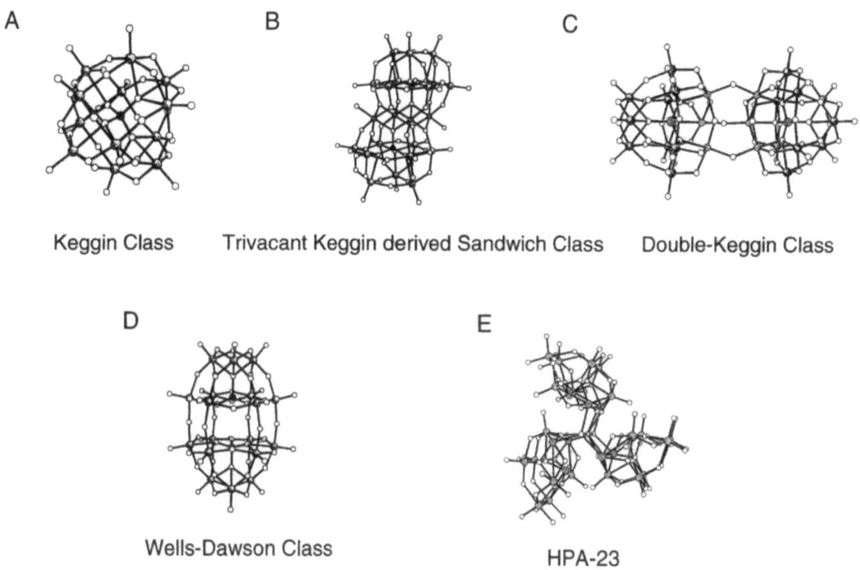

Fig. 1. Ball-and-stick drawings of representative structural families of POMs with antiviral activity. **A** Common Keggin structure $[XW_{12}O_{40}]^{x-}$ (the charge, x, depends on the heteroatom, X) (most antiviral POMs are of this structural class); **B** trivacant Keggin-derived sandwich complex, $[(M^{II})_4(PW_9O_{34})_2]^{10-}$ (M = a first-row transition metal ion); **C** double-Keggin structure, $[\{A\text{-}\alpha\text{-}SiO_4W_9O_{30}(OH)_3Nb_3\}_2(OH)_3]^{11-}$ (some of the most effective antiviral agents are of this class); **D** Wells–Dawson structure, $[X_2W_{18}O_{62}]^{x-}$ (x depends on X); **E** HPA-23, $[NaSb_9W_{21}O_{86}]^{18-}$. Atom designations: O (*open circle*); W (*dotted*); d-block ions (*parallel lines*); P (*crossed hatched*)

The application of POMs in antiviral chemotherapy has been a topic of recent research and interest [7]. A key advantage of POMs, especially in relation to their application as metallopharmaceuticals, is their remarkable alterability. Research in the last few years has led to reproducible methods for the synthesis of POMs of widely varying charge distribution, shape, acidity, polarity, hydrolytic stability and redox potential [7]. Methods are also now available for the replacement or insertion of d- or f-block metals into various POM structures, as well as attachment of organic groups [8]. Given these points, the number of synthetically accessible POMs with potential antiviral activity is truly substantial. In this chapter, the general attributes and chemistry of POMs that impact their anti-HIV activity will be discussed.

1.2 In Vitro and In Vivo Studies

Unlike most therapeutic agents, the data now available are consistent with POMs exhibiting multiple modes of action against HIV: inhibition of the viral enzymes reverse transcriptase (HIV RT) or protease (HIVP) and/or interference with surface viral proteins including gp120, a 120-kDa HIV glycoprotein that is present on the surfaces of both HIV-1 infected cells and on mature virions. Certain POMs inhibit HIV RT, while others inhibit HIVP in cell free assays, but this does not establish that POMs manifest their anti-HIV activity by these modes in cell culture or in mammals. In order to establish the potential relevance of viral enzyme inhibition by POMs, it must first be established that these molecules penetrate cellular membranes and localize intracellularly. We first summarize the experimental evidence that certain POMs do enter cells and then we address the possible modes of POM inhibition.

Although the penetration of cellular membranes by POMs may have been surprising at the outset of this research given the significant sizes (1–5 nm) and molecular charges (3 to 40 or more) of these compounds, there are several lines of evidence consistent with such penetration and cellular entry. A scanning proton microprobe was utilized by Cholewa et al. in two different studies to provide evidence for POM incorporation into cells. In the first study, the POM $[Co_4(H_2O)_2(PW_9O_{34})_2]^{10-}$ was incubated with peripheral blood mononuclear cells (PBMC) and two interesting observations were noted: the intracellular W/Co ratio was equivalent to the W/Co ratio for the free POM; and both W and Co were localized in the same cellular region [9]. In the second study, Cholewa and his co-workers made a number of assumptions to calculate the concentration of POMs and other compounds inside cells after the saturation point. The cellular metal concentration was found to be one order of magnitude lower than that of the cell medium [10]. While both of these reports do not unequivocally demonstrate that POMs are taken into cells and remain intact once inside, they do show that POMs may be incorporated initially.

Along these same lines, Cibert and Jasmin used Raman laser spectroscopy to detect HPA-23 $\{(NH_4)_{18}[NaSb_9W_{21}O_{86}]\}$ inside cells by detecting the characteristic W-O-W stretch of the POM intracellularly [11]. Further analysis of the cells by photonic microscopy revealed hexagonal-shaped precipitates similar in morphology to HPA-23 crystals. Examination of these precipitates by X-ray fluorescence revealed the presence of tungsten.

In a similar experiment, Berry and Galle utilized an electron probe to analyze cells removed from rats that had been treated with HPA-23 [12]. Analysis of these cells revealed cellular precipitates with the same W/Sb ratio as present in free HPA-23. Examination of the nucleus failed to reveal any precipitates.

Finally, Ni et al. used fluorescence microscopy to examine the increase in vacuoles due to POM uptake in cells [13]. Monitoring ^{125}I-labeled acetyl-low-density lipoprotein uptake, Ni et al. proposed that POMs may be moved into the cell via endocytosis by binding to the scavenger receptor of macrophages. Examination of these cells showed electron-dense cellular precipitates resembling those reported by Cibert and Jasmin [11].

Evidence to support the inhibition of viral proteins by POMs has emerged from our group and a few others. Inouye et al. noted in 1992 that POMs may inhibit HIV RT, but this inhibition did not correlate with HIV inhibition in cell culture [14]. Weeks et al. showed that organic-derivatized POMs were more active against HIV RT than cellular DNA polymerase [15]. This preference was the first indication that POMs are selective inhibitors of viral proteins. Weeks et al. noted that some mechanism of viral inhibition other than HIV RT inhibition must be operable since the IC_{50} value against HIV RT (the concentration of chemotherapeutic agent resulting in a 50% inhibition of enzymatic activity) was lower than the EC_{50} value (the concentration of chemotherapeutic agent resulting in a 50% inhibition of virus activity in cell culture). This mode of anti-HIV action was strengthened by data from Yamamoto et al. Using enzymatic studies, this group showed POMs inhibit HIV RT in a concentration dependent manner [16]. Like the initial work presented by Inouye et al., however, the correlation between inhibition of HIV RT and inhibition of HIV in cells was not comparable, again suggesting the presence of additional modes of inhibition intracellularly. Kim et al. also noted the ability of peroxyniobium POMs to inhibit HIV RT selectively over DNA polymerase [17], but the degree of RT inhibition did not account for the cellular inhibition. Similar results have been obtained in the Hill/Schinazi research groups using a variety of POMs [18–20].

Considerable recent experimental evidence from our research group and others indicates that POMs function in part by blocking or interrupting viral absorption. This mechanism of antiviral activity was first proposed in 1973 by Jasmin et al. [21] and since then this has been hypothesized to be important in POM antiviral activity. POMs were postulated to be effective in blocking viral adsorption because of their polyanionic nature. Previous research had established that polyanions, and most notably dextran sulfate and heparin, can act on the cell surface, interrupting viral adsorption [22,23]. The first experimental evidence for blockage of viral adsorption/fusion by POMs was provided by the Hill/Schinazi collaborative group [23]. Fusion of infected cells to uninfected cells was blocked when the two cells were incubated for 24 or 72 h with 50 and 150 μM solutions of POM. POM size and/or charge density correlated to some extent with antiviral activity: smaller, less charged POMs were less effective inhibitors than larger, more charged POMs [23].

Subsequently, Yamamoto et al. reported that heteropolytungstates were effective in blocking cellular adsorption of HIV in MT-4 cells, using time-of-addition studies [16]. These results correlated with those reported by Hill et al., that the majority of anti-HIV action was a result of viral absorption interruption and not direct inhibition of HIV [16,24].

Kim et al. evaluated the ability of POMs to interfere in the gp120 and the CD4 (T-helper membrane molecule that recognizes class II MHC molecule) binding process using an immunosorbent assay [17]. Data from this study indicate both inhibition of viral fusion as well as inhibition of HIV RT by the POMs.

In two separate publications, Inouye et al. provided strong evidence to support blockage of HIV-induced cellular fusion by POMs. In the first study, published in 1993, a Eu-containing POM was shown, by time-of-addition studies, to block viral adsorption and the fusion of HIV-infected cells with uninfected ones. Competition studies utilizing CD4 and the POM in a co-culture of HIV-infected cells showed that the POM prevented syncytium formation. Inhibition of HIV RT was also noted, but could not account for the total antiviral activity of the POM [25]. The second study, published in 1995, provided additional support for POM inhibition of syncytium formation using a syncytium formation assay [26]. Inouye et al. also showed that POM structure, shape or size, as well as viral strain and cell-line selection, may influence the ability of the POM to block viral adsorption.

Collective consideration of the data to date would indicate that POMs exhibit multiple modes of action against HIV. In addition, there seems to be a weak correlation between the size and charge density of POMs and their ability to effectively inhibit HIV. Further experimental evidence in this arena is needed. Judd et al. has presented work that addresses the effect of POM size and charge density POM-protein interactions [27]. Several classes of POMs were tested for their ability to block the binding of virion-derived gp120 to CD4. POMs with molecular weights above 3800 were found to completely inhibit the gp120–CD4 interaction. Keggin class POMs were generally less effective than the larger POMs or organic-derivatized POMs [27].

The promising in vitro results of the first-generation POMs did not translate into in vivo effectiveness. The first trial, conducted in 1985, involved administration of HPA-23 to four patients affected with AIDS [28]. After administration, a reduction in surrogate markers for HIV was demonstrated, but no other therapeutic effect was observed. Bolstered by this work, two separate clinical trials were undertaken in the US [29] and France [30]. Both these subsequent studies failed to show any reduction in the amount of virus present as indicated by HIV p24 antigemia. The marked toxicity of HPA-23 (hepatotoxicity, renal toxicity, and thromobocytopenia) and the apparent lack of efficacy terminated these clinical trials. It is unfortunate that these trials preceded most of the studies of POM biological chemistry and virology. It is clear today that HPA-23 is one of the most toxic POMs known. Indeed, other POMs prepared subsequent to the clinical trials have proven to be effectively non-toxic in high doses in mammals (vide infra).

1.3 Modeling Studies

The complex electronic structure and size of POMs have limited computational studies of their interactions with proteins. However, the last few years has seen some interesting developments as technology (software, hardware, and the availability of high resolution POM and protein X-ray crystal structures) has become more sophisticated. The research group of Sarafianos et al. recently reported the antiviral activity of two POMs, $[(O_3POPO_3)_4W_{12}O_{36}]^{16-}$ and $[(O_3PCH_2PO_3)_4W_{12}O_{36}]^{16-}$, against HIV RT [31]. Enzyme kinetic studies suggested inhibition of the HIV RT involved POM binding to the DNA docking domain of the enzyme. Confirmation of this mode of inhibition was accomplished utilizing radiolabeled DNA template-primer. Sarafianos' group then investigated the feasibility of this interaction by computationally addressing the binding energetics of one of these POMs. Using SYBYL 6.1, POM binding to the HIV RT active site was found to be energetically favorable overall (-1046 kJ mol^{-1}). Favorable electrostatic interactions between the anionic POM and the cationic area of the protein (-1155 kJ mol^{-1}) dominated some unfavorable steric interactions (109 kJ mol^{-1}) [31].

A second modeling study on two Nb-containing POMs, $[NbP_2W_{17}O_{62}]^{7-}$ and $[(NbO_2)P_2W_{17}O_{61}]^{7-}$, was undertaken by our group after we had found these two compounds were potent HIVP inhibitors (IC_{50} ~1 μM) [20]. In this study, the charge surfaces of the POMs and HIVP were analyzed to find the likely site of POM action. The study concluded that it was extremely unlikely on steric grounds that the POMs would bind to HIVP in the normal mode observed in many X-ray structures of inhibitor bound HIVP, namely with the protein flaps closed over the active site. Further investigation revealed that the binding of the POMs to the "flaps-open" form of the active site is favorable despite the anionic character of both protease and the POMs. Refined calculations using the DOCK3 program indicated that one binding site on HIVP is more favorable than the "flaps-open" form of the active site, namely the cationic "hinge" region where the flaps join the bulk of the HIVP structure. POM binding here is favorable both sterically and electrostatically [20].

1.4 Stability and Pharmacokinetics

A meaningful evaluation of activity and toxicity of a chemotherapeutic agent requires both knowledge of its stability and speciation under physiological conditions and its pharmacokinetics (distribution, metabolism and excretion in vivo). A key issue with POMs is that many are not hydrolytically stable at physiological pH. They decompose into POMs of lower nuclearity or simple inorganic products, and this limits the interpretability of biological evaluations [24]. In context with these concerns, the influence of buffer and cations on the POM drug should also be assessed.

The first report addressing the pharmacokinetics of POMs was contained in Bountiff's 1982 doctoral dissertation in which ^{185}W-labeled HPA-23 was monitored as it distributed through rats [32]. This structurally unusual POM (see Fig. 1) was found to accumulate in the liver, kidney, spleen and the brain, and limited quantities were excreted in the urine. Once the dosing regimen was terminated, HPA-23 cleared from all areas except the brain and spleen. This evidence is consistent with the reported toxic effects of HPA-23 when administered to humans [29,30].

The collaborative group of Boudinot, Schinazi and Hill reported the most comprehensive treatment of POM pharmacokinetics [33]. Three antivirally active POMs were injected into rats and monitored in the urine, plasma, faeces, and solubilized organ samples via atomic emission spectroscopy. At lower doses, the compounds bound preferentially to serum proteins. This led to the proposition of two protein binding sites: one with a low affinity and high capacity; and another with a high affinity and low capacity. Tissue samples taken one week after dosing revealed that significant concentrations of the POMs were still present, suggesting that the POMs are slow to move from tissues into the blood for clearance. Closer examination of clearance rates showed that, after one week, total POM recovery from faeces, urine and solubilized tissues accounted for only about half of the initial POM concentration. Contrary to the report by Bountiff [32], no POM was detected within the brain; thus, the POM must redistribute to accumulate in other tissues such as skin, bone and muscle [33].

In a subsequent study, Ni and Boudinot examined the renal and biliary clearances of the same three POMs [34]. The concentration of POMs present after 168 h was determined by examination of faeces, urine and plasma samples through atomic emission spectroscopy. The time dependence of biliary excretion was not similar to the time dependence of the concentration of unbound POM in the plasma, suggesting nonlinear biliary clearance (concentration dependent biliary excretion). This mode of excretion was supported when plots of biliary excretion rate versus unbound plasma POM concentration were examined. In contrast to the early clinical trials of HPA-23, these mammalian pharmacokinetic studies indicated little, if any, toxicity of these second-generation POMs.

1.5 Conclusions and Future Work for Anti-HIV Polyoxometalates

The high therapeutic indices of POMs coupled with the myriad of potentially accessible POMs define these inorganic cluster compounds to be of significant interest as anti-HIV agents. Nonetheless, several issues still need attention.

It might be defensible to expand the definition of "pharmaceutical" to legitimately include compounds other than the strictly organic, low molecular weight ones with high clearance rates that currently dominate research and development in the pharmaceutical industry. Progress in shifting this paradigm has been made with the advent of highly effective commercial metallopharmaceuticals such as cisplatin. In addition, there are significant and rapidly increasing demands for effective antiviral agents in third-world countries where the few anti-HIV agents approved for clinical use in the US are economically prohibitive. POMs still suffer from the stigma of the failed early clinical trials on HPA-23; however, some POMs including $[Si_2Nb_6W_{18}O_{77}]^{8-}$, one of the POMs whose pharmacokinetics were recently examined, are not only effective in vivo but apparently non-toxic in mammals even at doses of 10 mg kg^{-1} per day for two weeks [35].

The low rate of clearance of some POMs poses concern. This issue could be addressed by studying the long-term effects of accumulated POMs in mammals and their protein binding properties. The slow rate of clearance is also a potential advantage as lower and less frequent doses may be needed for efficacy.

Future work in POM antiviral chemotherapy must elucidate the specific modes and loci of action that give rise to some remarkably high biological activities. The specific peptide domains (the number and nature of POM–peptide bonds) involved in the inhibition of viral enzymes and surface proteins by POMs need to be fully elucidated. Such POM–protein interactions need to be optimized as a function of

POM properties (e.g. size and shape, charge density, redox potential, etc.). Finally, more pharmacokinetic investigations of lead POMs in mammals will be required.

2 Fullerenes as Anti-HIV Agents

2.1 Introduction to Fullerenes: Basic Chemistry and Synthesis of Water-Soluble Derivatives

The research activity on fullerenes continues to be intense with fullerene papers averaging the highest citation impact in the chemical literature in the 1990s. While the biological properties of fullerenes have garnered attention recently, the research activity in this sub-area of fullerene science has not matched that in other sub-areas for one simple reason: fullerenes are very hydrophobic and insoluble in water.

Several successful efforts have now been made to render fullerenes water soluble. Lead examples include: (1) treatment of C_{60} with fuming sulfuric acid to make polyarylsulfonated (FC4 S) and polyhydroxylated derivatives [36–38], (2) covalent binding of C_{60} to water-soluble polymers such as poly(ethylene glycol) (PEG), poly(propionylethyleneimine-co-ethyleneimine) and α-amino-ω-methoxypolyethylene glycol (MeO-PEG-NH_2) [39–44], (3) reaction of free amine forms of amino acids and peptides with C_{60} to form fullerene amino acids and peptides [40,45–49], and (4) condensations between porphyrins and fullerenes to make porphyrin–fullerene conjugates [50]. These relatively new water-soluble fullerenes, not surprisingly, have a tendency to aggregate in water or polar solvents by hydrophobic fullerene–fullerene interactions. Such aggregation phenomena make interpretation of their biological activities and toxicities less than straightforward [51].

While the properties of tractable new fullerene derivatives have been assessed in context with antitumoral, cell toxicity, and antiviral studies, and this literature was reviewed three years ago [52], this article focuses solely on anti-HIV activity and in vitro toxicity. Some biological studies of general interest or of peripheral relevance to anti-HIV chemotherapy the reader may wish to consult, however, are as follows: (1) quenching of hydroxyl and superoxide radicals in vitro [53,54], (2) photodynamic tumor therapy using PEG-modified C_{60} as a photosensitizer [55], (3) liposome and lipid bilayer penetration [56], and (4) photochemical cleavage of DNA [57].

2.2 In Vitro and In Vivo Studies

The limited water solubility of even the "water-soluble" fullerenes has necessitated the use of DMSO/water or pyridine solvent systems to prepare the fullerenes in these studies. Pyridine is, of course, toxic and the evalations requiring that solvent were conducted at levels non-toxic to cells. The results of the in vitro assays involving different strains of HIV in various cell lines and the enzymes HIV-1 RT and DNA polymerase α are summarized in Table 1. The table is organized in such a manner as

Table 1. Antiviral activity of fullerenes

Fullerene and reference(s)	HIV or polymerase	Cell line	EC_{50} (μM)	IC_{50} (μM)	Comments
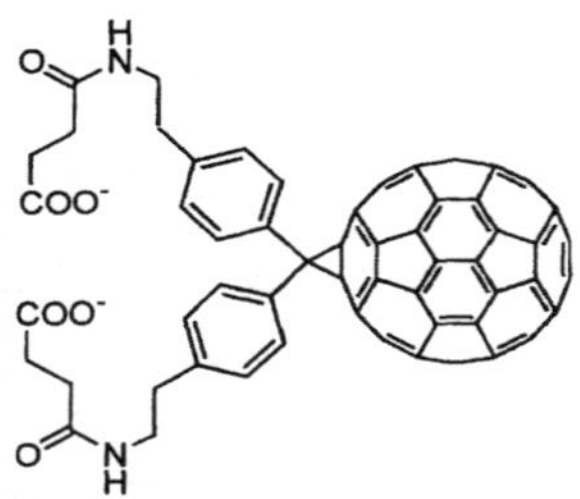	HIV-1_{LAI}	PBMC	7.3	>100	toxicity determined at day 6
	HIV-2_{ROD}	PBMC	5.5	>100	toxicity determined at day 6
	HIV-1_{IIIB}	H9	10.8	>100	toxicity determined at day 6
		Vero		>100	toxicity determined at day 3
		CEM		>100	toxicity determined at day 6
	HIV-1_{H112-2}	PBMC	2.8		AZT-susceptible HIV
	HIV-$1_{G9610-6}$	PBMC	2.7		AZT-resistant HIV
	HIV-1 RT			4.6	
	DNA Poly α			4.9	
	HIVP		2.0		
[58,60,66,76]	HIVP		5.3*		* K_i value from kinetic study
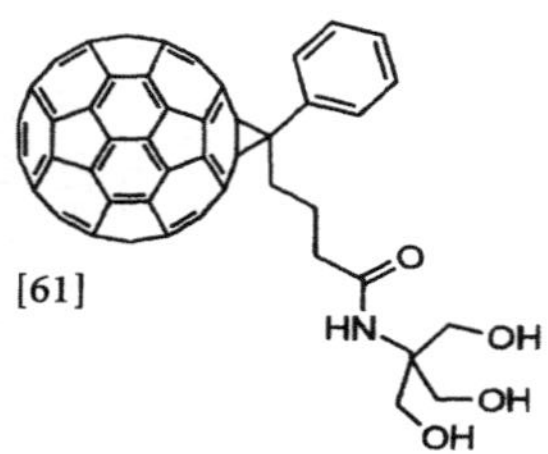	HIV-1_{LAI}	PBMC	2.5	>100	
		Vero		>100	
[61]		CEM		>100	
$C_{60}(OH)_n$ [63]	HIVP		8.3		

$C_{60}(OH)_n$ n=24–26 [66]	HIV-1_{LAI}	PBMC	15.0	>100*	* tritiated thymidine uptake
	HIV-1_{LAI}	H9	>100		
	HIV-1_{LAI}	Vero	36.7	68.2**	** cell proliferation assay
		CEM		8.3**	
$C_{60}(OH)_n$ n=18–20 [66]	HIV-1_{LAI}	PBMC	19.0	>100*	* tritiated thymidine uptake
	HIV-1_{LAI}	H9	>100		
	HIV-1_{LAI}	Vero	38.2	>100**	** cell proliferation assay
		CEM		3.36**	
$C_{60}(PCS)_n$ PCS=polycyclosulfate [66]	HIV-1_{LAI}	PBMC	42.0	>100*	* tritiated thymidine uptake
	HIV-1_{LAI}	H9	>100		
	HIV-1_{LAI}	Vero	>100	>100**	** cell proliferation assay
		CEM		~115**	
$C_{60}(OH)_n$ n=≥16 [66]	HIV-1_{LAI}	PBMC	>100	>100*	* tritiated thymidine uptake
	HIV-1_{LAI}	H9			
	HIV-1_{LAI}	Vero		77.9**	** cell proliferation assay
		CEM		>100**	
$C_{70}(OH)_n$ [63]	HIVP		0.46		
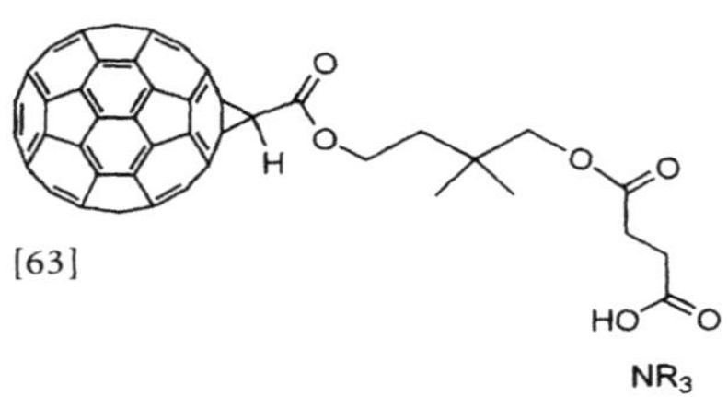 [63]	HIVP		0.24		
	HIVP		0.32*		* K_i value from kinetic study

Table 1. (*Contd.*)

Fullerene and reference(s)	HIV or polymerase	Cell line	EC_{50} (μM)	IC_{50} (μM)	Comments
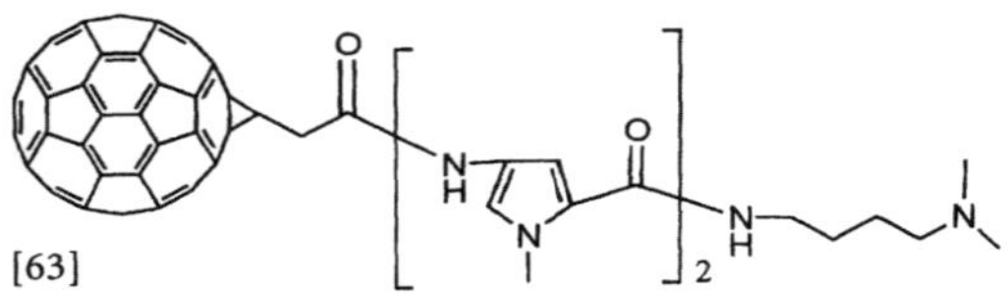 [63]	HIVP		0.59		
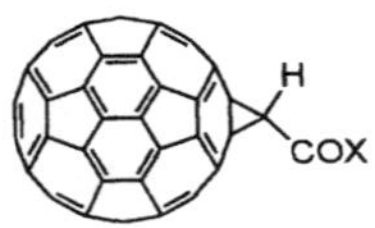 [77]	HIVP		100		X=peptide T [4–8]
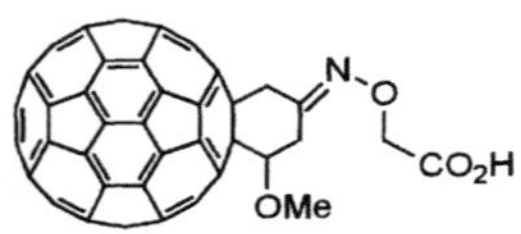 [62]	HIV-1_{LAI}	PBMC Vero	0.9	>100 >100	
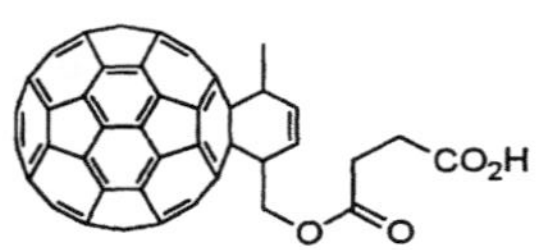 [62]	HIV-1_{LAI}	PBMC Vero	2.2	>100 >100	

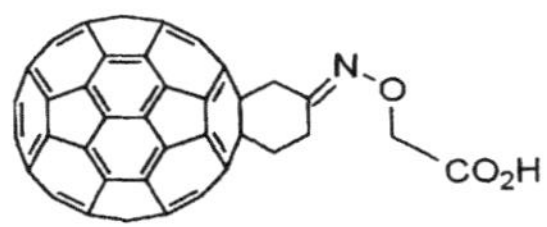 [62]	HIV-1_{LAI}	PBMC Vero	2.9	>100 >100
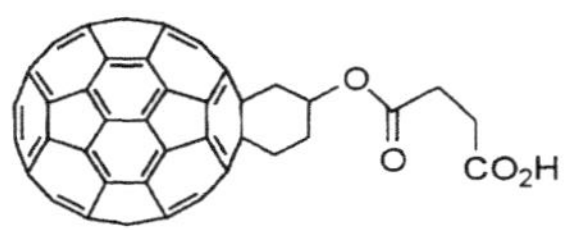 [62]	HIV-1_{LAI}	PBMC Vero	6.3	>100 >100
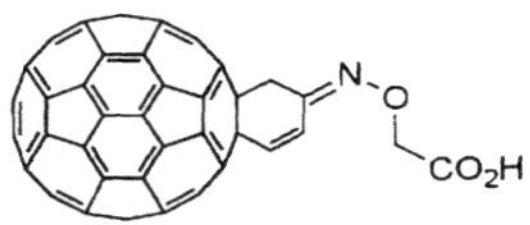 [62]	HIV-1_{LAI}	PBMC Vero	7.3	>100 33.9
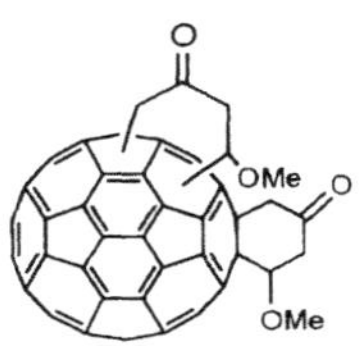 isomer A [62]	HIV-1_{LAI}	PBMC Vero	>100	>100 >100

Table 1. (*Contd.*)

Fullerene and reference(s)	HIV or polymerase	Cell line	EC_{50} (μM)	IC_{50} (μM)	Comments
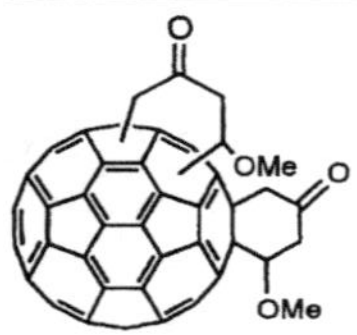 isomer B [62]	$HIV\text{-}1_{LAI}$	PBMC Vero	7.7	>100 >100	
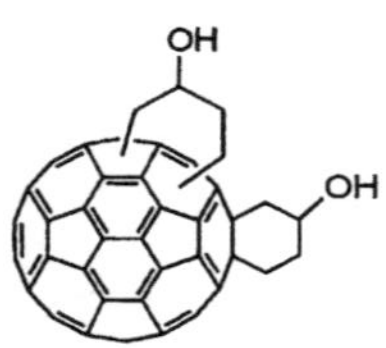 [62]	$HIV\text{-}1_{LAI}$	PBMC Vero	17.6	>100 >100	
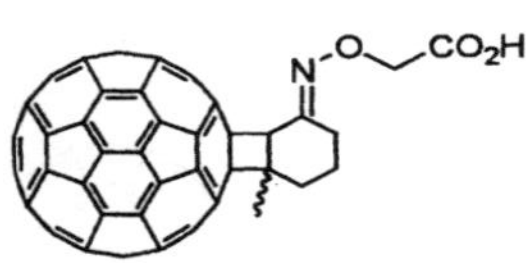 [62]	$HIV\text{-}1_{LAI}$	PBMC Vero	21.7	>100 >100	

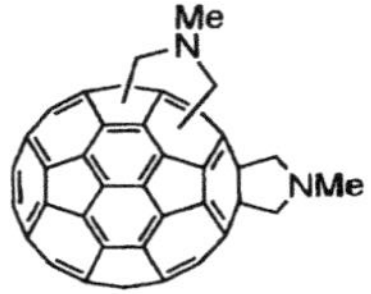 [62]	HIV-1_{LAI}	PBMC	72.7	>100	
		Vero		≥100	
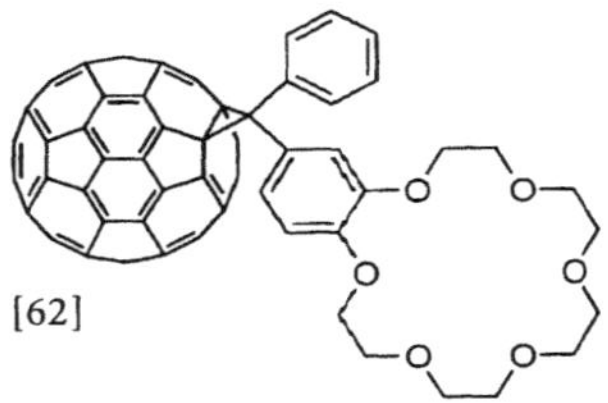 [62]	HIV-1_{LAI}	PBMC	137	>100	
		Vero		>100	
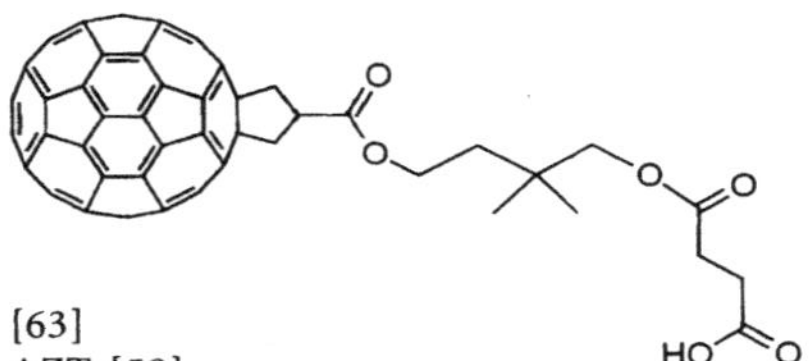 [63]	HIV RT			0	IC_{50} value at 10 μM
AZT [58]	HIV-1_{LAI}	PBMC	0.004	>100	
	HIV-1_{ROD}	PBMC	0.003		
	HIV-1_{IIB}	PBMC	>100		
		H9		60	
		Vero		23	
		CEM		13	
	HIV-1 RT			0.04	
	DNA Poly α			>100	

to allow efficient examination of the data: column 1 provides a structural drawing of the fullerene derivative, along with the applicable reference to the original work; column 2 lists the strain of HIV or specific polymerase type against which the fullerene was evaluated; column 3 provides the cell line used in the assay; columns 4 and 5 provide the activity, EC_{50} and toxicity, IC_{50} values in µM, respectively; and column 6 gives pertinent comments.

The first definitive in vitro assays of fullerenes were conducted in 1993 [58] on the bis(monosuccinimide) derivative of p,p′-bis(2-aminoethyl)diphenyl-C_{60} [59]. This methanofullerene is soluble and thermally stable at pH = 7, making it suitable for biological evaluation. This fullerene exhibited moderate activity against HIV strains in both acutely and chronically infected cells, with no apparent cytotoxicity. Of particular note was the effectiveness of the compound against AZT (3′-azido-3′-deoxythymidine)-resistant strains. When the fullerene was assayed in PBMC against the AZT-resistant HIV-1_{G910-6}, the EC_{50} value was 2.7 µM; when tested against the AZT-susceptible HIV-1_{H112-2}, the EC_{50} value was 2.8 µM [58]. These data suggest that this fullerene and AZT could be used in combination chemotherapy.

This methanofullerene exhibited EC_{50} values of 4.6 and 4.9 µM against HIV RT and DNA polymerase in cell-free systems, respectively. The similarity in EC_{50} values for polymerase inhibition indicates the fullerene is non-selective with respect to HIV RT. The demonstrated ability of this fullerene to non-selectively inhibit DNA polymerase α and HIV RT, combined with its non-toxicity in different cell lines, suggests this compound may function by a virucidal mode of action, since it is logical to assume that if it is transported into the cells there would be some degree of toxicity. The same group also reported an IC_{50} value against HIVP of 2.0 µM in a cell-free system [58].

In 1994, additional evidence was reported that the methanofullerene had anti-HIV activity, namely that the compound reduced the infectivity of cell-free HIV-1 by more than 95% [60]. In this experiment, the fullerene was incubated with cell-free HIV-1 for 2 h. The virus was then inoculated into mitogen-stimulated PBMCs. This group also examined the ability of the fullerene to interact with the CD4 receptor on lymphocytes and interfere with gp120-CD4 binding. In CEM (human T-cell lymphoma) cells, the fullerene blocked the binding of the anti-CD4 antibody by only 23% at 100 µM. In PBMC, there was no evidence of CD4 receptor blockage at methanofullerene concentrations up to 100 µM. In related experiments, the fullerene did not inhibit fusion between CEM and TF228.1.16 (gp160 transfected B-cell lymphoma) cells. These data suggest the fullerene does not interfere in the gp120-CD4 fusion process by blocking the CD4 receptor [60].

In 1995, the collaborative group of Schinazi, Hill and Wudl continued their evaluations of fullerene in vitro activity by examining an *N*-tris(hydroxymethyl)propylamido methanofullerene (C_{60}) derivative [61]. As seen from Table 1, this tris(hydroxymethyl) fullerene exhibits a higher activity against HIV than the bis(monosuccinimide) fullerene (EC_{50} = 2.5 µM vs. 7.5 µM, respectively) with no increase in toxicity in several cell lines [61]. The *N*-tris(hydroxymethyl)propylamido methanofullerene was found to be more soluble in pyridine than DMSO/water. The assays were conducted using stock solutions of the fullerene in pyridine or DMSO/water emulsions, at levels non-toxic to cells [61].

In 1996, Schuster et al. used new synthetic routes to prepare 11 new fullerenes for evaluation against HIV-1 [62]. Again, limited water solubility required the use of DMSO/water emulsions in the studies. Unlike the fullerenes addressed above, most of these new derivatives were chiral. It is interesting to note the marked difference in activity of two closely related isomers (cf. isomer A and isomer B, Table 1): EC_{50} = 7.7 and >100 μM, respectively [62].

Nakamura et al. investigated two classes of water-soluble fullerene derivatives, detergent- and sphere-type [63]. The detergent-type is amphiphilic and the sphere-type has a random distribution of polar groups on the fullerene surface. Both types of compounds were reported to be highly soluble in water, precluding the need for DMSO/water emulsions. Both detergent- and sphere-type derivatives inhibited HIVP. However, a detergent-type fullerene did not inhibit HIV [63].

In 1997, Schinazi, Hill, Chiang and co-workers reported the activities of four polyhydroxy fullerenes prepared by different synthetic procedures [36,64–67]. All four of these fullerenes (each a complex mixture of non-separable isomers) exhibited some degree of activity against HIV in acutely infected PBMCs, while none showed activity in chronically infected H9 (human T-cell lymphoma) cells, implying that these fullerenes did not inhibit HIVP. The polyhydroxy fullerenes were evaluated in DMSO/water or DMF as they were not sufficiently water soluble.

2.3
Fullerene Toxicity

The first examination of fullerene toxicity was provided by Nelson et al. [68]. This group examined the carcinogenicity of C_{60} administered topically to mice. Repeated application of the C_{60} for 24 weeks did not result in formation of benign or malignant skin tumors. These researchers also noted that 72 h post-treatment, neither epidermal DNA synthesis nor ornithine decarboxylase activity were affected [68].

In 1994, Schinazi et al. reported the first systemic toxicity studies of fullerenes in mammals [60]. Using random-bred Swiss CFW mice, a 2% bis(monosuccinimide) fullerene in DMSO/phosphate buffered solution was injected intraperitoneally at doses of 15, 25 and 50 mg kg^{-1} per day for six days. Those in the test group showed a slight decrease in weight after the first dose, which was subsequently reversed. Doses up to 50 mg kg^{-1} per day were tolerated in the animals with no apparent side effects (i.e. hair or weight loss, ruffled fur or death). After 2 months of monitoring, no animals had died [60]. A subsequent study by the groups of Wudl, Schinazi and Boudinot further examined the toxicity and pharmacokinetics of this fullerene derivative [69]. Intravenous administration of the fullerene (15 mg kg^{-1} dissolved in DMSO) was followed by the plasma concentration declining in a bi- or triexponential manner with more than 99% of the fullerene bound to the protein. The rate of clearance was determined to be 0.19 l h^{-1} kg^{-1}. Examination of urine samples 24 h post-injection failed to detect any of the fullerene, suggesting that renal clearance was not a mode of drug elimination. A single intravenous dose of 25 mg kg^{-1} resulted in spasms, shortness of breath, and death within 5 min post-injection.

The uptake of fullerenes by human keratinocytes was investigated by Scrivens et al. using ^{14}C-labeled fullerenes solubilized as a fine, aqueous suspension of small C_{60} molecules [70]. Uptake of the fullerenes was rapid with more than 50% of associated radioactivity occurring within 6 h. No further accumulation occurred after 9 h. The fullerenes remained associated with the cell based on simple cell washing experiments. The fullerenes did not affect thymidine incorporation into human fibroblasts or keratinocytes [70].

The genotoxicity of fullerene compounds was examined in *E. coli* and larvae by Zakharenko and co-workers [71]. The lack of mutations present after treatment of the cells and larvae with the fullerenes indicated the compounds are non-mutagenic [71].

Yamago et al. reported the first fullerene pharmacokinetics in mammals using a ^{14}C-labeled derivative [72]. This group found that when the fullerene was administered orally in rats, the vast majority of the compound was excreted in the faeces with little or no absorption in tissue. Intravenous injection resulted in the fullerene being transported to the tissues within 1 h. After 1 week, most of the fullerene was distributed to the skeletal muscle and hair. Less then 2% remained in the tissues and approximately 5% was faecally excreted [72]. This group also reported that single, intraperitoneal doses of 200–500 mg kg^{-1} did not prove fatal to mice after one week, although there were some initial side-effects, including weight loss [72].

Cellular penetration by unfunctionalized C_{60} was confirmed by Moussa et al. [73]. Injection of large amounts of C_{60} into Swiss mice led to the observation of C_{60} deposits inside liver and spleen cells, proving that C_{60} is able to cross some cellular membranes [73].

Conflicting assertions exist regarding the in vivo behavior of fullerenes. The most serious point of contention is whether or not fullerenes cross the blood-brain barrier. In the review by Jensen et al. opposing views are presented by Yamago et al. and Bullard-Dillard et al. [52]. Another significant unknown is whether or not fullerene cages, and in particular the relatively stable C_{60} skeleton, are susceptible to metabolic attack. Recent evidence presented by Hamamo et al. indicates that fullerenes are resistant to cytochrome P-450 catalyzed oxidative degradation in rat liver microsomes [74]. Clearly, these issues warrant further investigation. More detailed pharmacokinetic studies of fullerene derivatives are needed.

2.4 Modeling Studies

Despite the fact that recent advances in hardware and software are making computational investigation of fullerenes, like those of POMs, increasingly viable and productive, there is only one published study on the modeling of fullerene–biomacromolecule interactions, that involving C_{60} derivatives and HIVP. The realization that HIVP inhibition via fullerene binding at the active site might lead to an effective therapeutic approach, led Friedman et al. to model this interaction [75].

These investigators rationalized that fullerenes would be good protease inhibitors based on the complementarity of their size and hydrophobicity with that of the HIVP active site. The energetics of binding of various C_{60} derivatives in the "flaps-open" form of the HIVP active site were calculated using the DOCK3 program. The loss of hydrophobic surface area accessible to the solvent was calculated. With the fullerene docked, 92% of the hydrophobic surface accessible to the solvent was blocked, translating to a ΔG_{bind} of 8–12 kcal mol^{-1} [75]. These modeling studies led this group to propose that another synthetically accessible, water-soluble fullerene derivative, a bis(phenethylaminosuccinate) C_{60}, would also be a strong protease inhibitor. This compound was synthesized and did, in fact, inhibit HIVP (cf. Table 1). Friedman and co-workers argued this DOCK3 method could be incorporated into a modeling-based methodology for structure-based design of fullerene HIVP inhibitors [76].

2.5
Conclusions and Future Work for Anti-HIV Fullerenes

The work to date on fullerenes and their derivatives as anti-HIV agents is provocative but sufficiently limited so that little can be concluded at this time. The unique size, shape and chemical nature of fullerenes make them of potential interest in antiviral chemotherapy in part as many biological targets including the active site of HIVP are of similar dimensions. Better and more general synthetic routes to water-soluble fullerenes are needed. While modeling studies and in vitro assays point to protease inhibition (vide supra) as the primary mode of viral inhibition, there is not yet sufficient information available to confirm this is the mechanism of anti-HIV activity. The role enantioselectivity and isomerism play in the inhibition of the protease must be further explored and established.

3
Conclusions

Both POMs and fullerenes are cluster compounds with varied physicochemical properties, and in particular water solubilities. Yet these two classes of compounds have been shown to have selective anti-protease activity. Other mechanisms for this antiviral activity are possible and will be explored in the near future. A particularly important research area is their interaction with human macrophages and their immunological properties. The design of new compounds coupling the properties of POMs with fullerenes are underway. Clearly, the versatility of their chemical composition, structure, stereochemistry and physicochemical properties should provide a rich resource of potent compounds with selective antiviral activity.

Acknowledgement. We thank the National Institutes of Health (R01 AI32903-04A1) and the Molecular Design Institute (Office of Naval Research, grant N00014-95-1-1116) for support.

References

1. Hill CL (ed) (1998) Special Issue: Polyoxometalates. Chem Rev 98
2. Pope MT (1983) Heteropoly and isopoly oxometalates. Springer, Berlin Heidelberg New York Tokyo
3. Pope MT, Müller A (ed) (1993) Polyoxometalates: from platonic solids to anti-retroviral activity. Kluwer Academic Publishers, Dordrecht, Netherlands

4. Raybaud N, Jasmin C, Huppert J, Chermann JC, Mathé G, Raynaud M (1972) Rev Eur Etud Clin Biol 17:295
5. Barnard DL, Hill CL, Gage T, Matheson JE, Huffman JH, Sidwell RW, Otto MI, Schinazi RF (1997) Antiviral Res 34:27
6. Huffman JH, Sidwell RW, Barnard DL, Morrison A, Otto MJ, Hill CL, Schinazi RF (1997) Antiviral Chem Chemother 8:75
7. Rhule JT, Hill CL, Judd DA, Schinazi RF (1998) Chem Rev 98:327
8. Gouzerh P, Proust A (1998) Chem Rev 98:77
9. Cholewa M, Legge GJF, Weigold H, Holan G, Birch CJ (1994) Life Sci 54:1607
10. Cholewa M, Turnbull IF, Legge GJF, Weigold H, Marcuccio SM, Holan G, Tomlinson E, Wright PJ, Dillon CT, Lay PA, Bonin AM (1995) Nucl Instrum Methods Phys Res Sect B 104:317
11. Cibert C, Jasmin C (1982) Biochem Biophys Res Commun 108:1424
12. Berry JP, Galle P (1990) Exp Mol Path 53:255
13. Ni L, Greenspan P, Gutman R, Kelloes C, Farmer MA, Boudinot FD (1995) Antiviral Res 32:141
14. Inouye Y, Tokutake Y, Kunihara J, Yoshida T, Yamase Y, Nakata A, Nakamura S (1992) Chem Pharm Bull 40:805
15. Weeks MS, Hill CL, Schinazi RF (1992) J Med Chem 35:1216
16. Yamamoto N, Schols D, De Clercq E, Debyser Z, Pauwels R, Balzarini J, Nakashima H, Baba M, Hosoya M, Snoeck R, Neyts J, Andrei G, Murrer BA, Theobald B, Bossard G, Henson G, Abrams M, Picker D (1992) Mol Pharmacol 42:1109
17. Kim G-S, Judd DA, Hill CL, Schinazi RF (1994) J Med Chem 37:816
18. Judd DA, Chen Q, Campana CF, Hill CL (1997) J Am Chem Soc 119:5461
19. Hill CL, Judd DA, Tang J, Nettles J, Schinazi RF (1997) Antiviral Res 34:A43
20. Judd DA, Nettles J, Nevins N, Tang J, Schinazi RF, Hill CL submitted for publication
21. Jasmin C, Raybaud N, Chermann JC, Haapala D, Sinoussi F, Loustau CB, Bonissol C, Kona P, Raynaud M (1973) Biomedicine 18:319
22. De Clercq E (1995) J Med Chem 38:2491
23. Hill CL, Hartnup M, Faraj M, Weeks M, Prosser-McCartha CM, Brown JRB, Kadkhodayan M, Sommadossi J-P, Schinazi RF (1990) Polyoxometalates as inorganic anti-HIV-1 compounds. Structure-activity relationships. In: Diasio RB, Sommadossi J-P (eds) Advances in Chemotherapy of AIDS. Pergamon Press, New York, p 33
24. Hill CL, Weeks MS, Schinazi RF (1990) J Med Chem 33:2767
25. Inouye Y, Tokutake Y, Yoshida T, Seto Y, Hujita H, Dan K, Yamamoto A, Nishiya S, Yamase T, Nakamura S (1993) Antiviral Res 20:317
26. Inouye Y, Fujimoto Y, Sugiyama M, Yoshida T, Yamase T (1995) Biol Pharm Bull 18:996
27. Judd DA, Schinazi RF, Hill CL (1994) Antiviral Chem Chemother 5:410
28. Rozenbaum W, Dormont D, Spire B, Vilmer E, Gentilini M, Griscelli C, Montagnier L, Barre-Sinoussi F, Chermann J-C (1985) Lancet 450
29. Moskovitz BL (1988) Antimicrob Agents Chemother 32:1300
30. Burgard M, Sansonetti P, Vittecoq D, Descamps P, Guetard D, Herson S, Rozenbaum W, Rouzioux C (1989) AIDS 3:665
31. Sarafianos SG, Kortz U, Pope MT, Modak MJ (1996) Biochem J 319:619
32. Bountiff L (1982) PhD thesis, University of Reading
33. Ni L, Boudinot FD, Boudinot SG, Henson GW, Bossard GE, Martellucci SA, Ash PW, Fricker SP, Darkes MC, Theobald BR, Hill CL, Schinazi RF (1994) Antimicrob Agents Chemother 38:504
34. Ni L, Boudinot FD (1995) Eur J Drug Metab Pharmacokinet 20:209
35. Ni L, Henson GW, Darkes MC, Bossard GE, Boudinot FD (1996) Pharm Sci 2:137
36. Chiang LY, Bhonsle JB, Wang L, Shu SF, Chang TM, Hwu JR (1996) Tetrahedron 52:4963
37. Sun D-Y, Liu Z-Y, Guo X-H, She Y-M, Zhou Y, Liu S-Y (1996) Changchun Inst Applied Chem, Chinese Acad Sci 17:19
38. Wang B-C, Cheng C-Y (1997) Theor Chem 391:179
39. Zgonnik VN, Vinogradova LV, Melenevskaya EY, Litvinova LS, Kever EE, Bykova EN, Klenin SI (1997) Zh Prikl Khim 70:1159
40. Sun Y-P, Liu B, Lawson GE (1997) Photochem Photobiol 66:301
41. Sun Y-P, Liu B, Moton DK (1996) Chem Commun 2699
42. Tabata Y, Murakami T, Ikada Y (1997) Fullerene Sci Technol 5:989

43. Tabata Y, Murakami Y, Ikada Y (1997) International Symposium for Controlled Release of Bioactive Materials
44. Nakajima N, Nishi C, Li F-M, Ikada Y (1996) Fullerene Sci Technol 4:1
45. Vol'pin ME, Romanova VS, Parens ZN (1996) Mol Cryst Liq Cryst Sci Technol, Sect C 7:53
46. Timofeeva GI, Kuleshova EF, Romanova VS (1997) Mendeleev Commun 37
47. Timofeeva GI, Kuleshova EF, Romanova VS (1997) Russ Chem Bull 46:472
48. Timofeeva GI, Romanova VS, Lopanova LA, Nesmeyanov AN (1996) Izv Akad Nauk, Ser Khim 879
49. Kotelnikova RA, Kotelnikov AI, Bogdanov GN, Romanova VS, Kuleshova EF, Parnes ZN, Vol'pin ME (1996) FEBS Lett 389:111
50. Baran PS, Monaco RR, Khan AU, Schuster DI, Wolson SR (1997) J Am Chem Soc 119:8363
51. Guldi DM (1997) J Phys Chem A 101:3895
52. Jensen AW, Wilson SR, Schuster DI (1996) Bioorg Med Chem 4:767
53. Yu C, Bhonsle JB, Wang LY, Lin JG, Chen B-J, Chiang LY (1997) Fullerene Sci Technol 5:1407
54. Dayong Z, Yanshi L, Ziyang L, Liu G, Guo C, Zhan R, Liu S (1997) Chin Sci Bull 42:748
55. Kaesermann F, Kempf C (1997) Antiviral Res 34:65
56. Tien HT, Wang LG, Wang X, Ottova AL (1997) Bioelectrochem Bioenerg 42:161
57. Tokuyama H, Yamago S, Nakamura E, Shiraki T, Sugiura Y (1993) J Am Chem Soc 115:7918
58. Schinazi RF, Sijbesma R, Srdanov G, Hill CL, Wudl F (1993) Antimicrob Agents Chemother 37:1707
59. Sijbesma R, Srdanov G, Wudl F, Castoro JA, Wilkins C, Friedman SH, DeCamp DL, Kenyon GL (1993) J Am Chem Soc 115:6510
60. Schinazi RF, McMillian A, Juodawlkis AS, Pharr J, Sijbesma R, Srdanov G, Hummelen J-C, Boudinot FD, Hill CL, Wudl F (1994) Recent Advances in the Chemistry and Physics of Fullerenes and Related Materials. San Francisco, USA
61. Schinazi RF, Bellavia C, Gonzalez R, Hill CL, Wudl F (1995) Recent Advances in the Chemistry and Physics of Fullerenes and Related Materials. Reno, USA
62. Schuster DI, Wilson SR, Schinazi RF (1996) Bioorg Med Chem Lett 6:1253
63. Nakamura E, Tokuyama H, Yamago S, Shiraki T, Sugiura Y (1996) Bull Chem Soc Jpn 69:2143
64. Chiang LY, Wang L-Y, Swirczewski JW, Soled S, Cameron S (1994) J Org Chem 59:3960
65. Chiang LY, Upasani RB, Swirczewski JW (1992) J Am Chem Soc 114:10154
66. Schinazi RF, Chiang LY, Wilson LJ, Cagle DW, Hill CL (1997) Recent Advances in the Chemistry and Physics of Fullerenes and Related Materials. Montreal, Canada
67. Li J, Takeuchi A, Ozawa M, Li X, Saigo K, Kitazawa K (1993) J Chem Soc Chem Commun 1784
68. Nelson MA, Domann FE, Bowden GT, Hooser SB, Fernando Q, Carter DE (1993) Toxicol Ind Health 9:623
69. Rajagopalan P, Wudl F, Schinazi RF, Boudinot FD (1996) Antimicrob Agents Chemother 40:2262
70. Scrivens WA, Tour JM, Creek KE, Pirisi L (1994) J Am Chem Soc 116:4517
71. Zakharenko LP, Zakharov IK, Lunegov SN, Nikiforov AA (1994) Dokl Akad Nauk 335:261
72. Yamago S, Tokuyama H, Nakamura E, Kikuchi K, Kananishi S, Sueki K, Nakahara H, Enomoto S, Ambe F (1995) Chem Biol 2:385
73. Moussa F, Trivin F, Céolin R, Hadchouel M, Sizaret P-Y, Greugny V, Fabre C, Rassat A, Szwarc H (1996) Fullerene Sci Technol 91:259
74. Hamamo T, Mashino T, Nagano T, Arakana K, Ryu R, Mashiko S, Hirobe M (1995) Pacifichem-96. Honolulu, Hawaii
75. Friedman SH, DeCamp DL, Sijbesma RP, Srdanov G, Wudl F, Kenyon GL (1993) J Am Chem Soc 115:6506
76. Friedman SH, Wudl F, Rubin Y, Kenyon GL (1994) Recent Advances in the Chemistry and Physics of Fullerenes and Related Materials. San Francisco, USA
77. Toniolo C, Bianco A, Maggini M, Scorrano G, Prato M, Marastoni M, Tomatis R, Spisani S, Palú G, Blair ED (1994) J Med Chem 37:4558

Vanadium-Containing Insulin Biomimetic Drugs

K.H. Thompson[1], J.H. McNeill[2], C. Orvig[3]

Medicinal Inorganic Chemistry Group, Department of Chemistry and Faculty of Pharmaceutical Sciences, University of British Columbia, B.C., Canada V6T 1Z1
E-mail: [1]*kthompso@chem.ubc.ca,* [2]*jmcneill@unixg.ubc.ca,* [3]*orvig@chem.ubc.ca*

Vanadium-containing metallocomplexes are being widely investigated as potential insulin-enhancing or replacement pharmaceutical agents. Experimental evidence of inorganic vanadium's insulin-mimetic effects, with the immensely appealing feature of oral bioavailability, has prompted design and development of novel organic ligands. Desirable features include low molecular weight, neutral charge, thermodynamic and hydrolytic stability, and a balance between aqueous solubility and lipophilicity. From the initial development of bis(maltolato)oxovanadium(IV), containing the ligand, maltol, which is an approved food additive, to more recent development of metallocomplexes such as imidazoleoxobisperoxovanadate(V), which seek to mimic the active site of vanado-enzymes, such as bromperoxidase, researchers have examined a variety of compounds as solutions to the problems of improved absorption, coupled with minimal toxicity. Also detailed are 5-, 6-, and 7-coordinated compounds, with V-N, V-O, V-S, and temporary complexation, as well as the peroxovanadates currently in development as insulin-mimetic agents.

Keywords. Vanadium, Insulin-mimetic, Diabetes, Peroxovanadate

List of Abbreviations and Symbols

STZ	Streptozotocin
BMOV	Bis(maltolato)oxovanadium(IV)
VPA	Bis(picolinato)oxovanadium(IV)
VO-MPA	Bis(methylpicolinato)oxovanadium(IV)
V-P	Bis(pyrrolidine-*N*-carbodithioato)oxovanadium(IV)
IRK	Insulin receptor kinase
PTPase	Phosphotyrosine phosphatase
IGF-II	Insulin-mediated growth factor II
bpV(phen)	Potassium oxodiperoxol(1,10-phenanthroline)vanadate(V) trihydrate
mpV(2,6-pdc)	Pyridine-2,6-dicarboxylatooxoperoxovanadate
mpV(pic)	Picolinatooxoperoxovanadate
$mpV(ox)_2$	Bisoxalatooxoperoxovanadate
IDDM	Insulin dependent diabetes mellitus
NIDDM	Non-insulin dependent diabetes mellitus

1 Introduction

1.1 Initial Interest in Vanadium as an Insulin Mimetic

The first recorded use of vanadium as a therapeutic agent for diabetes was in France in 1899 [1]. Sodium vanadate, given orally, was reported to have an insulin-like effect in two of three diabetic patients, as shown by lower concentrations of glucose in their urine. However, it was not until 1985, when McNeill and co-workers showed that sodium vanadate given orally had insulin-like effects in a diabetic rat model, that this ultratrace element was seriously considered in this regard [2]. Since then, because insulin is not orally active, there has been a burgeoning interest in the potential of oral vanadium to replace, at least partly, insulin therapy, as evidenced by several recent review volumes [3–5] devoted to this topic.

Here, we summarize current knowledge of coordination complexes of vanadium which are being or have been investigated as insulin biomimetics. Strategies for design of an effective hormone replacement metallocomplex in general will be reviewed, and salient features of the earliest proposed vanadium insulin-mimetic drugs detailed. New synthetic approaches will be outlined, and an overview of the drug testing process presented.

1.2 Strategies for Design of an Effective Hormone Replacement Metallocomplex

1.2.1 *Physical Properties*

For metallocomplexes to be useful as biomimetic drugs, they must be able to cross biological membranes, generally by passive diffusion, because for most metal ions active or facilitated transport mechanisms are absent. Thus, a potential agent should preferably have low molecular weight, neutral charge, and some aqueous solubility. The lipophilicity of the complex should be balanced with the hydrophilicity (i.e., water solubility) of the molecule. Moreover, the metal-ligand complex should be thermodynamically and hydrolytically stable in water [6]. Many pharmacologically useful metals, including vanadium, readily undergo hydrolysis in an aqueous environment. This is especially so in a biological system if the metal ligand binding constants of the metal complex are not high enough [7], and where the concentrations are usually low ($\mu mol\ l^{-1}$ to $mmol\ l^{-1}$). The kinetics of complexation/decomplexation should also be considered; a chelating ligand may be specifically designed to remove a metal from, or to deliver a metal to, an active site, in which case the kinetics are of primary importance and may be faster or slower, respectively [8].

1.2.2
Biological Targeting

Biological targeting of a proposed metallocomplex requires that ligand functionalization be considered. A biologically directing portion of the molecule such as a steroid or enzyme substrate may be used to functionalize the ligand in order to direct biologically the complex towards an active site, or target organ. Patterns of uptake and distribution which result in metal accumulation in target tissues, as opposed to wide and undifferentiated perfusion, may serve as selection criteria [6].

1.2.3
Pharmacological Testing

Before being considered for human therapeutic use, a new candidate drug must undergo extensive pharmacological testing, both in vitro and in vivo. In vitro studies of the insulin-mimetic actions of vanadium compounds are usually undertaken with phosphatases and/or kinases related to the insulin signaling cascade, or with enzymes involved in glucose and lipid metabolism, in adipocytes or other tissues [9,10], since these systems appear most relevant to the mechanism of action of vanadium in alleviation of diabetic symptomatology [11]. An important factor in determining the utility of a candidate compound is oral activity, because insulin itself is not orally bioavailable and must be taken by injection. In vivo efficacy testing may be followed by tissue distribution experiments, toxicity testing, and ultimately, if the results are still favorable, by testing in human volunteers.

1.3
Current Research Directions

Some commercially-available vanadium salts (vanadate, vanadyl) have already been tested in clinical trials [12–15]. Chelated vanadium complexes are still in the animal testing stage, although some tissue retention and toxicity testing have also been carried out in selected lead compounds [16,17]. Of major research importance is to learn more about the mechanism of action of vanadium and vanadium metallocomplexes [18–20]. Unfortunately, in trying to deduce the mechanism of action, there are a myriad of sites at which vanadium can interact [9,21]. Recent evidence suggests that vanadyl may be acting to stimulate a cytosolic protein tyrosine kinase [19], distinct from the usual insulin receptor mechanism, and also unlike vanadate's mechanism of action [10,22]. Much work remains in order to elucidate clearly the mechanism – one major problem with diabetes is the great difficulty in picking specific in vitro systems which actually relate to the in vivo results, such as those reported here.

2
Animal Models of Diabetes

2.1
STZ-Diabetic Rat Model of Insulin-Dependent Diabetes

For in vivo testing, a number of experimental models of diabetes in rodents have been used. Most widely accepted is the streptozotocin (STZ)-induced diabetic rat [23]. STZ is an antibiotic that specifically attacks the insulin secreting β-cells in the pancreas in a dose-responsive fashion [24]. The STZ-diabetic rat model of diabetes is obtained by administering intravenous streptozotocin (STZ) to rats, usually at doses of 45–75 mg kg^{-1} body weight. This results in greatly reduced insulin secretory capacity of the rat pancreas and hence the development of diabetic characteristics (reduced insulin levels, elevated levels of glucose in blood and urine). STZ-treated rats are insulinopenic, hyperphagic, and catabolic. The model does not completely parallel type I diabetes in humans, in that STZ-diabetic rats can survive without administration of exogenous insulin (providing an untreated control group for comparison as an experimental benefit); however, it is relatively simple, inexpensive, reproducible, and reliable.

2.2
Genetic (Spontaneous) Animal Models of Diabetes

Other animal models of diabetes are especially developed rodent strains that are spontaneously diabetic, whether insulin-dependent or not [23]. The key end-point of interest is usually blood glucose-lowering, which is known to be correlated closely with plasma lipid-lowering in diabetic animals. The BioBreeding (BB) Wistar rat is a spontaneous model of diabetes that closely resembles type I diabetes [25]. It is characterized by rapid onset of hyperglycemia at between 60 and 120 days of age and by loss of pancreatic insulin production. Insulin resistance leading to overt diabetes despite hyperinsulinemia is a hallmark of *ob/ob* diabetic mice, while *db/db* mice are obese and glucose intolerant [26]. The *fa/fa* Zucker rat is considered a genetically determined model of obesity and mild glucose intolerance [27].

3
Inorganic Vanadium Salts

3.1
Vanadate

The earliest experiments with vanadium as a pharmacological agent used sodium metavanadate [1]. In the late 1950s and early 1960s, a number of experiments with human volunteers used diammonium vanadotartrate, 80–420 μmol V day^{-1}, as a potential cholesterol-lowering drug. Most investigators reported gastrointestinal distress in a majority of the experimental subjects at the highest doses, but there were neither other toxic symptoms nor a consistent cholesterol-lowering therapeutic effect in these early human trials [28–31].

3.1.1
In Vitro Stimulation of Glucose Uptake, Inhibition of Lipolysis

Characterization of the effects of inorganic vanadium on cell homogenates led to the first recognition of its insulin-mimetic effects. Sodium vanadate stimulated glucose uptake and glucose oxidation in rat adipocytes, stimulated glycogen synthesis in rat diaphragm and liver, and inhibited hepatic gluconeogenesis, usually at millimolar concentrations of added vanadate or vanadyl [32,33]. Lipid and glycolytic pathways were affected in specific tissues, mimicking insulin's effects, in most cases. For instance, vanadate inhibited lipolysis and stimulated lipogenesis in adipocytes [34,35]. Glucose uptake and transport were enhanced by vanadate in other tissue types: mouse brain [36], rat skeletal muscle [37], and isolated villus cells [38]. Conversely, vanadate inhibited glucose transport in rat intestine when added to the perfusate of an everted gut sac preparation [39], or when added as a supplement to the drinking water of diabetic and non-diabetic rats at concentrations up to 0.08 mmol l^{-1} for 2–4 weeks [40]. Stimulated translocation of the GLUT-4 glucose transporter (which is regulated by insulin) to the plasma membrane was demonstrated in rat adipocytes [41], and increased glucose transporter expression in the presence of vanadate was shown in vitro in NIH 3T3 mouse fibroblasts [42] and in vivo in rat skeletal muscle [43]. Vanadate increased sensitivity to insulin in activating glucose transport in normal and insulin-resistant rat adipocytes [44] (but not human adipocytes [45]), via enhanced insulin binding. Some insulin-like effects of vanadium are concentration-dependent. For example, in perfused rat liver, glucose output was inhibited by the addition of vanadate in the range of 0.5–1.0 mmol l^{-1}, but was increased by concentrations greater that 25 mmol l^{-1} [46,47]. In other experiments, in vitro and in vivo effects were found to be contradictory; for instance, when using a concentration range of 0.1–5 mmol l^{-1} vanadate in rat hepatocytes, both non-insulin-like glycogenolytic (glycogen breakdown) effects [48] and insulin-like glycolytic effects [49] were demonstrated. Furthermore, these apparently opposing effects were also demonstrable in hepatocytes isolated from diabetic rats [50,51].

3.1.2
Discovery of in Vivo Efficacy in Lowering of Blood Glucose

The first definitive study to demonstrate in vivo glucose-lowering and anti-diabetic effects of inorganic vanadium was by McNeill and co-workers [2]. Addition of sodium orthovanadate (0.39–0.54 mmol kg^{-1} day^{-1}) to the drinking water of STZ-diabetic rats over a period of 6 weeks dramatically lowered plasma glucose levels without increasing depressed plasma insulin levels. Subsequent studies have shown that the glucose-lowering effect with orally administered vanadate or vanadyl solutions is persistent, relatively non-toxic, and accompanied by other desirable effects in plasma parameters, such as triglyceride and cholesterol lowering, and restoration of thyroxine levels to normal [52–56]. Although overall mean body weight in diabetic rats is not improved, the treatment consistently normalizes the elevated intake of food (hyperphagia) and fluids (polydipsia), both characteristic symptoms of STZ-diabetes [57]. Vanadate treatment of diabetic animals partially or completely restores liver and muscle activities of enzymes involved in glycolysis and glycogenesis, such as glucokinase [58], phosphoenolpyruvate carboxykinase [59,60], pyruvate kinase [61,62], and glycogen synthase [63], independent of normalized plasma glucose levels [64]. In vanadate-treated diabetic rats having plasma concentrations of 17.5–22.0 $\mu mol\ l^{-1}$ V, liver glycogen increased, and enzymes involved in glycogen metabolism were affected in an insulin-like manner [54].

Vanadium treatment in vivo resulted in an apparently enhanced peripheral insulin sensitivity and a lowered insulin demand [53]. The insulin-lowering effect of vanadium has particular relevance to type II, or non-insulin-dependent diabetes mellitus (NIDDM), in which hyperinsulinemia is a frequent concomitant, and may be a contributing factor in the development of secondary complications of the disorder [65].

That peripheral insulin resistance is also reversed by vanadium treatment has been shown by glucose clamp studies in which in-dwelling venous and arterial catheters permit infusion of tritiated glucose to maintain plasma glucose levels in the presence of high insulin levels (hyperinsulinemia). Using this method, hepatic glucose production and glucose utilization in response to submaximal or maximal insulin levels have been shown to be improved with vanadium treatment in STZ-diabetic rats [55,66] and in partially pancreatectomized rats [67]. In contrast to vanadium's effects on STZ-diabetic rats, no effect on plasma triglycerides has been seen in models of type II diabetes. Glucose tolerance of genetically obese *(db/db)* diabetic mice [68] and insulin resistance of *ob/ob* mice [69,70] were improved by vanadate treatment. In the *fa/fa* Zucker rat (genetically obese and mildly glucose intolerant), vanadate treatment reduced food and fluid intake, reduced weight gain, lowered hyperinsulinemia towards normal levels, and improved glucose tolerance, whereas pair feeding only partially reversed these parameters [71]. Vanadate treatment improved insulin-mediated glucose utilization, without increasing glucose transporter levels in muscle [72] of *fa/fa* rats.

3.2 Vanadyl Sulfate

3.2.1 *Lower Toxicity, Less Risk of Dehydration*

By comparison with the earliest studies of sodium vanadate in STZ-diabetic rats [2], later experience with vanadyl sulfate suggested that this form of vanadium was less toxic and had a lower risk of dehydration [73–75]. For most investigators, this has been the inorganic vanadium salt of first choice [19,74,76]. Vanadyl sulfate treatment of the genetically diabetic BB rat reduces, but does not completely eliminate, its exogenous insulin requirement [53,77].

3.2.2 *Prolonged Effect*

An intriguing discovery was that euglycemia and improved glucose tolerance, along with plasma, cardiac, and adipose tissue abnormalities, were maintained for a long period (up to 30 weeks) after vanadyl treatment was withdrawn [56,73,74]. This may be attributed partly to an accumulation of vanadium in various tissue stores [78] or to the preservation or improvement of pancreatic β-cell insulin content in vanadium treated rats, leading to an indefinite period of near-normal glucose homeostasis in the fed state in these animals [56]. Overall, moderate to good diabetic control has been obtained in the STZ-diabetic rat, with inorganic vanadium, at oral doses of between 0.1 and 0.7 mmol kg^{-1} day^{-1} [3,52,59,74,79,80]. The dose of vanadium required to achieve good diabetic control varies with the initial diabetic state of the animal [80], addition of other trace elements [81], and as yet undetermined individual factors [82,83].

4 Chelated Oxovanadium(IV) Complexes

Chelated vanadyl complexes are a more recent development [84,85] and offer a wide scope for deliberate design improvements. Two types of chelated vanadium complexes have been studied for their insulin-mimetic effect: these comprise (1) oxovanadium(IV) and (V) complexes [84,85] and (2) combinations of vanadate and hydrogen peroxide, called peroxovanadates [86]. Development of various ligands designed to improve the absorption, tissue uptake, and intracellular mobility of vanadium compounds, thereby reducing the dose required for optimal insulin mimesis, exemplifies one strategy. With the exception of the peroxovanadium(V) complexes, ligands have been chosen to chelate vanadyl; these ligands have generally been chosen to impart specific features to the resulting vanadium complexes – improved lipophilicity (vanadyl cysteine methyl ester, naglivan), improved oral absorption by passive diffusion (BMOV), potentiation of in vitro insulin-mimetic effect (the monoperoxo- and diperoxovanadates), or facilitation of transmembranal ion uptake (RL-252 and analogs). These are discussed below.

4.1 Bis(maltolato)oxovanadium(IV)

An oxovanadium(IV) complex which has been synthesized in our laboratories (Fig. 1), and has undergone extensive testing over the last few years is bis(maltolato)oxovanadium(IV) (BMOV) [87]. BMOV can be readily synthesized [88] by combining vanadyl sulfate and maltol, an approved food additive in both Canada and the USA. Potentially useful properties of BMOV include significant water solubility, neutral charge, and lipophilicity, a combination designed to enhance gastrointestinal absorption. It is relatively hydrolytically and thermodynamically stable

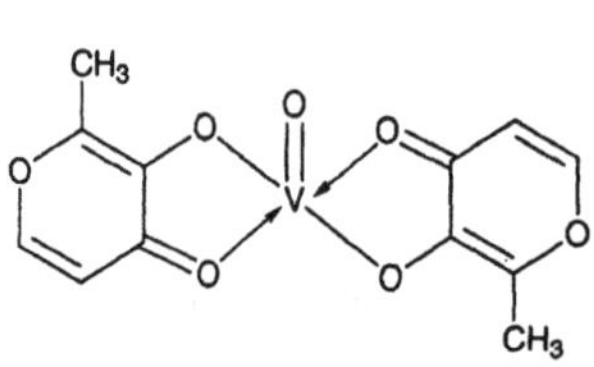

Fig. 1. Bis(maltolato)oxovanadium(IV), BMOV

[89,90]. BMOV has a molecular weight of 317 and is soluble (mmol l^{-1} scale) in a number of organic solvents and in water [88]. Stability constants for the binding of 1 and 2 maltolato ligands to vanadyl are log K_1 = 8.80, and log K_2 = 7.51, for the bis(ligand) complex, log β_2 = 16.31 [88]. The geometry around the vanadium in $VO(ma)_2$ is square pyramidal [88]. BMOV has one unpaired electron, characteristic of the vanadyl unit, and a fairly high V = O stretching frequency in the infrared spectrum (995 cm^{-1}), suggesting that there is no ligand (or a weakly bound solvent) in the sixth position.

BMOV is effective in lowering blood glucose at a lower dose than vanadyl sulfate, and does not show evidence of toxicity over a six-month period of administration in STZ-diabetic rats [91,92]. Both oral gavage and intraperitoneal administrations indicate that BMOV is two to three times more potent than its parent compound, vanadyl sulfate, in bringing about acute glucose lowering [93].

4.1.1
Redox Chemistry

Redox chemistry between the vanadium(IV) and vanadium(V) oxidation states is of fundamental importance in elucidating the mechanism of action of BMOV. In methanol, or in any alcoholic solvent, BMOV oxidizes to form an alkoxobis(maltolato)oxovanadium(V) complex, *cis*-$VO(OR)(ma)_2$ [88], the oxidation kinetics being second order, a function of the concentrations of both complex and molecular oxygen [90]. The reaction between BMOV and molecular oxygen, in a 4:1 ratio, gives the vanadium(V) species, consistent with the fact that BMOV undergoes a 1-electron oxidation and O_2 is a 4-electron oxidant. The observed rate constant is directly proportional to the molecular oxygen concentration, consistent with this stoichiometry and the overall rate at 25 °C. Two pathways, aquo and hydroxo, give the dioxoanion, *cis*-$[VO_2(ma)_2]^-$, for oxidation of BMOV with O_2 in water [90].

The stomach has a pH of 2–3, depending on its contents [94], which may preclude passage of the complex through the acidic environment of the stomach without some dissociation. BMOV is partially hydrolyzed at pH levels below 3; however, sufficient biologically active BMOV remains intact in vivo to yield distinct advantages over inorganic vanadium compounds in terms of its orally active insulin-mimetic properties, ease of administration, low toxicity, enhanced tissue uptake, and favorable tissue localization [88].

4.1.2
Tissue Distribution

Experiments undertaken to investigate tissue distribution of BMOV [16] involved oral or intraperitoneal administration of ^{48}V in a carrier-added form. From these studies, the absorption of vanadium from an oral dose of ^{48}V-BMOV was determined to be about twice as high as that from an equivalent dose of ^{48}V-vanadyl sulfate; and tissue distribution between the two compounds differed. Compartmental analysis of the results showed that the proportion of vanadium taken up by liver following BMOV treatment was almost four times higher than with $VOSO_4$ treatment, whereas that taken up by kidney was less than 50% higher, and that by bone (at 24 h) was almost three times higher. The ratio of vanadium predicted by the model in bone:kidney:liver was 8:3:2 for BMOV and 6:4:1 for $VOSO_4$, demonstrating a different pattern of tissue uptake for inorganic vanadium compared with the organically chelated vanadium complex [16]. The increase in uptake into liver, kidney, and bone was, on average, 2.7 times higher for BMOV compared to $VOSO_4$, consistent with the two to three times increased potency of BMOV compared to $VOSO_4$ seen in acute testing [91,93]. Taking into account total tissue weight, the principal uptake of ^{48}V from an oral dose of BMOV vs $VOSO_4$ was into bone (8.62 vs 2.99% administered dose, AD), blood (3.55 vs 2.73% AD), muscle (1.14 vs 0.67% AD), liver (0.82 vs 0.21% AD), and kidney (0.23 vs 0.17% AD), based on model-predicted compartmental masses at 24 h following gavage [16].

4.2
V-N and V-S Coordination Vanadyl Complexes

Bidentate ligands with one ionizable proton can be used to form neutral bis(ligand) metal complexes with vanadyl. Ligands that also contain a ring oxygen (e.g., maltol) also tend to be water soluble. These properties together (neutral charge and aqueous solubility) contribute to high oral bioavailability [95].

Bis(picolinato)oxovanadium(IV) (VPA) has a VN_2O_2 coordination mode [96]. VPA (Fig. 2) is slightly soluble in water and is gradually oxidized by air in solution. It is stable under inert gas in the solid state, or in 5% acacia suspension. In the latter form, at a dose of 2.2–2.5 mg V day^{-1} (approximately 10 mg V kg^{-1} day^{-1}), STZ-diabetic rats became normoglycemic within 7 days and, after another 7 days of daily oral gavage, remained nearly normoglycemic for an additional 30 days, with no treatment during that period. VPA administered at a higher dose (5.3–6.2 mg V day^{-1}) by oral gavage resulted in 8 out of 11 STZ-diabetic rats becoming normoglycemic within 12 h; however, this faster glucose-lowering was accompanied by diarrhea. VPA, given in the drinking water, 0.75 mg ml^{-1} (approximately 9 mg V day^{-1}) led to significant glucose lowering (to 14.2 ± 2.0 mmol l^{-1} in treated diabetic rats, compared to 18.3 ± 0.6 mmol l^{-1} in untreated diabetic animals) without toxic side effects [97].

Bis(methylpicolinato)oxovanadium(IV) (VO-MPA) (Fig. 2) has also recently been synthesized, characterized, and tested, both in vitro and in vivo, for insulin-mimetic activity [98]. VO-MPA, 10 mg V kg^{-1} day^{-1}, given orally by gavage as a 5% acacia gum suspension, led to sustained glucose-lowering over an 80-day period following cessation of treatment. The decreased body weight gain and increased bilirubin levels, seen in the early stages of treatment, were correctable by lowering the dose to 5 mg V kg^{-1} day^{-1} [98].

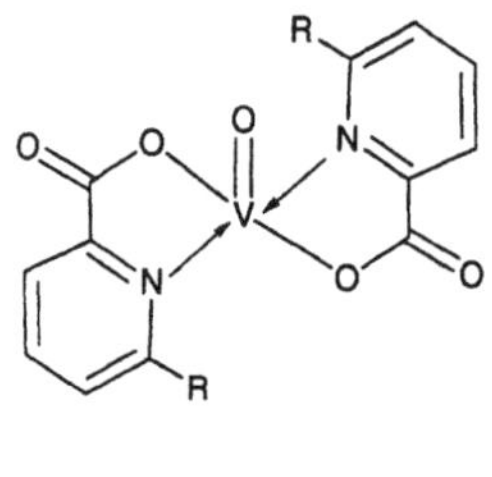

Fig. 2. Bis(picolinato)oxovanadium(IV), VPA, R = H; bis(methylpicolinato)oxovanadium(IV), VO-MPA, R = CH_3

Two new V-N coordinated vanadyl complexes, both having distorted octahedral geometries and having the characteristics of VO(XeX′) type complexes have been synthesized [99]. Vanadyl-*N,N′*-ethylene bis-glycine (ethylenediamine *N,N′*-diacetic acid) proved to be more effective than vanadyl-*N,N′*-ethylene bis(*S*)-methionine (and much more effective than vanadyl sulfate, used as a control) in stimulating free fatty acid release from rat adipocytes in vitro. No in vivo studies have yet been reported.

A vanadyl bis(cysteine methyl ester) complex (Fig. 3) of vanadium(IV) [100] at a dose of 10 mg V kg^{-1} day^{-1} in 5% acacia gum, by oral gavage, was slightly more effective than vanadyl complexes containing other ligands (such as malonate, oxalate, salicylaldehyde, and tartrate) in normalizing plasma glucose levels within 24 h of administration. There was no obvious toxicity at this dose; however, at ten times the glucose-lowering dose, all the test animals died of diarrhea within 4 days, suggesting the necessity for careful ligand design if the concept of chelated vanadyl is to prove worthwhile [84].

V-P [bis(pyrrolidine-*N*-carbodithioato)oxovanadium(IV)] (Fig. 4) was initially tested as an in vitro insulin-mimetic by inhibition of free fatty acid release from rat adipocytes [101]. It was administered orally to STZ diabetic rats at an initial dose (for 2 days) of 10 mg V kg^{-1} day^{-1} (0.2 mmol kg^{-1} day^{-1}) to achieve normoglycemia, followed by a maintenance dose of 5 mg V kg^{-1} day^{-1} (0.1 mmol kg^{-1} day^{-1}). Intraperitoneal administration of this compound proved to be more effective than oral treatment but both achieved significant glucose-lowering.

The water-insoluble vanadyl complex naglivan [bis(*N*-octylcysteineamido)-oxovanadium(IV)] (Fig. 5) has been given to STZ-diabetic rats in a suspension of 3% acacia gum by oral gavage. Naglivan doses of 5–15 mg V kg^{-1} day^{-1} (0.1–0.3 mmol kg^{-1} day^{-1}) effectively lowered blood glucose levels to near normal, although the onset of action was significantly slower than with inorganic vanadate or vanadyl [73,102]. Naglivan treatment of both control and experimentally diabetic animals was not accompanied by weight loss or a reduction in food or fluid intake over an eight-week period; however, neither was there diarrhea associated with naglivan treatment.

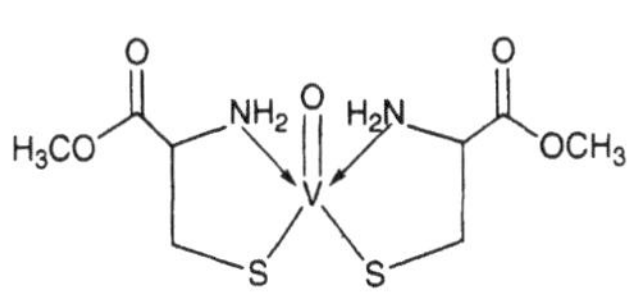

Fig. 3. Vanadyl bis(cysteine methyl ester)

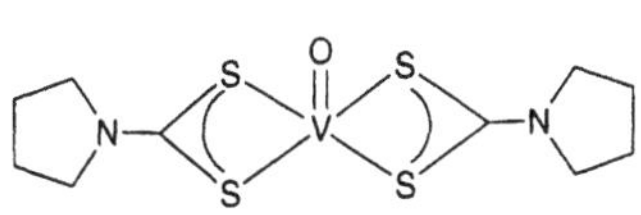

Fig. 4. Bis(pyrrolidine-*N*-carbodithioato)oxovanadium(IV), V-P

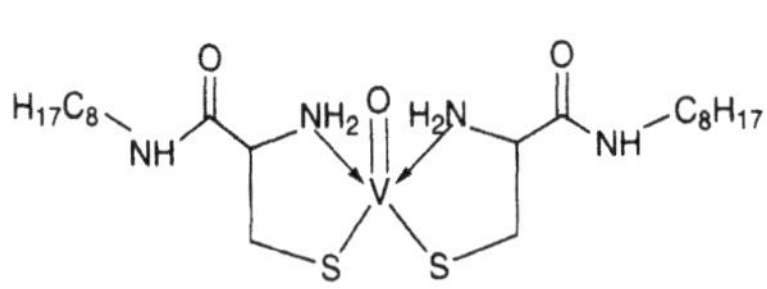

Fig. 5. Naglivan

4.3
Temporary Complexation (Carrier Ligands)

A series of dihydroxamic acid chelators have been designed as hydrophobic carriers of vanadyl [103]. In an in vitro assay of lipogenic stimulation in rat adipocytes, RL-252 (Fig. 6) was maximally effective at molar ratios of 10:1 vanadyl sulfate:chelator, suggesting a shuttle mechanism of action. These compounds were electrically neutral, lipid-soluble, and optically chiral; they released the bound metal ion when treated with aqueous glutathione solutions [103].

5
Peroxovanadates

Vanadate and hydrogen peroxide alone each have insulin-mimetic activity [104, 105]. In combination (as peroxovanadate, a poorly characterized mixture), the effect has been shown to be strongly synergistic for stimulating insulin-mediated growth factor II (IGF-II) translocation and insulin receptor tyrosine kinase activation in rat fat cells [106]. Complexation of vanadium(V) with hydrogen peroxide increased phosphorylation of the β-subunit [106–108], stimulated lipogenesis, inhibited lipolysis, and promoted protein synthesis in rat adipocytes at micromolar concentrations of vanadium [108,109]. Peroxovanadates also increased the phosphorylation of several proteins, including those of the insulin receptor, when injected in the portal vein of rat livers [110], as well as lowering blood glucose in diabetic rats when administered intraperitoneally, but were (at least initially) ineffective when administered orally [111]. An observed synergism when vanadate and hydrogen peroxide were co-administered was suggested to be due to the formation of aqueous peroxovanadium complexes with peroxide ion enclosed within the coordination sphere of the vanadate [106].

Because formation of fairly random peroxovanadates appeared to potentiate vanadium's insulin mimetic effect [112,113], it was speculated that design of specific peroxovanadate compounds might have a markedly potent insulin biomimetic effect [114]. It was also hoped that judicious choice of specific ligands would solve the problem of instability of the spontaneously formed peroxovanadate mixtures, especially in aqueous solutions [112,115]. Although the bisperoxo species appear to

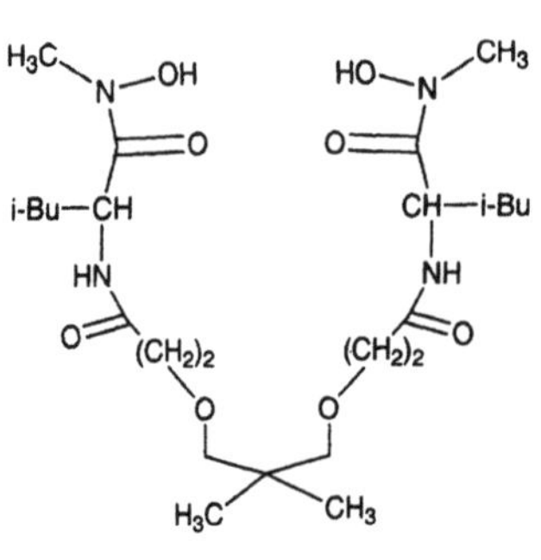

Fig. 6. RL-252

predominate at the physiologically most relevant neutral pH, some monoperoxo species have also been synthesized.

5.1 Ligandoxobis(peroxo)vanadate(V)

Two of the earliest discrete diperoxovanadate(V) compounds, potassium oxodiperoxo(pyridine-2-carboxylato)vanadate(V) and potassium oxodiperoxo(3-hydroxypyridine-carboxylato)vanadate(V) [112], were direct descendants of analogous chromium complexes [116,117], which had been shown to improve membrane fluidity and increase the rate of insulin uptake. Although these peroxovanadates (Fig. 7) are stable indefinitely in the solid state, they are prone to decomposition in aqueous solutions. Both have distorted bipyramidal ligand geometry around the vanadium ion [112], and have been shown to be effective in stimulating insulin receptor kinase (IRK) activity in hepatoma cells and inhibiting phosphotyrosine phosphatase (PTPase) activity in rat liver endosomes [114]. In the latter study, altogether 12 different bisperoxovanadium compounds of the $[VO(O_2)_2L\text{-}L']^{n-}$ type [114] were compared. The most stable, potassium oxodiperoxo(1,10-phenanthroline)vanadate(V) trihydrate, [bpV(phen)], was tested in vivo in fasted female Sprague-Dawley rats, by intrajugular injection of 6 μmol kg^{-1}, a dose which resulted in glucose-lowering equivalent to 15 μg kg^{-1} insulin administered by the same method. At doses of 0.75 to 6 μmol kg^{-1}, bpV(pic), bpV(phen), and bpV(Me_2phen) were effective in lowering plasma glucose in BB rats whether given intravenously, intraperitoneally, or subcutaneously [118]. Only bpV(phen), 20–200 μmol kg^{-1}, was shown to be effective when administered by oral gavage [119].

Demonstrably increased insulin binding to intact rat adipocytes [120] (but not human NIDDM fat cells [121]) in the presence of 0.5 mmol l^{-1} bpV(pic), possibly due to increased insulin receptor affinity, suggests an insulin-sparing mechanism of action for peroxovanadates of this type [120]. Effective concentrations of all the peroxovanadates tested for in vitro inhibition of phosphotyrosine phosphatase (PTPase) and stimulation of insulin receptor tyrosine kinase was in the 5–80 mmol l^{-1} vanadium range [114]. The nature of the ancillary ligand has profound effects on the specificity and potency of the complex, with increasing bulkiness, whether polar [e.g., carboxylation in bpV(bipy)] or nonpolar [e.g., methylation in bpV(phen)], decreasing IRK stimulation, and increasing methylation of the phenanthroline ligand, thus reducing PTPase inhibition [114].

Synthesis and characterization of a large number of peroxovanadate heteroligands, both mono- and polydentate [122] revealed a range of stabilities toward decomposition in aqueous solution, depending on the nature of the heteroligand.

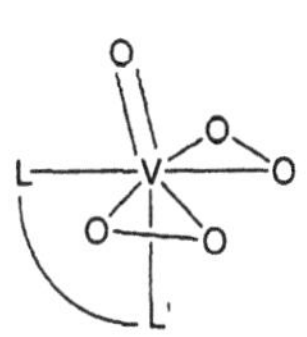

$[VO(O_2)_2(L\text{–}L')]^{n-}$ n=1-3 **Fig. 7.** Ligandoxobis(peroxo)vanadate(V)

The fact that none of the peroxovanadates formed can be considered hydrolytically stable, and all are subject to redox processes which ultimately result in radical formation with the potential for increased intracellular oxidative stress [123], may limit biomedical utility [122].

5.2 Imidazoleoxobisperoxovanadate(V)

A 6-coordinate bisperoxovanadium imidazole compound (Fig. 8) has been synthesized and shown to enhance insulin receptor autophosphorylation in human liver cell culture, as well as to increase glucose transport in rat adipocytes, at concentrations ranging from 1 μmol l^{-1} to 1 mmol l^{-1} [124]. The coordination of vanadium(V) to imidazole presents structural analogies to the coordination of vanadium to histidine residues in vanadium-containing haloperoxidases [125] and some phosphorylases [126].

5.3 Ligandoxoperoxovanadate(V)

A variety of well-characterized monoperoxovanadate(V) complexes (Fig. 9) have also recently been prepared and tested in vitro and by intraperitoneal and/or subcutaneous injection in vivo [118,127]. Picolinatooxoperoxovanadate [mpV(pic)] was effective in achieving a 20% decrease in plasma glucose in STZ-diabetic Sprague Dawley and BB rats at a lowest effective dose (LED) of 0.4 μmol kg^{-1} while the lowest dose producing mortality was more than 15 times higher. By contrast, pyridine-2,6-dicarboxylatooxoperoxovanadate [mpV(2,6-pdc)] had an LED of 24 μmol kg^{-1} which was twice the lowest dose producing mortality [118]. In each case, O or N donor atoms were bound to two or more sites in the V(V) trigonal bipyramidal coordination sphere, usually with a −1 or -2 charge [118]. Bisoxalatooxoperoxov-

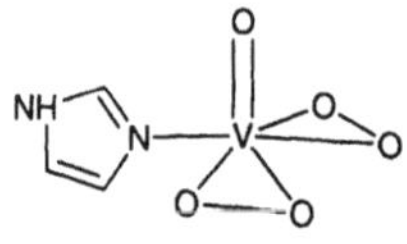

$[VO(O_2)_2(HIm)]^-$

Fig. 8. Imidazoleoxobisperoxovanadate(V)

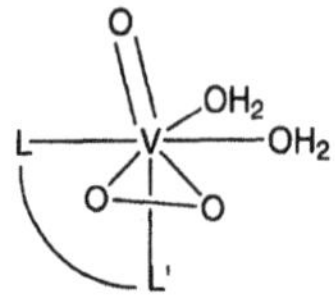

$[VO(O_2)(H_2O)_2(L\text{–}L')]^{n-}$ n=0,1

Fig. 9. Ligandoxoperoxovanadium(V)

anadate [mpV(ox)$_2$] was only minimally stimulatory on [^{14}C]glucose incorporation into diaphragm glycogen, unlike mpV(pic) which resulted in a nine-fold stimulation over control levels [127].

6 Miscellaneous Other Vanadium Complexes

In addition to the vanadyl, vanadate, and peroxovanadate complexes mentioned above, a variety of bis(ligand)vanadyl complexes have been reported by Sakurai and co-workers [84]. This series of bis(ligand)oxovanadium(IV) complexes was tested on STZ-diabetic rats [84]; ligands included salicylate, oxalate, malate, and tartrate. None of these compounds was as effective as the cysteinyl methyl ester oxovanadium(IV) complex (see above).

Bis-salicylidine ethylenediiminato oxovanadium(IV) was found to be stable, water-soluble, and effective in acute glucose-lowering [128]. Blood glucose levels in alloxan-induced diabetic rats decreased from hyperglycemic to hypoglycemic during oral intubation with 7.5 mg V kg^{-1} day^{-1} for a period of 30 days. Withdrawal of treatment brought an immediate reversion to hyperglycemia [128].

7 In Vitro and in Vivo Testing: Steps in the Drug Development Process

Ideally, vanadium complexes perceived as having insulin-mimetic potential would be screened initially using an in vitro method, similar to methods developed for screening chromium complexes for biological activity [129]. Both phosphotyrosine phosphatase inhibition [119] and activation of insulin receptor kinase [130] have been suggested as methods of comparing various oxovanadium complexes. Insofar as the mechanism of vanadium's actions as an insulin-mimetic are seen as deriving from its ability to stimulate nonreceptor protein tyrosine kinases, it should be possible to measure insulin-mimetic potential according to degree of inhibition by the specific inhibitor of cytosolic tyrosine kinases, staurosporine, in the presence of varying concentrations of the new complexes [131]. However, the differential effects observed for PTPase inhibition vs tyrosine kinase activation [113,119], as well as specific target tissue effects [127], renders these methods of evaluation somewhat questionable. Comparative studies of cellular uptake kinetics and PTPase inhibition [17] and of lymphocyte signal transduction vs PTPase inhibition [123,132], have shown conclusively that even structurally similar oxovanadium species can have markedly different biopotencies and bioavailabilities, and that the particular in vitro method employed is an important factor in determinations of relative efficacy.

For insulin-mimetic efficacy, the simplest relevant systems appear to be isolated, intact cells [133]. Studies of this type have demonstrated profound effects of ligand variations on biopotency [134], with steric hindrance playing a critical role in terms of cellular uptake [17]. The simplest and most reliable tests of insulin-mimetic effect remain those that are carried out in vivo [84,93,135], with the relevant functional end-points being plasma and lipid glucose lowering in experimentally diabetic animals following oral administration.

An important feature of potential new drug testing in animals is establishment of an index of toxicity (LD_{50}). Other determinations include effective dose (ED_{50}), absorption characteristics under fasted and non-fasted conditions, accumulation under conditions of chronic use, and profile of uptake, distribution and excretion [16,133].

Recently, clinical trials of vanadium compounds have been initiated on human type I (IDDM) and type II (NIDDM) diabetic subjects. Treatment with sodium metavanadate (52 mg V day^{-1}) for 2 weeks resulted in significant increases in mean rates of glucose metabolism during a euglycemic clamp in two out of five subjects with IDDM and in five out of five subjects with NIDDM [13,14]. There were decreased insulin requirements in the IDDM subjects, and lowered serum cholesterol levels in all subjects. Treatment of NIDDM subjects with vanadyl sulfate (24 mg V day^{-1}) for 3 weeks caused an improved insulin sensitivity, a reduction in hepatic glucose production, and an increased rate of glucose disposal, all of which were sustained for 2 weeks after treatment was withdrawn [12]. In both of these studies, there were reported incidences of mild gastrointestinal intolerance.

Vanadyl therapy for 3 weeks at 24 mg V day^{-1} resulted in improved insulin sensitivity in NIDDM, but not in obese non-diabetic, subjects [15], with an insulin-enhancing inhibition of lipolysis. Peak serum vanadium at 50 mg V day^{-1} was 82.4 ± 43.2 ng ml^{-1}. At 25 mg V day^{-1} there was no change in glucose and lipid metabolic parameters and the peak serum vanadium was 18.2 ± 5.3 ng ml^{-1}. There was no increase in thiobarbituric acid reactive substances (an indicator of in vivo lipid peroxidation) at these doses [14].

8 Summary and Conclusions

The potential of vanadium compounds to mimic insulin when administered orally has stimulated renewed interest in the pharmacological effects of these unique and enigmatic compounds. A variety of coordination compounds of vanadium have been shown to lower blood glucose and ameliorate other diabetic symptoms in a variety of animal models of both insulin-dependent and non-insulin-dependent diabetes. Opportunities exist for lowering the effective dose of vanadium by appropriate chelation, taking into account those properties of a metal complex most conducive to passive diffusion – namely, neutral charge, low molecular weight, amphiphilicity, and low redox activity. Hydrolytic and thermodynamic stability are also desirable. The development of BMOV is an example of how the biological properties that need to be incorporated into a complex were included in its chemical synthesis and design. Other methods of achieving the desired insulin mimesis in vivo without accompanying toxicity are in an earlier development stage, but may benefit from our increasingly detailed knowledge of the insulin signaling cascade and of the biochemical role of vanadium in mammalian systems. Therapeutic benefit will derive from optimization of ligand design to improve bioavailability and consistency of biological response.

References

1. Lyonnet BM, Martz Martin E (1899) La Presse Medicale 7:191
2. Heyliger CE, Tahiliani AG, McNeill JH (1985) Science 227:1474
3. Chasteen ND (ed) (1990) Vanadium in biological systems. Kluwer, Dordrecht
4. Sigel H, Sigel A (eds) (1995) Vanadium and its role in life. Dekker, New York
5. Srivastava AK, Chiasson J-L (eds) (1995) Vanadium compounds: biochemical and therapeutic applications. Kluwer, Dordrecht (Mol Cell Biochem 153)
6. Thompson KH, Yuen VG, McNeill JH, Orvig C (1998) Chemical and pharmacological studies of a new class of antidiabetic vanadium complexes. In: Tracey AS, Crans DC (eds) Vanadium compounds. Chemistry, biochemistry, and therapeutic applications. American Chemical Society, Washington, DC, chap 26
7. Nelson WO, Karpishin TB, Rettig SJ, Orvig C (1988) Inorg Chem 27:1045
8. Thompson KH, McNeill JH, Orvig C (1997) Rev Port Quim 4:3
9. Shechter Y, Li J, Meyerovitch J, Gefel D, Bruck R, Elberg G, Miller DS, Shisheva A (1995) Mol Cell Biochem 153:39
10. Elberg G, He Z, Li J, Sekar N, Shechter Y (1997) Diabetes 46:1684
11. Orvig C, Thompson KH, Battell M, McNeill JH (1995) Vanadium compounds as insulin mimics. In: Sigel H, Sigel A (eds) Vanadium and its role in life. Dekker, New York, chap 17
12. Cohen N, Halberstam M, Shlimovich P, Chang CJ, Shamoon H, Rossetti L (1995) J Clin Invest 95:2501
13. Goldfine AB, Simonson DC, Folli F, Patti M-E, Kahn CR (1995) J Clin Endocrinol Metab 80:3311
14. Goldfine AB, Willsky G, Kahn CR (1998) Vanadium salts in the treatment of human diabetes mellitus. In: Tracey AS, Crans DC (eds) Vanadium compounds. Chemistry, biochemistry, and therapeutic applications. American Chemical Society, Washington, DC, chap 28
15. Halberstam M, Cohen N, Shlimovich P, Chang CJ, Shamoon H, Rossetti L (1996) J Clin Invest 45:659
16. Setyawati IA, Thompson KH, Sun Y, Lyster DM, Vo C, Yuen VG, Battell M, McNeill JH, Ruth TJ, Zeisler S, Orvig C (1998) J Appl Physiol 84:569
17. Reuland DJ, Cao ZX, Rosenhein L, Poehlein S, McCoy L, Turner J, Durham T, Ganguli S (1998) Horm Metab Res 30:29
18. Brichard SM, Henquin J-C (1995) Trends Pharmacol Sci 16:265
19. Li J, Elberg G, Crans DC, Shechter Y (1996) Biochemistry 35:8314
20. Henquin J-C, Brichard SM (1992) La Presse Medicale 21:1100
21. Sekar N, Li J, Shechter Y (1996) Crit Rev Biochem Mol Biol 31:339
22. Zhang M, Zhou M, Van Etten RL, Stauffacher CV (1997) Biochemistry 36:15
23. Bell RH, Jr, Hye RJ (1983) J Surg Res 35:433
24. Rakieten N (1963) Cancer Chemother Rep 29:91
25. Marliss E (1983) Metab 32 [Suppl 1]:1
26. Mordes JP, Rossini AA (1981) Am J Med 70:353
27. Penicaud L, Ferre P, Terretaz J, Kinebanyan MF, Leturque A, Dore E, Girard J, Jeanrenaud B, Picon L (1987) Diabetes 36:626
28. Curran GL, Azarnoff DL, Bolinger RE (1959) J Clin Invest 38:1251
29. Schroeder HA, Balassa JJ, Tipton IH (1963) J Chron Dis 16:1047
30. Somerville J, Davies B (1962) Am Heart J 64:54
31. Dimond EG, Caravaca J, Benchimol A (1963) Am J Clin Nutr 12:49
32. Shechter Y, Karlish SJD (1980) Nature (Lond) 284:556
33. Dubyak GR, Kleinzeller A (1980) J Biol Chem 255:5306
34. Duckworth WC, Solomon SS, Liepnieks J, Hamel FG, Hand S, Peavy DE (1988) Endocrinology 122:2285
35. Degani H, Gochin M, Karlish SJD, Shechter Y (1981) Biochemistry 20:5795
36. Amir S, Meyerovitch J, Shechter Y (1987) Brain Res 419:392

37. Okumura N, Shimazu TJ (1992) Biochemistry (Tokyo) 112:107
38. Hajjar JJ, Dobish MP, Tomicic TK (1989) Proc Soc Exp Biol Med 190:35
39. Kellett GL, Barker ED (1989) Biochim Biophys Acta 979:311
40. Madsen KL, Porter VM, Fedorak RN (1993) Diabetes 42:1126
41. Paquet MR, Romanek RJ, Sargeant RJ (1992) Mol Cell Biochem 109:149
42. Mountjoy KG, Flier JS (1990) Endocrinology 127:2025
43. Strout HV, Vicario PP, Biswas C, Saperstein R, Brady EJ, Pilch PF, Berger J (1990) Endocrinology 126:2728
44. Eriksson JW, Lonnroth P, Smith U (1992) Diabetologia 35:510
45. Lonnroth P, Eriksson JW, Posner BI, Smith U (1993) Diabetologia 36:113
46. Bruck R, Prigozin H, Krepel Z, Rotenberg P, Shechter Y, Bar-Meir S (1991) Hepatology 14:540
47. Roden M, Liener K, Fürnsinn C, Prskavec M, Nowotny P, Steffan I, Vierhapper H, Waldhäusl W (1993) Diabetologia 36:602
48. Bosch F, Arino J, Gomez-Foix A, Giunovart JJ (1987) J Biol Chem 262:218
49. Gomez-Foix AM, Rodriguez-Gil JE, Fillat C, Guinovart JJ, Bosch F (1988) Biochem J 255:507
50. Rodriguez-Gil JE, Gomez-Foix AM, Fillat C, Bosch F, Guinovart JJ (1991) Diabetes 40:1355
51. Rodriguez-Gil JE, Gomez-Foix AM, Arino J, Guinovart JJ, Bosch F (1989) Diabetes 38:793
52. Meyerovitch J, Farfel Z, Sack J, Shechter Y (1987) J Biol Chem 262:6658
53. Ramanadham S, Cros GH, Mongold JJ, Serrano JJ, McNeill JH (1990) Can J Physiol Pharmacol 68:486
54. Pugazhenthi S, Khandelwal RI (1990) Diabetes 39:821
55. Venkatesan N, Avidan A, Davidson MB (1991) Diabetes 40:492
56. Ramanadham S, Brownsey RW, Cros GH, Mongold JJ, McNeill JH (1989) Metabolism 38:1022
57. Meyerovitch J, Shechter Y, Amir S (1989) Physiol Behavior 45:1113
58. Gil J, Miralpeix M, Carreras J, Bartrons R (1988) J Biol Chem 263:1868
59. Blondel O, Simon J, Chevalier B, Portha B (1990) Am J Physiol 258:E459
60. Valera A, Rodriguez-Gil JE, Bosch F (1993) J Clin Invest 92:4
61. Saxena AK, Srivastava P, Baquer NZ (1992) Eur J Pharmacol 216:123
62. Brichard SM, Desbuquois B, Girard J (1993) Mol Cell Endocrinol 91:91
63. Bollen M, Miralpeix M, Ventura F, Toth B, Bartrons R, Stalmans W (1990) Biochem J 267:269
64. Rossetti L, Laughlin MR (1989) J Clin Invest 84:892
65. Reaven GM (1993) Ann Rev Med 44:121
66. Blondel O, Bailbe D, Portha B (1989) Diabetologia 32:185
67. Cordera R, Andraghetti G, DeFronzo RA, Rossetti L (1990) Endocrinology 126:2177
68. Pugazhenthi S, Angel JF, Khandelwal RL (1991) Metabolism 40:941
69. Brichard SM, Bailey CJ, Henquin J-C (1990) Diabetes 39:1326
70. Meyerovitch J, Rothenberg P, Shechter Y, Bonner-Weir S, Kahn CR (1991) J Clin Invest 87:1286
71. Brichard SM, Pottier AM, Henquin J-C (1989) Endocrinology 125:2510
72. Brichard SM, Ongemba LN, Henquin J-C (1992) Diabetologia 35:522
73. Cam MC, Faun J, McNeill JH (1994) Metabolism 44:332
74. Pederson RA, Ramanadham S, Buchan AMJ, McNeill JH (1989) Diabetes 38:1390
75. Brichard SM, Lederer J, Henquin J-C (1991) Diabetes Metab (Paris) 17:435
76. Dai S, Thompson KH, McNeill JH (1994) Pharmacol Toxicol 74:101
77. Battell ML, Yuen VG, McNeill JH (1992) Pharmacol Commun 1:291
78. Parker RDR, Sharma RP (1978) J Environ Pathol Toxicol 2:235
79. Sekar N, Kanthasamy A, Williams S, Subramanian S, Govindasamy S (1990) Pharmacol Res 22:207
80. Thompson KH, Leichter J, McNeill JH (1993) Biochem Biophys Res Comm 197:1549
81. Srivastava P, Saxena AK, Kale RK, Baquer NZ (1993) Res Comm Chem Path Pharm 80:283
82. Cam MC, Pederson RA, Brownsey RW, McNeill JH (1993) Diabetologia 36:218
83. Bendayan M, Gingras D (1989) Diabetologia 32:561
84. Sakurai H, Tsuchiya K, Nukatsuka M, Kawada J, Ishikawa S, Yoshida H, Komatsu M (1990) J Clin Biochem Nutr 8:193
85. McNeill JH, Yuen VG, Dai S, Orvig C (1995) Mol Cell Biochem 153:175
86. Posner BI, Shaver A, Fantus IG (1990) Insulin mimetic agents: vanadium and peroxovanadium compounds. In: Bailey CJ, Flatt PR (eds) New antidiabetic drugs. Smith Gordon, London, p 107
87. McNeill JH, Yuen VG, Hoveyda HR, Orvig C (1992) J Med Chem 35:1489

88. Caravan P, Gelmini L, Glover N, Herring FG, Li H, McNeill JH, Rettig SJ, Setyawati IA, Shuter E, Sun Y, Tracey AS, Yuen VG, Orvig C (1995) J Am Chem Soc 117:12,759
89. Hanson GR, Sun Y, Orvig C (1996) Inorg Chem 35:6507
90. Sun Y, James BR, Rettig SJ, Orvig C (1996) Inorg Chem 35:1667
91. Yuen VG, Orvig C, McNeill JH (1993) Can J Physiol Pharm 71:263
92. Yuen VG, Orvig C, Thompson KH, McNeill JH (1993) Can J Physiol Pharm 71:270
93. Yuen VG, Orvig C, McNeill JH (1995) Can J Physiol Pharm 73:55
94. Waldron-Edward D (1971) Effects of pH and counter-ion on absorption of metal ions. In: Skoryna SC, Waldron-Edward D (eds) Intestinal absorption of metal ions, trace elements and radionuclides. Pergamon, Toronto, p 373
95. Orvig C, Thompson KH, Cam MC, McNeill JH (in press) Vanadium compounds as possible insulin modifiers. In: Farrell ND (ed) Uses of inorganic chemistry in medicine. Royal Society of Chemistry, London
96. Sakurai-H, Fujii K, Watanabe H, Tamura H (1995) Biochem Biophys Res Comm 214:1095
97. Melchior M, Thompson KH, Jong JM, Rettig SJ, Shuter E, Yuen VG, Zhou Y, McNeill JH, Orvig C (1999) Inorg Chem 38:2288
98. Fujimoto S, Fujii K, Yasui H, Matsushita R, Takada J, Sakurai H (1997) J Clin Biochem Nutr 23:113
99. Kawabe K, Tadokoro M, Kojima Y, Fujisawa Y, Sakurai H (1998) Chem Lett 1998:9
100. Sakurai H, Taira Z, Sakai N (1988) Inorg Chim Acta 151:85
101. Watanabe H, Nakai M, Komazawa K, Sakurai H (1994) J Med Chem 37:876
102. Cam MC, Cros GH, Serrano J-J, Lazar R, McNeill JH (1993) Diab Res Clin Practice 20:111
103. Shechter Y, Shisheva A, Lazar R, Libman J, Shanzer A (1992) Biochemistry 31:2063
104. Czech MP, Lawrence JCJ, Lynn WS (1974) Proc Natl Acad Sci USA 71:4173
105. May JM, de Haën C (1979) J Biol Chem 254:9017
106 Kadota S, Fantus IG, Deragon G, Guyda HJ, Posner BI (1987) J Biol Chem 262:8252
107. Kadota S, Fantus G, Deragon G, Guyda HJ, Hersh B, Posner BI (1987) Biochem Biophys Res Comm 147:259
108. Fantus IG, Kadota S, Deragon G, Foster B, Posner BI (1989) Biochemistry 28:8864
109. Shisheva A, Shechter Y (1993) Endocrinology 133:1562
110. Hadari YR, Geiger B, Nadiv O, Sabanay I, Roberts CTJ, LeRoith D, Zick Y (1993) Mol Cell Endocrinol 97:9
111. Shisheva A, Ikonomov O, Shechter Y (1994) Endocrinology 134:507
112 Shaver A, Ng JB, Hall DA, Soo Lum B, Posner BI (1993) Inorg Chem 32:3109
113. Bevan AP, Drake PG, Yale J-F, Shaver A, Posner BI (1995) Mol Cell Biochem 153:49
114. Posner BI, Faure R, Burgess JW, Bevan AP, Lachance D, Zhang-Sun G, Fantus IG, Ng JB, Hall DA, Soo Lum B, Shaver A (1994) J Biol Chem 269:4596
115. Eriksson JW, Lönroth P, Posner BI, Shaver A, Wesslau C, Smith UPG (1996) Diabetologia 39:235
116. Stearns DM, Armstrong WH (1992) Inorg Chem 31:5178
117. Evans GW, Bowman TD (1992) J Inorg Biochem 46:243
118. Yale J-F, Lachance D, Bevan AP, Vigeant C, Shaver A, Posner BI (1995) Diabetes 44:1274
119. Yale J-F, Vigeant C, Nardolillo C, Chu Q, Yu J-Z, Shaver A, Posner BI (1995) Mol Cell Biochem 153:181
120. Yu Z-W, Posner BI, Smith U, Eriksson JW (1996) Biochim Biophys Acta 1310:103
121. Yu Z-W, Jansson P-A, Posner B1, Smith U, Eriksson JW (1997) Diabetologia 40:1197
122. Djordjevic C, Vuletic N, Reslo ML, Puryear BC, Alimard R (1995) Mol Cell Biochem 153:25
123. Krejsa CM, Nadler SG, Esselstyn JM, Kavanagh TJ, Ledbetter JA, Schieven GL (1997) J Biol Chem 272:11,541
124. Crans DC, Keramida AD, Hoover-Litty H, Anderson OP, Miller MM, Lemoine LM, Pleasic-Williams S, Vandenberg M, Rossomoando AJ, Sweet LJ (1997) J Am Chem Soc 119:5447
125. Messerschmidt A, Wever R (1996) Proc Natl Acad Sci USA 93:392
126. Cornman CR, Geiser-Bush KM, Rowley SR, Boyle PD (1997) Inorg Chem 36:6401
127. Bevan AP, Burgess JW, Yale J-F, Drake PG, LaChance D, Baquiran G, Shaver A, Posner BI (1995) Am J Physiol 268:E60
128. Durai N, Saminathan G (1997) J Clin Biochem Nutr 22:31
129. Anderson RA, Brantner JH, Polansky MM (1978) J Agric Food Chem 26:1219

130. Band CJ, Posner BI, Dumas V, Contreres J-O (1997) Mol Endocrinol 11:1899
131. Elberg G, Li J, Shechter Y (1994) J Biol Chem 269:9521
132. Schieven GL, Wahl AF, Myrdal S, Grosmaire L, Ledbetter JA (1995) J Biol Chem 270:20,824
133. Shechter Y, Meyerovitch J, Gefel D, Bruck R, Shisheva A, Elberg G, Libman J, Shanzer A (1993) Frontiers Endocrinol 1:51
134. Liu J, Reuland DJ, Rosenhein L, Cao ZX, Franklin LA, Ganguli S (1997) Metab Clin Exp 46:565
135. Yuen VG, Caravan P, Gelmini L, Glover N, McNeill JH, Setyawati IA, Shuter E, Zhou Y, Orvig C (1997) J Inorg Biochem 68:109

Bismuth Antiulcer Complexes

Hongzhe Sun, Peter J. Sadler[1]

Department of Chemistry, University of Edinburgh, King's Buildings, West Mains Road, Edinburgh EH9 3JJ, UK
E-mail: [1]*p.j.sadler@ed.ac.uk*

Bismuth compounds have been widely used in medicine for more than 200 years, and new bismuth-containing drugs are now being developed. However, the biological chemistry of bismuth is poorly understood. In this review, the use of Bi(III) in antiulcer and antibacterial agents is described, as well as in anti-HIV and radiotherapeutic agents. Bi(III) exhibits a highly variable coordination number and coordination geometry, and alkoxide ligands can induce a strong stereochemical 'lone-pair effect'. The chemistry of Bi(III) carboxylates and aminocarboxylates is dominated by intermolecular interactions which lead to polymeric structures. Bi(III) binds strongly to the thiolate sulfur of the tripeptide glutathione, but these adducts are also kinetically labile which allows rapid translocation of Bi(III) inside cells. The major biological target for Bi(III) appears to be proteins and not DNA. Bi(III) can bind to both Zn(II) sites (e.g. metallothionein) and Fe(III) sites (e.g. transferrin and lactoferrin) in proteins and enzymes. The mechanism of its antibacterial and antimicrobial activity may therefore involve interference with both Fe(III) and Zn(II) pathways. Further work is needed to understand the mechanism of action of bismuth and provide a basis for the design of more effective drugs.

Keywords. Antiulcer, Bismuth, *Helicobacter pylori*, Metallodrugs, Bioinorganic chemistry

List of Abbreviations

ADH	alcohol dehydrogenase
[15]aneN_4	1,4,8,12-tetraazacyclopentadecane
BSS	bismuth subsalicylate
CBS	colloidal bismuth subcitrate
dien	diethylenetriamine
dota	1,4,7,10-tetraazacylododecane *N,N′,N″,N‴*-tetraacetate
dtpa	diethylenetriamine pentaacetate
edta	ethylenediamine tetraacetate
egta	ethyleneglycol- *O,O′*-bis(2-aminoethyl)-*N,N,N′,N′*-tetraethanoate
gsh	glutathione (γ-L-Glu-L-Cys-Gly)
hbdtta	*N,N′*-bis(2-hydroxybenzyl)diethylenetriamine-*N,N′,N″*-triacetate
himda	*N*-(2-hydroxyethyl)imino-diethanoate
H. pylori	*Helicobacter pylori*
hTF	human serum transferrin
ida	iminodiethanoate
IL-2	interleukin-2
MIC	minimum inhibitory concentration
MT	metallothionein
nac	*N*-acetyl-L-cysteine
nta	nitrilotriacetate
RBC	ranitidine bismuth citrate
trien	triethylenetetramine

1
Bismuth in Medicine

Bismuth has long been associated with medicine. The earliest use of bismuth in medicine appears to have been in the Middle Ages as bismuth subnitrate, a white pigment which was also used in beauty care and painting. The first full account of the internal administration of a bismuth compound was in 1786 by Louis Odier for the treatment of dyspepsia [1].

In the UK, bismuth seems to have entered firmly into medicinal use as early as the 19th century via Alexander Marcet at Guy's Hospital, London (1805) [2], where he used bismuth subnitrate to treat spasms of the stomach. In the late 19th century, several new bismuth compounds were introduced into medicine specifically for the treatment of intestinal and gastric disorders. These included derivatives of phenol, bromophenol, pyrogallol, naphthol and salicylate.

In 1889, Felix Balzer first discovered that bismuth might be useful as an anti-syphilitic agent [3]. Even after the introduction of penicillin as a safe and rapid treatment for treponemal infections, bismuth was still found to be useful. Its gradual action was thought to be particularly valuable in the tertiary and quaternary stages of syphilis. In addition, bismuth nitrate in combination with morphine was a constituent of Ferrier's snuff, an inhalation for nasal catarrh. Bismuth and iodoform were even widely advocated as a surgical wound dressing due to their antimicrobial and antibacterial effects [4].

During this century, various bismuth complexes (subnitrate, subgallate, subcitrate, tartrate, subcarbonate and subsalicylate) have been used to treat syphilis, hypertension, infections, skin conditions and gastrointestinal disorders (Table 1) [5,6]. Since the 1970s, two bismuth compounds have been most commonly used worldwide - bismuth subsalicylate (BSS, Pepto-Bismol; Procter & Gamble Co., Cincinnati, Ohio, USA) for the prevention and treatment of diarrhoea and dyspepsia, and colloidal bismuth subcitrate (CBS, De-Nol; Gist Brocades, Delft, The Netherlands, launched in 1976) for the treatment of peptic ulcers. The latter (CBS) has been successfully used in the treatment of both gastric and duodenal ulcer disease. It is said to be as effective as histamine H_2-antagonists such as cimetidine. CBS differs from previous bismuth compounds in that it is highly water soluble, giving a colloidal solution in contrast to bismuth subnitrate and subsalicylate. Attempts to investigate the solid-structure of CBS have been reported [7,8]. Recently, the effectiveness of bismuth has been attributed to its bactericidal action against *Helicobacter pylori* (an organism which was first discovered in 1983 named as *Campylobacter pyloridis*, later amended to *H. pylori*). Several bismuth compounds alone or in combination with antibiotics have been found to be effective against *H. pylori*

Table 1. Use of bismuth compounds for the treatment of various diseases [5,6]

Diseases	Description
Internal diseases	Gastric disorders, gastric and duodenal ulcers, *H. pylori*, traveller's diarrhoea (prevention), dyspepsia, gastrointestinal disorders; hypertension; edema; lupus erythematodes; upper respiratory infections
Ear diseases	Inflammation of the middle ear
Eye diseases	Ceratoconjunctivitis, corneal abscesses
Gynaecological disorders	Local infections
Skin diseases	Burns, eczema, warts and wounds
Infectious diseases	Malaria, herpes, zoster
Parasitic disease	Scabies, threadworms
Venereal diseases	Syphilis, gonorrhea

in vitro, and the longer remission times achieved with bismuth therapy are probably due to clearance of the organism by bismuth [9]. Clinical studies with CBS and BSS show that patients treated with bismuth alone experience a slower relapse than patients treated with other ulcer-healing agents [10], due to bactericidal action of these two complexes against *H. pylori.*

In the 1970s there was an outbreak of bismuth-induced encephalopathy in France and Australia which led to the withdrawal of bismuth drugs in these countries [11]. However, there have been few such side effects reported in the UK and the USA where there is widespread use of bismuth drugs (200 and 507 tons per annum in the UK and USA, respectively). The mechanism of action of bismuth drugs needs to be investigated, as toxic side effects may be related to bismuth interactions with certain molecules in the body. As with other metallodrugs, such as gold antiarthritic and platinum anticancer agents, new methods (i.e. different from those used routinely for organic drugs) need to be developed in clinical chemistry laboratories to determine both the total bismuth contents of biological samples and to investigate speciation. Such studies will lead to their safer use.

New bismuth-containing drugs are currently being developed. Recently, a new ranitidine bismuth citrate compound (made by GlaxoWellcome plc., Tritec and Pylorid) has been approved for marketing and is on sale in a number of countries around the world. It combines the antisecretory action of ranitidine with the mucosal protectant and the bacteriocidal properties of bismuth [12,13].

Another use of bismuth in medicine is in radiotherapy. ^{212}Bi is a strong α-particle emitter, has a short half-life (1 h) [14], and can be obtained in large quantities from a ^{224}Ra generator. This isotope can be used as a targeted radiotherapeutic agent for cancer therapy when attached via complexing ligands such as dtpa (diethylenetriaminepentaacetate) and dota (1,4,7,10-tetraazacylododecane *N,N′,N″,N‴*-tetraacetate) to monoclonal antibodies [15]. Using the method developed by Krejcarek and Tucker [16], ^{212}Bi has been successfully conjugated to anti-Tac, a monoclonal antibody directed towards the human interleukin-2 (IL-2) receptor. Recently, ^{212}Bi-anti-Tac has shown potential for eliminating alloreactive T-cells in graft-versus-host disease, and in organ allograft settings, where inflamed tissues are more rapidly targeted because of favourable circulatory distribution and vascular leakage [17].

Bismuth complexes such as $Na_2[BiO(mp)_3]\cdot 3H_2O$ and $[Bi(tgn)_3(H_2O)]\cdot 3.5H_2O$ (where mp = 6-mercaptopurine; tgn = thioguanine) have been shown to exhibit anticancer activity [18]. Skinner et al. found that $Na_2[BiO(mp)_3]\cdot 3H_2O$ was significantly effective in treating Dunning ascitic leukemia in rats [19]. Recently, it has been found that some organometallic bismuth(III) thiolate complexes (Scheme 1) exhibit an optimum cure rate of 100% against fluid Ehrlich ascites tumor with a therapeutic index of 3.2 to 5.0 [20]. Some bismuth complexes have also recently been shown to inhibit HIV-1 virus production from chronically infected H9 cells with a selectivity index of ca. 5 [21].

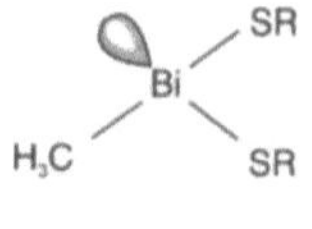

Scheme 1. Organometallic bismuth(III) thiolate complexes which exhibit antitumor activity [20]. R=CH_3, *p*-$PhNH_2$, *p*-$PhN^+H_2CH_3$

2
The Chemistry of Bismuth

2.1
Properties of the Element

Bismuth was first discovered as early as the 15th century, and was established as an element in 1739 by Potts and Bergmann [22]. It is most commonly found as the oxide (Bi_2O_3), carbonate [$(BiO)_2CO_3$] and sulfide (Bi_2S_3), and is also obtained as a by-product of lead, zinc and copper mining. It is the heaviest stable element in the periodic table, and is in group 15 together with nitrogen, phosphorus, arsenic and antimony. The main features of bismuth are shown in Table 2.

2.2
Oxidation States and Redox Chemistry

There are two major oxidation states of bismuth (III and V), with the 3+ state being the most common and most stable form, in contrast to arsenic and antimony. The electronic configurations of bismuth(III) and bismuth(V) are:

Bi(III): $(Xe)4f^{14}5d^{10}6s^2$
Bi(V): $(Xe)4f^{14}5d^{10}$

Bi(V) is a powerful oxidant in aqueous solution, with a Bi(V)/Bi(III) potential $E^o = 2.03$ V, although recently two Bi(V) complexes with benzenoid and non-benzenoid arenes as ligands have been reported to be stable even in aqueous media [23]. Little redox data for either Bi(III) or Bi(V) complexes appear to have been reported.

2.3
Bismuth(III) Compounds

Bismuth is widely used in medicine [24] but is also of current interest in high T_c superconducting materials and sol-gel and vapor deposition precursor compounds [25]. However, very little is known about its solution and solid-state chemistry. According to Pearson's HSAB (Hard-soft Acid-base) theory [26], Bi(III) is a border-

Table 2. Main features of bismuth

Atomic number	83
Electronic structure	$(Xe)4f^{14}5d^{10}6s^26p^3$
Occurrence in earth's crust	0.2 ppm
Natural abundance	100% ^{209}Bi
Nuclear spin quantum number	9/2
Radio isotope	^{212}Bi (α-emission with half-life 1 h and max. energy 6.09 and 8.78 MeV) [14]
Atomic weight	208.9804 (relative to ^{12}C=12,000)
Radii: Bi(III) (CN: 6)	1.03 Å
Bi(V) (CN: 6)	0.76 Å

line or soft metal ion [27]. Recently, Bi(III) has been found to have a high affinity for both oxygen and nitrogen ligands in aqueous solution. It binds to nitrogen donor macrocycles even in strongly acidic solutions (pH ca. 0) [28]. The stability constants of Bi(III) with a series of nitrogen donor ligands have been determined via differential pulse polarography. Separate peaks can be observed in differential pulse polarograms for free Bi(III) and complexed Bi(III) which greatly simplify the determinations. Bi(III) binding to ligands is usually fast with equilibration within a few minutes. An exception is the macrocycle 1,4,8,12-tetraazacyclopentadecane ([15]aneN_4), for which equilibration required two to three weeks. Some Bi(III) binding constants can be predicted from the following empirical equation [28a]:

$$\log K_1 \text{ (polyamine)} = 1.152 \log \beta_n(NH_3) + (n-1)\log 55.5 \quad (1)$$

where β_n refers to the stability of the complex containing n NH_3 analogues of the polyamine. This equation has been extended to polyaminocarboxylates such as nta and edta [29].

Stability constants for the binding of Bi(III) to some common ligands are listed in Table 3 [30]. It has been found that there is a good linear free-energy relationship between log K_1 values for Bi(III) and Pb(II) complexes, and between log K_1 values for Bi(III) and Fe(III) complexes (Fig. 1).

In general, the structures of Bi(III) compounds are similar to those of As and Sb compounds, but more complicated. The coordination number of Bi varies from 3 to 9 [31,32]. Compared to As and Sb, more structures of Bi complexes are known. Table 4 lists the range of stereochemistries and coordination numbers of Bi(III) compounds.

The structure of the aqua complex of $[Bi(H_2O)_9]^{3+}$, which is the same as those of the lanthanide complexes $[Ln(H_2O)_9](SO_3CF_3)_3]$, was solved only recently (Fig. 2) [32]. Bismuth coordinates to nine water molecules with a tricapped trigonal prismatic structure (CN: 9) without recognisable stereochemical activity of the lone pair of electrons. The oxide Bi_2O_3 is strictly basic, whereas As_2O_3 and Sb_2O_3 are acidic and amphoteric. Bi_2O_3 is soluble in mineral acids to give salts. The hydroxide, $Bi(OH)_3$, can be precipitated by the addition of base. The first pK_a for a Bi(III) aqua

Table 3. Binding constants for Bi(III) with common ligands [30]

Ligand	log K_1	Ligand	log K_1
hydroxide (OH^-)	12.49	2-mercaptoethanol	13.6
ammonia	5.1	glycine	10.0
dien	17.4	aspartate	10.5
trien	21.9	glutamate	10.5
[15]-aneN_4	23.5	nta	17.5
oxalate	7.7	ida	12.9
L-malate	9.9	hda	12.5
citrate	13.5	edta	27.8
ascorbate (vitamin C)	25.3	egta	23.8
succinate	8.8	cdta	31.9
fumarate	6.9	dtpa	35.6

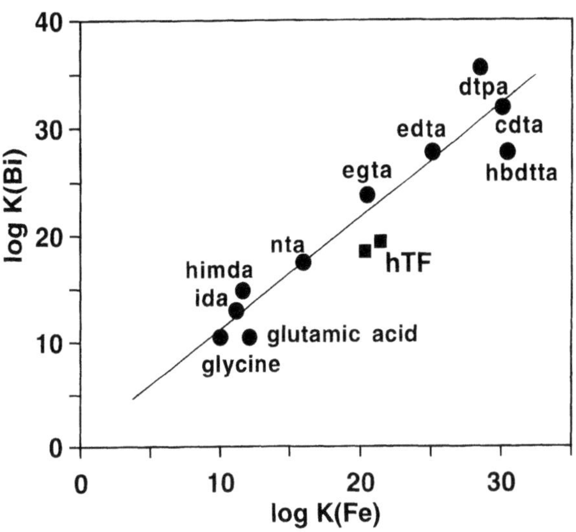

Fig. 1. Linear free-energy relationship for the complexation of Bi(III) and Fe(III) with oxygen and nitrogen donor ligands. Data taken from [30]. The points for the two binding constants of human transferrin are shown as *squares*

Table 4. Structural data for Bi(III) compounds [31]

Coord. number	Geometry	Examples	Ref.
3	Pyramidal	$[Bi(SAr)_3]$	38
4	Trigonal bipyramidal	$[Bi\{OP(NMe_2)_3\}_2\{Fe(CO)_2(\eta\text{-}C_5H_5)_5\}_2][PF_6]$	39
5	Square-based pyramidal	$Na_2[Bi(SC_6F_5)_5](THF)_4$	40
5	Trigonal antiprism	$\{Bi(NO_3)$bis[1-azepanyl-4-(2-thienyl)-2,3-diazapenta-1,3-diene-1-thiolato-N^3,S]}	67
6	Octahedral	$[(RS)_3Bi(SR')_3]$, $[Bi_6O_4(OH)_4]^{6+}$	33
7	Trigonal dodecahedron	$\{Bi(NO_3)$bis[1-azepanyl-4-(2-pyridyl)-2,3-diazapenta-1,3-diene-1-thiolato-N',N^3,S]}	67
8	Bicapped trigonal prism	$[Bi(nta)(H_2O)_2]$ and $[Bi(Hedta)]\cdot 2H_2O$	55
9	Tricapped trigonal prism	$[Bi(H_2O)_9](CF_3SO_3)_3$ (Fig. 2)	31
9	Monocapped-square antiprism	$(guanidinium)_2[Bi(dtpa)]\cdot 4H_2O$	55

ligand is 1.51 [30]. At high pH, $[Bi(O)]^+$ forms, and polymeric cations such as $[Bi_6O_4(OH)_4]^{6+}$ always form in solution before precipitation of the hydroxide complex. The X-ray crystal structure of this cation shows that six Bi atoms are at the apices of an octahedron at non-bonding separations. The octahedron is face-capped by O and OH, forming Bi–O–Bi bridges [33]. The lone pair of electrons on Bi is directed away from the cage [34]. Bi(III) also shows a high affinity for other oxygen donor ligands. Bi(III) complexation with polyethylene glycols and crown ethers has been reported recently [35]. Most of the Bi(III) polyethylene glycol complexes are dimeric, with bridging by two alkoxide oxygen atoms. Coordination of each Bi(III) is completed by a nitrate anion. The Bi–alkoxide bonds are short (2.24 Å), indicative of strong covalent character, and this strong bond appears to give rise to a stereochemical role for the lone pair of electrons. The crown ether complexes do not exhibit such strong covalent interactions, and do not display lone-pair effects. Bis-

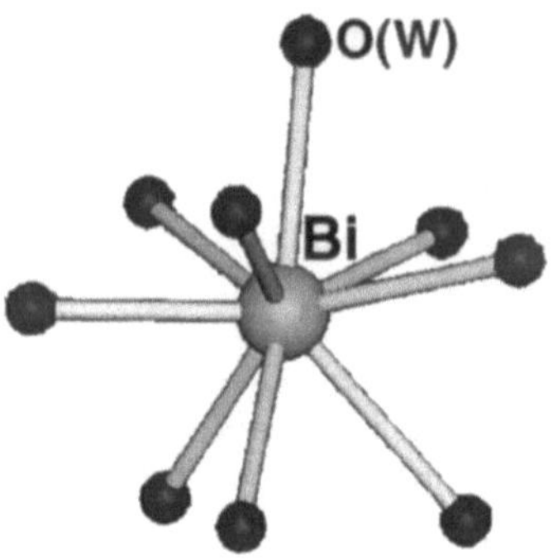

Fig. 2. Structure of $[Bi(III)(H_2O)_9]^{3+}$[32]. The hydrogen atoms are omitted. The Bi–O lengths are 2.5–2.6 Å

muth alkoxide complexes (e.g. $[Bi(OC_2H_5)_3]$ and $[Bi(O\text{-}iC_3H_7)_3]$) show potential as precursors for new superconductor formulations [36].

Another class of complexes which may be relevant to the biological chemistry of bismuth is that with thiolate-containing ligands [37–41]. In these complexes, bismuth exhibits coordination numbers of three, four, five and six with thiolate ligands. Overall the structures are either monomers or loose dimers as shown in Scheme 2. The Bi–S bond distance is ca. 2.5–2.7 Å for intramolecular interactions, and ca. 3.1–3.4 Å for intermolecular (long-range) interactions.

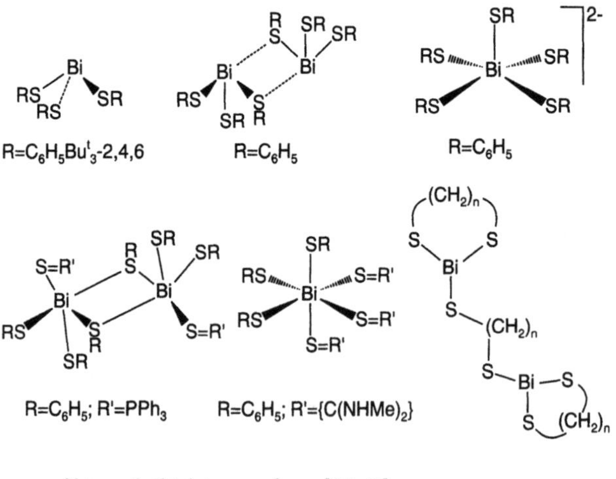

Scheme 2. Structures of bismuth thiolate complexes [37–41]

Bi–Bi bonds are also observed in some complexes. The most significant one is tetraphenyldibismuth, Bi_2Ph_4, which shows a staggered trans conformation (Scheme 3). The Bi-Bi bond length is 2.990 Å [42]. The first stable Bi=Bi double bond was observed in a dibismuthene complex [TbtBi=BiTbt], where Tbt is 2,4,6-tris[bis(trimethylsilyl)methyl]phenyl. The X-ray crystal structure reveals that the Bi=Bi double bond (2.821 Å) is 6% shorter than a typical Bi–Bi single bond, and that the Bi–Bi–C angle is 100.5° which deviates greatly from the ideal sp^2 hybridized bond angle [43]. Other bismuth organometallic complexes have been extensively reviewed [44].

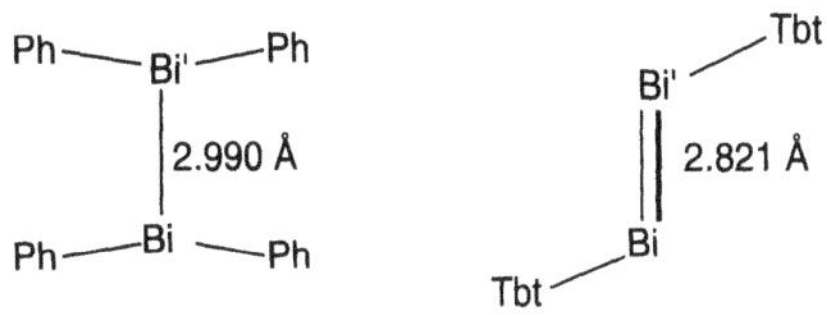

Scheme 3. Bi_2Ph_4 and TbtBi=BiTbt showing the Bi–Bi single and double bonds [42,43]

2.4 Bismuth(V) Compounds

Most of the Bi(V) complexes prepared to date are five-coordinate, although Bi(V) in triarylbismuthate complexes such as $Ar_3Bi(OCHO)_2$ (where Ar=Ph, *p*-Tol) [45], is six-coordinate. Almost all of these Bi(V) complexes are not stable. However, seven-coordinate complexes tri(aryl)tropolonatobismuth(V) (where R=H, NO_2; R′=H, CH_3, Scheme 4) have recently been reported to be unusually stable, probably due to the steric shielding of the bismuth ion [23].

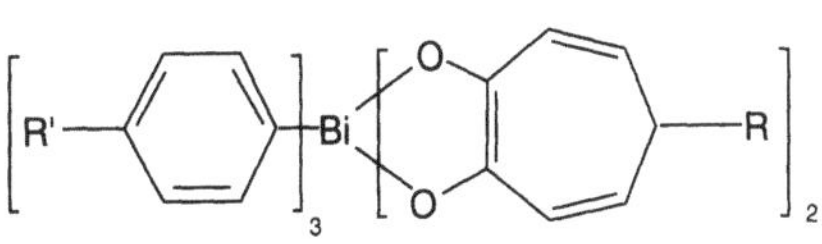

Scheme 4. Tri(aryl)tropolonatobismuth(V)

3 Physical Methods for the Study of Bismuth Compounds

Table 5 summarizes the various physical methods which are used for the study of bismuth compounds. There are few informative methods for the direct probing of Bi in bismuth complexes. ^{209}Bi NMR has not been widely used because it has a large electric quadruple moment (-0.37×10^{-28} m^2), thus leading to broad resonances. Resonances for ^{209}Bi can be detected only for highly symmetrical environments, and only two ^{209}Bi NMR spectra have been reported; those of $Bi(NO_3)_3$ (dissolved in nitric acid) with a linewidth of 3.2 kHz (chemical shift of −24 ppm) and $[BiF_6]^-$, with a linewidth of 44 Hz, chemical shift 0.0 ppm, and $J_{Bi\text{-}F}$ 3823 ± 3 Hz [46], and no useful chemical studies have been carried out. NMR studies of bismuth complexes have therefore relied on observations of ligand nuclei (e.g. ^1H,^{13}C). Nuclear quadruple resonance has also not been useful for studies of bismuth. Single crystal X-ray diffraction gives useful structural information if suitable crystals can be obtained, and since bismuth is a very heavy element, very small crystals are used to avoid absorption effects. X-ray absorption near edge structure (XANES) and extended X-ray absorption fine structure (EXAFS, e.g. Bi L_{III} edge) are likely to be useful in the study of bismuth coordination spheres but there appear to be no previous reports of

Table 5. Physical methods for the study of bismuth complexes

Technique	Information	Comments
NMR spectroscopy	Types and number of coordinated ligands, conformation, kinetics and dynamics (ligand exchange)	Only two reports of ^{209}Bi NMR, probing ligand rather than Bi, usually ^{1}H, ^{13}C, ^{15}N and ^{31}
Infrared and Raman spectroscopy	Ligand and Bi-ligand vibration frequencies	Can be difficult to assign Bi–S or Bi–O vibrational frequencies. Usually compare free ligand and complex
UV/Vis and circular dichroism spectroscopy	Electronic energy levels and symmetry	Absorptions are usually weak in the visibleregion, charge-transfer absorptions can beobserved in the UV region
Mass spectrometry	Molecular mass of parent and fragments	Compound must be introduced into the ion source, fragmentation pattern can be obtained
Atomic absorption spectroscopy	Total Bi, detection limit 0.09 ppb (μg/l)	Need to digest the sample, desalt biological samples
ICP-MS	Total Bi, detection limit 0.01 ppb (μg/l), direct observation of ^{209}Bi isotope	Samples only need dilution by 10% HNO_3, less interference (than AAS), suitable for biological samples
X-ray crystallography	3D structure (bond lengths, angles)	Need single crystals
X-ray spectroscopy (XAS, XANES and EXAFS)	Number and type of bonded atoms, interatomic distances	Requires synchrotron radiation, can be used for both solution and amorphous solid, difficult to distinguish N and O, and P, S and Cl
Differential pulse polarography	Half-wave potentials and concentrations, Detection limit, ca. 10^{-7} M	Distinguishes between Bi(III) and Bi(V), speciation, Bi binding at very low concentration (μM), wide pH range

this. Using EXAFS, we have recently been able to determine the coordination spheres of Bi in glutathione, metallothionein and transferrin complexes [47].

The most common method of determining the total bismuth content of a sample is atomic absorption spectroscopy (AAS), and, more recently, inductively coupled plasma mass spectrometry (ICP-MS) has been used. The former requires digestion of the sample, and large interferences from high concentrations of salts retard the application to biological samples. ICP-MS seems to be a suitable method for determining bismuth in all kinds of materials. Since bismuth is a very heavy element (atomic mass 208.98), there is little interference from other elements, and the detection limit can be as low as 0.01 ppb (μg/l, 4.8×10^{-9} M) [48]. Using direct inject nebulization (DIN) instead of pneumatic nebulization (PN), memory effects can be dramatically suppressed and less sample volume is required [48]. By combining HPLC (FPLC) and ICP-MS, it is possible to study the speciation of Bi in biological systems. Recently, the Bi content of blood plasma has been determined by ICP-MS before, during and after the intake of the bismuth antiulcer drug CBS [49].

4
Helicobacter pylori

Helicobacter pylori was discovered about 100 years ago by German pathologists, and was isolated by two Australian doctors, Warren and Marshall, in 1983 [50]. The presence of this bacterium in gastric mucosa is associated with chronic active gastritis, and has been implicated in more severe gastric mucosal conditions, including gastric atrophy, peptic ulceration, mucosa-associated lymphoid tissue lymphomas, and even gastric cancer [51].

Helicobacter pylori is a micro-aerophilic, slow-growing, gram-negative spiral-shaped bacterium. It has a smooth outer coat, with several whip-like tails (flagellae) at one end (Fig. 3). Due to the presence of the flagellae, it can penetrate the mucus layer of the mucosa. The curved structure of *H. pylori*, as well as specific adhesions and its flagellae mechanism, enable it to travel within the sanctuary zone. It produces the multisubunit, nickel-containing enzyme urease [52]. This enzyme catalyzes the hydrolysis of urea to form ammonia and carbon dioxide [Eq. (2)]. It is commonly thought that the ammonia produced from urea in the stomach catalyzed by urease gives rise to severe cytotoxic effects on mammalian cells within the gastric epithelium neutralizes the micro- and macro-environment of bacteria and, therefore, aids the survival of the bacteria under the acidic conditions of the gastric lumen and mucosa (pH ~2).

$$H_2N - CO - NH_2 + H_2O \xrightarrow{\text{urease}} [H_2N - COO^- + NH_4^+] \rightarrow 2NH_3 + CO_2 \qquad (2)$$

Another important enzyme in *H. pylori* is catalase, a manganese-containing enzyme, which protects *H. pylori* from being destroyed by neutrophils, part of the natural defense (immune) system of the body. Similar to all other bacteria, *H. pylori* requires

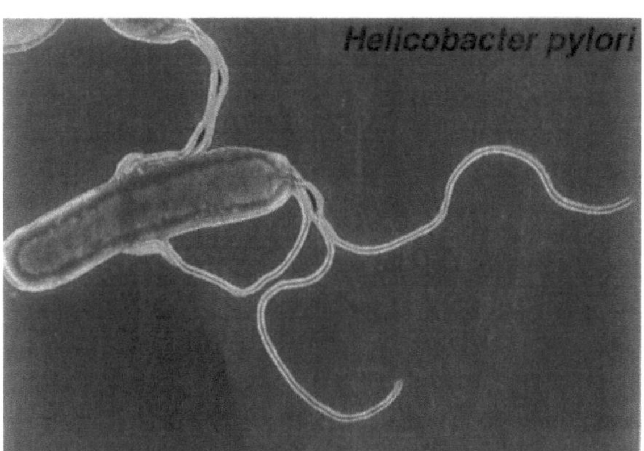

Fig. 3. *Helicobacter pylori*

an iron-scavenging system for survival in the host. The bacterium possesses several systems for iron-uptake, such as the siderophore-mediated iron-uptake *fec* system of *E. coli*, the anaerobic *feo* (ferrous iron) system of *E. coli* and receptors for the iron-binding protein lactoferrin, which is abundant in mucus [53,54].

5
Antibacterial and Antiulcer Bismuth Complexes

Although it has been known for a long time that bismuth complexes exhibit antibacterial and antiulcer activity, the structures of several of them have only recently been determined. $[Bi(nta)(H_2O)_2]$, $[Bi(Hedta)]\cdot 2H_2O$ and $(guanidinum)_2[Bi(dtpa)]\cdot 4H_2O$ all exhibit antibacterial activities, and their structures were determined in 1994 by Palenik et al. [55]. The least soluble complex, $[Bi(nta)(H_2O)_2]$ (solubility in aqueous solution ca. 2 mM), was found to be the most active. In the neutral complex $[Bi(nta)(H_2O)_2]$, bismuth is eight-coordinate, bound to three oxygens and one nitrogen from nta, and two carboxylate oxygen atoms from a neighbouring $[Bi(nta)(H_2O)_2]$ molecule, with the coordination sphere being completed by two water molecules, forming a bicapped trigonal prism. In $[Bi(Hedta)]\cdot 2H_2O$ (solubility ca. 30 mM, pH 7.0), the bismuth atom is also eight-coordinate with a bicapped trigonal prism arrangement, but there is no water coordinated to bismuth. The edta ligand coordinates in hexadentate mode via the two nitrogen atoms and four oxygen atoms of the carboxylates; two carboxylate oxygen atoms from another edta ligand complete the coordination sphere. In $(guanidinum)_2[Bi(dtpa)]\cdot 4H_2O$, bismuth is coordinated in a monocapped-square antiprism arrangement.

The empirical formula of colloidal bismuth subcitrate (CBS), one of the clinically used antiulcer drugs, is often given as: $K_3(NH_4)_2[Bi_6O_3(OH)_5(Hcit)_4]$ [56]. By variation of the pH and the ratios of bismuth and citrate, nine different bismuth citrate adducts have been isolated and characterized by X-ray crystallography. Most of them contain the stable dinuclear unit $[Bi(cit)_2Bi]^{2-}$ with additional O^{2-}, OH^- and H_2O ligands as shown in Fig. 4a. This dinuclear unit can aggregate further via citrate bridging to form channels and also form sheets via H-bonding; the high solubility of CBS is probably due to the formation of such channels and sheets. The coordination number of bismuth in these complexes is usually high, from 6 to 9, and there are short Bi–alkoxide (C–O^- of citrate) bonds of ca. 2.13 Å. The stereochemical role played by the lone pair of electrons of Bi(III) in these compounds is particularly notable. All of the bound atoms lie on one side of the Bi coordination sphere, and the vacant axial site is occupied by the lone pair of electrons. However none of these complexes has exactly the same composition as CBS, and, therefore, the solid-state and solution structures of CBS are still unknown, although it seems likely that it consists of dinuclear units $[Bi(cit)_2Bi]^{2-}$ aggregated through citrate bridges and H-bonding. It is clear that bismuth citrate complexes can have complicated structures, being dependent on pH, concentration, the [Bi/cit] ratio and the counter cations. At neutral pH, multinuclear clusters such as $[Bi_{12}O_8(cit^{4-})_8]^{12-}$ (Fig. 4c) are also formed by citrate bridging [57]. The latter cluster can further aggregate to form Bi_{24} in concentrated solutions, and dissociates into smaller clusters such as $[Bi_6O_4(cit^{4-})_4]^{6-}$, with some release of citrate when diluted [57].

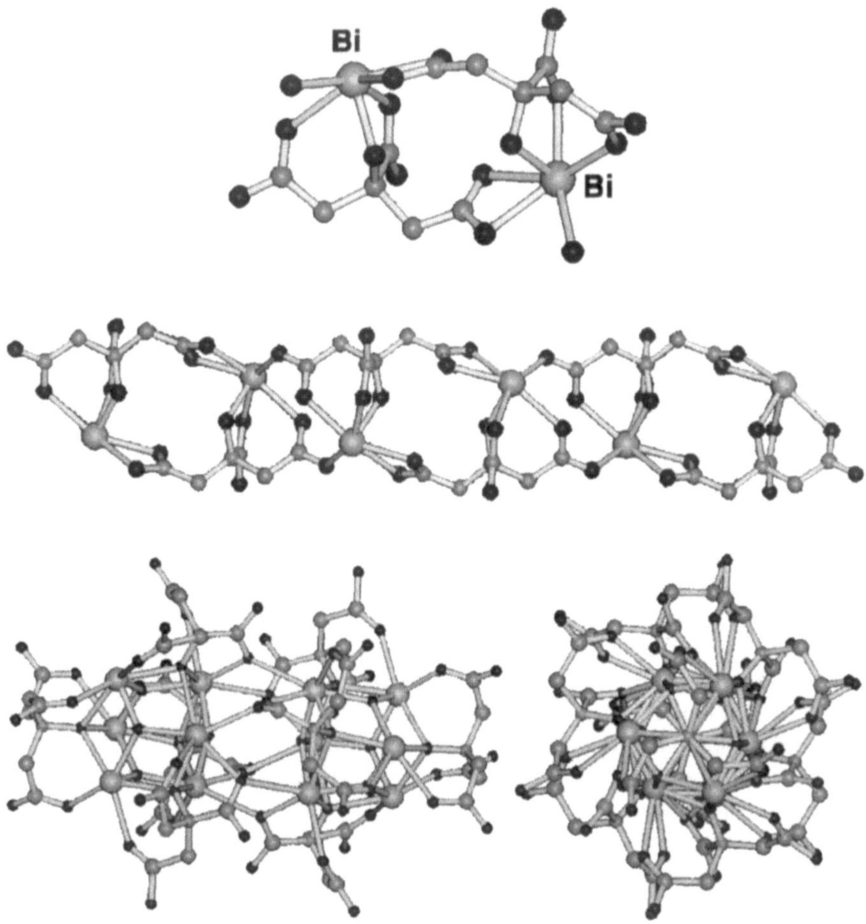

Fig. 4. Structures of Bi(III) citrate complexes. *a* The basic bismuth citrate dimeric unit, the single O atom bonded to each Bi can be from H_2O or from a neighbouring carboxylate; *b* polyanionic chain of $[\{Bi(cit)_2Bi\}_n]^{2n-}$; *c* dodecanuclear cluster of $[Bi_{12}O_8(cit)_8]^{12-}$ with different views

The new antiulcer drug, ranitidine bismuth citrate (RBC, Tritec, Pylorid), is highly soluble in water (ca. 1.0 g/ml), giving a pH of 4.6. It is highly active against *H. pylori* (Table 6) [58]. Possible solution and solid-state structures of RBC have been discussed [59,60]. In aqueous solution, different bismuth citrate species are in fast exchange on the NMR time scale; at pH* >6.2 the exchange rate decreases and different species are observed [59,61]. By means of diffusion-ordered 2D $[^1H,^{13}C]$ HSQC NMR spectroscopy, together with isotope labelling of the citrate ($^{13}C2/^{13}C4$), a wide range of types of bound citrate in aqueous solution of ranitidine bismuth citrate at pH* 7.4 can readily be detected at low Bi concentrations [62]. There appears to be an equilibrium between the free citrate $[Bi(cit)_2Bi]_n^{2n-}$ and multinuclear bismuth citrate clusters at physiological pH values.

Table 6. In vitro activity of bismuth compounds against *H. pylori* [58]

Compound	Formula	Solubility in water	MIC_{50}/nM (ng/l)[a]
Bismuth subnitrate	$Bi(O)NO_3$	Insoluble	0.45 (128)
Bismuth subgallate	$Bi(OH)C_7H_4O_5$	Insoluble	0.04 (16)
Bismuth subsalicylate	$BiOC_7H_5O_3$[b]	Very low	0.09 (32)
CBS	$K_3(NH_4)_2[Bi_6O_3(OH)_5(C_6H_5O_7)_4]$[b]	Soluble	0.04 (8)
RBC	$(Ranitidine^+)Bi(cit)$[b] $C_{19}H_{27}N_4O_{10}SBi$	Soluble (>1 g/ml)	0.011 (8)

[a] MIC, minimum inhibitory concentration
[b] Approximate

In RBC, ranitidine is not coordinated to Bi, and appears to be involved in specific second-coordination sphere interactions with bismuth citrate via its $HNMe^+$ group. Complexation of Bi(III) to both citrate and ranitidine in acidic solutions (pH 2.5–3.0) has been detected by differential pulse polarography [59]. There is a rapid deprotonation equilibrium between at least two species in solutions containing Bi(III)/citrate in a 1:10 mol ratio. At pH >5.8, a second peak appears suggesting that at least two types of Bi(III) citrate complex exist under these conditions, in agreement with NMR data. The solid-state structure of ranitidine bismuth citrate may be closely related to that of $Na_2[Bi_2(citrate)_2]\cdot 7H_2O$ in view of the similarity in their chemical compositions (Bi/cit=1:1), crystallization conditions (ca. pH 4), and solid-state ^{13}C NMR spectra [60,61]. The complex $Na_2[Bi_2(citrate)_2]\cdot 7H_2O$ also contains the highly stable dimeric unit $[Bi(cit)_2Bi]^{2-}$ in which two Bi(III) ions are doubly bridged by citrate carboxylate oxygens as shown in Fig. 4a. These dimeric units are assembled into infinite polymeric chains via double *syn-anti* carboxylate bridges (Fig. 4d). Such polymers, formed into sheets, could be deposited on the ulcer crater forming a protective coating, or on bacteria in ulcers. Cleavage of the double bridges between dimeric units by citrate or water can be envisaged to give singly bridged polyanionic species. Various tiny crystalline species have been reported to be present in the ulcer craters of patients after treatment with colloidal bismuth subcitrate [63].

It is possible to envisage that dimeric bismuth citrate units could inhibit the uptake of Fe(III) into some types of bacteria since Fe(III) also forms a dimeric citrate complex, $[Fe(cit)_2(H_2O)_2]^{2-}$ [64]. In this Fe(III) complex, the double bridge is provided by alkoxide oxygens rather than carboxylates as is the case with Bi(III).

Both Bi(III) and Bi(V) tropolonate complexes have shown promising activity against *H. pylori* in an agar dilution test. Some of them even have higher in vitro activity. In these complexes, the tropolonates are bidentate with Bi–O bond lengths of 2.1–2.4 Å for Bi(III) and 2.3–2.6 Å for Bi(V) (Schemes 4 and 5) [65,66].

The complexation of bismuth to active ligands can greatly enhance the effectiveness against *H. pylori*. For example, thiosemicarbazones exhibit antibacterial and

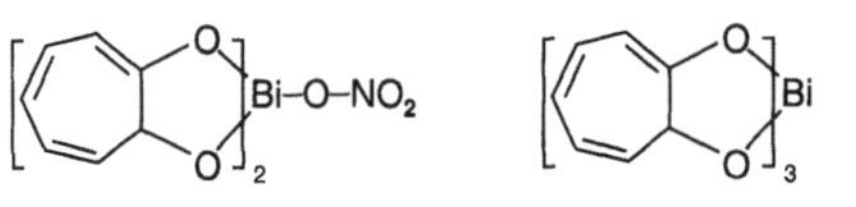

Scheme 5. Bi(III) tropolonato complexes which show anti *H. pylori* activity

other activity. Recently, a series of bismuth complexes with thiosemicarbazones and dithiocarbazonic acid methyl ester derivatives as ligands has been synthesized and characterized by spectroscopic methods and X-ray crystallography [67]. Bi(III) coordinates to these ligands in bi-, tri- or penta-dentate modes. These complexes were tested against ten *H. pylori* strains, and exhibited activity similar to that of clinically used bismuth compounds (bismuth citrate and bismuth salicylate).

Several thiol ligands (e.g. 2-mercaptoethanol) have shown high activity against microbial and plant urease [68,69]. Urease is a nickel enzyme which catalyzes the conversion of urea into ammonia. This allows *H. pylori*, for example, to survive under the highly acidic conditions in the stomach. The structure of 2-mercapto-ethanol-inhibited urease from *Bacillus pasteurii* shows that one 2-mercaptoethanol bridges the two nickel ions in the active site through the sulfur atom and chelates one Ni through the OH group. Another molecule of inhibitor forms a mixed disulfide bond with a nearby Cys residue, and hence seals the entrance to the active site cavity [70]. The crystal structures of bismuth complexes of 2-mercaptoethanol have been solved by two groups independently [71,72]. These complexes show structural flexibility for the bicyclic framework as shown in Scheme 6.

Surprisingly, all of these complexes have almost identical minimum inhibitory concentrations (MIC) against *H. pylori* in vitro, although their solid-state structures are different, indicating either that the antibacterial activity toward *H. pylori* is a general property of Bi(III) compounds or that all these complexes are effective delivery agents for mercaptoethanol. These complexes have enhanced activity against *H. pylori*-produced urease compared to the free ligand, 2-mercaptoethanol. Antibacterial activity of Bi(III) can be greatly enhanced by 25- to 300-fold by combining bismuth with lipophilic thiols [73]. The most active bismuth mono- and dithiol complexes are also the most soluble in butanol. The enhanced activity of bismuth complexes over simple bismuth salts is likely to be due to increased solubility, and involve a facilitated transport of bismuth into bacteria via thiols as carriers.

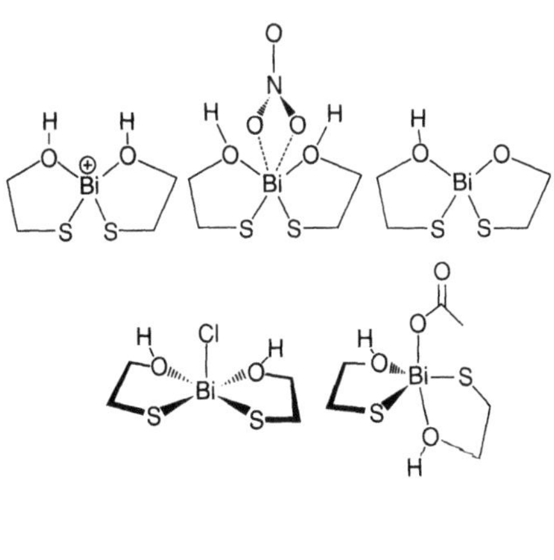

Scheme 6. Various types of Bi(III) mercaptoethanol complexes [71,72]

Several sialic acid derivatives exhibit antimicrobial activity, and one of these compounds is in phase III clinical trials [74,75]. Complexation of these sialic acid inhibitors with bismuth has been shown to give activity against two strains of

H. pylori comparable to that of bismuth citrate. These complexes may have an advantage over current bismuth compounds in that the sialic acid inhibitor can interfere with the adhesion of *H. pylori* to the gastric epithelium [76].

6 Bismuth Complexes with Biomolecules

6.1 Bismuth Binding to Oxygen-Containing Molecules

Bi(III) is readily hydrolyzed with log K_{OH} ca. 12.5, and also readily forms hydroxide- and oxide-bridged multinuclear clusters even at pH 1.5 [34]. The structures of some bismuth complexes with oxygen-containing biomolecules crystallised at acidic pH values have recently been solved by Herrmann and co-workers. These include D-lactate, L-(−)-malate, oxalate and L-(+)-tartrate complexes [77]. In these complexes, bismuth exhibits a high coordination number (e.g. 8 and 9), and all the complexes are polymeric via carboxylate oxygen bridging. In the bismuth oxalate complex, the bismuth is also bridged by chloride ions. Five-membered chelate rings are preferentially formed in these complexes probably due to the large size of Bi(III) [78]. It is not clear if these ligands are targets for bismuth in biological systems.

6.2 Bismuth Complexes with Thiolate Ligands

As Bi(III) is a border-line or soft metal ion, it is expected to bind strongly to thiolate sulfur. The amino acid cysteine and tripeptide glutathione (γ-L-Glu-L-Cys-Gly, gsh) can prevent the precipitation of CBS at pH 2.0, and animal studies have shown that simultaneous oral administration of bismuth salts and thiolates produces a significant rise in the bismuth concentration in blood plasma [79,80]. Infrared spectroscopic studies suggest that Bi(III) coordinates only to the thiolate group of cysteine [81], while the X-ray crystal structure of the complex $[\{SC(CH_3)_2CH(NH_2)CO_2\}BiCl]$ shows chelation of Bi(III) by sulfur, nitrogen and oxygen of tridentate D-penicillamine [82]. The interaction of bismuth with oxygen atoms of the carboxylate groups leads to a polymeric two-dimensional structure. Bismuth forms stable complexes with gsh and *N*-acetyl-L-cysteine (nac) with a stoichiometry $[Bi(H_{-1}gsh)_3]$ or $[Bi(H_{-1}nac)_3]$ (Scheme 7) [83], and the binding constants have been obtained from studies of competition reactions between edta and gsh (or nac) giving log K 29.6 and 31.4 (I = 0.1 M $NaNO_3$, 298 K) for $[Bi(H_{-1}gsh)_3]$ and $[Bi(H_{-1}nac)_3]$, respectively. Both NMR and EXAFS [61] suggest that the deprotonated thiolate group is the only strong binding site for Bi(III) in these ligands, and Bi(III) induces unusually large chemical shift changes of 1.4 ppm for the two βCH_2 Cys protons of gsh and nac. In spite of the extremely high thermodynamic stability of $[Bi(H_{-1}gsh)_3]$, bound gsh is kinetically labile and bound gs^- exchanges with free gsh slowly at pH 4 (3 s^{-1}, 298 K) but faster at biological pH (ca. 1500 s^{-1}, 298 K). Bismuth appears to pass through red cell membranes slowly ($t_{1/2}$ ca. 3 h), probably via a shuttle mechanism involving membrane proteins, and forms an intracellular complex with gsh, probably $[Bi(H_{-1}gsh)_3]$.

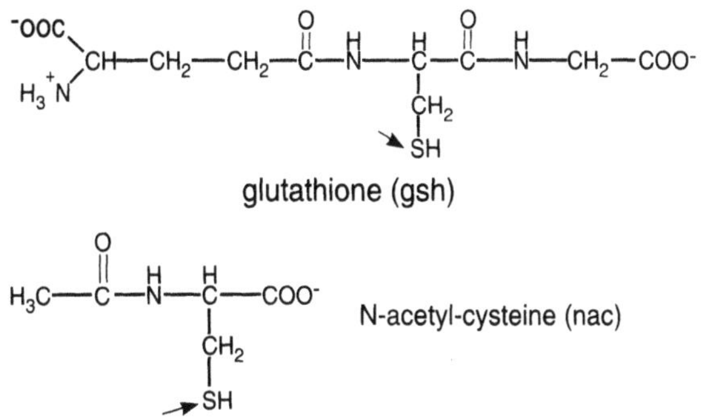

Scheme 7. Structures of glutathione and *N*-acetylcysteine

6.3 Bismuth(III) Complexation with Proteins and Enzymes

Systematic studies of the binding of high molecular mass ligands to Bi(III) have not been reported [84,85]. In the stomach, bismuth has been shown to bind strongly to connective tissue proteins, mucus glycoproteins and enzymes [86], but little is known about the binding mode or kinetic behavior. The most detailed studies on proteins are described below.

Metallothionein. Bismuth is known to be a potent inducer of renal metallothionein (MT) synthesis, and pretreatment with bismuth can protect from some of the toxic side effects induced by the anticancer drug cisplatin [87–89]. The protection probably involves Bi(III) induction of metallothionein synthesis. However, there appear to be few reported studies of the chemistry of Bi-metallothionein complexes. Strong binding of Bi(III) to metallothionein is expected, since 20 of its 61 amino acid residues are cysteines. The protein appears to be involved in the normal storage of up to seven Zn(II) ions or ten Cu(I) ions, and toxic ions such as Cd(II), Hg(II), and Au(I) from antiarthritic drugs [90]. An interaction has also been suggested between bismuth and a copper-binding protein different from metallothionein, as well as with the copper-containing blood plasma protein caeruloplasmin [91]. After administration of bismuth, a significant rise in both the copper content and the amount of MT-like proteins is observed in the kidney. Simultaneous administration of selenium has been reported to inhibit the binding of bismuth to a Bi-binding protein in the kidney, which is similar to the effect of Se on the binding of mercury to a Hg-binding protein [92,93].

Our recent work [94] has shown that Bi(III) binds very strongly to metallothionein with a stoichiometry Bi/MT=7:1 and can readily displace both Zn(II) and Cd(II) in biphasic processes. In contrast to Zn(II) and Cd(II), bismuth is still bound to the protein even in strongly acidic solutions (pH <2). ^1H NMR data suggest that both Zn(II) and Cd(II) in the β-domain (three metal cluster) of MT are displaced by Bi(III) much faster than from the α-domain (four metal cluster). Extended X-ray absorption fine structure (EXAFS) spectroscopy indicates that Bi(III) is coordinated to three sulfur atoms with average Bi–S distances of 2.55 Å [94].

Albumin. Serum albumin (66.5 kDa) is a single-chain protein of about 585 amino acids with three structurally homologous domains and is largely α-helical [95]. It is the most abundant protein in serum blood (ca. 0.65 mM) and is known to bind and transport various endogenous small molecules and metal ions [e.g. Zn(II)], as well as drugs and xenobiotics [96,97]. Feldman et al. have suggested that the main binding target for Bi(III) in blood plasma is albumin, since it contains a free thiolate group at Cys34 (with pK_a ca. 5) [84]. Binding of antiarthritic gold(I) complexes to Cys34 and redox reactions of disulfides with Cys34 can be readily detected by ^{1}H NMR [98,99] and we have applied the same methods to study Bi(III) binding. Our preliminary data [100] suggest that Bi(III) binds to albumin only weakly and non-specifically. The availability of Cys34 for binding appears to depend both on the metal ion and on the ligands bound to the metal. This residue is situated in a cleft in the protein and the factors which control cleft opening are not yet understood. Cd(II) (also a soft metal ion) appears to bind at oxygen- and nitrogen-containing sites rather than Cys34 [101]. Proteins other than albumin may bind to Bi(III) in blood plasma and investigations of the chemical forms of bismuth present in blood plasma are urgently needed [102].

Transferrin. Transferrin is an 80 kDa glycoprotein which transports Fe(III) in blood, and is recognised by cell surface receptors (proteins) when fully loaded with Fe(III) in both its binding sites. The diferric protein is internalised by cells, placed in vesicles (endosomes) where the pH is lowered to 5.5 and the Fe(III) is released [103]. In human blood, transferrin is only ca. 30% saturated with Fe(III), and hence there is capacity (ca. 50 μM) for binding to other metal ions that enter the human body. Similar proteins have been found in mucous (lactoferrin) and several gram-negative bacteria (ferric-ion-binding protein, FBP) [104]. Bismuth binds strongly to both N- and C-lobe iron binding sites of human serum transferrin (Scheme 8) [105]. The uptake of Bi(III) by apo-transferrin from bismuth citrate complexes is very slow (hours at 310 K) and occurs in at least two steps, whereas transfer from bismuth nitrilotriacetate is rapid (minutes).

Both UV and ^{13}C NMR data suggest that bismuth binds to transferrin along with carbonate (CO_3^{2-}) as a synergistic anion, which is similar to Fe(III). Bi(III) binds preferentially to the C-lobe, followed by the N-lobe. This order of lobe loading has been confirmed by 2D [^{1}H,^{13}C] heteronuclear multiple-quantum coherence (HMQC) NMR studies of recombinant ε-[^{13}C]Met-hTF (Fig. 5). The changes in shift of the ^{1}H,^{13}C NMR resonances can be used as fingerprints of conformational changes induced by Bi(III) and other metal ions. Bi(III), Fe(III), Ga(III) and Al(III) probably induce similar conformational changes in hTF, since the changes in shifts of resonances are almost identical [106].

Although Bi(III) binds to human serum transferrin relatively strongly (log K_1 19.42 and log K_2 18.58, at 310 K, 5 mM bicarbonate, pH 7.4), it can be displaced by Fe(III). Competition reactions between transferrin and citrate indicate that even in the presence of 100-fold excess of citrate, Bi(III) can still bind to transferrin. The strong binding of Bi(III) to transferrin can be correlated with its high acidity [107,108] (Fig. 6). The correlation allows prediction of the strong binding of other highly acidic metal ions such as Ti(IV) used in anticancer agents, which has now been verified [109], and suggests that the two tyrosine ligands play a dominant role in the strength of binding.

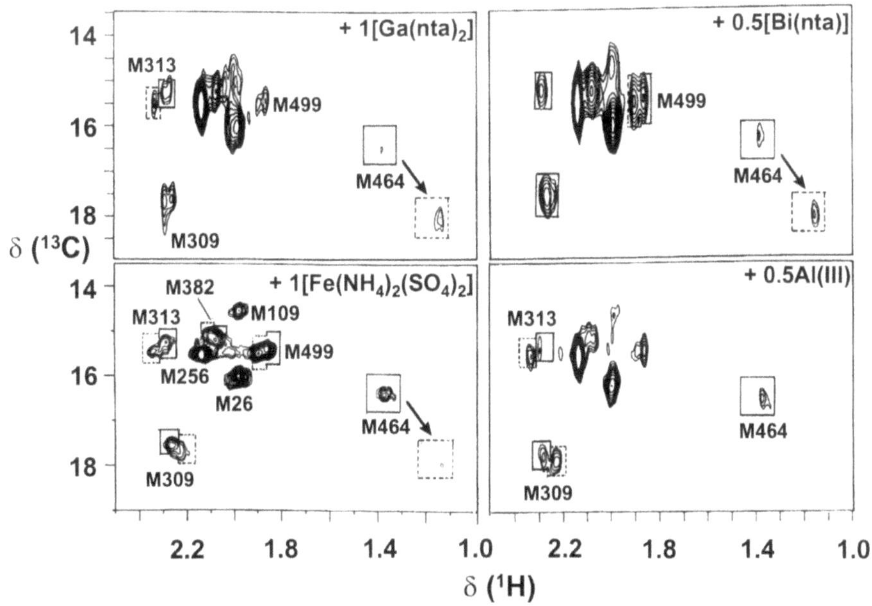

Fig. 5. NMR determination of the order of lobe-loading (N- and C-lobe metal-binding sites) of human serum transferrin with metal ions. 2D [^{1}H,^{13}C] HMQC spectra of ε-[^{13}C]Met-apo-hTF in the presence of 10 mM bicarbonate, pH* 7.4, except Al(III) at pH* 8.8, after addition of different metal complexes. The *solid boxes* indicate initial peaks and the *dotted boxes* show new peaks (adapted from [106]). A shift of the peak for Met464, which is close to the C-lobe iron site, indicates binding of the metal to the C-lobe, whereas shifts of Met309 and Met313, for example, indicate N-lobe binding. Under these conditions, Bi(III) preferentially loads the C-lobe, Al(III) loads the N-lobe, and Fe(III) [as $Fe(NH_4)_2(SO_4)_2$] and Ga(III) load both lobes

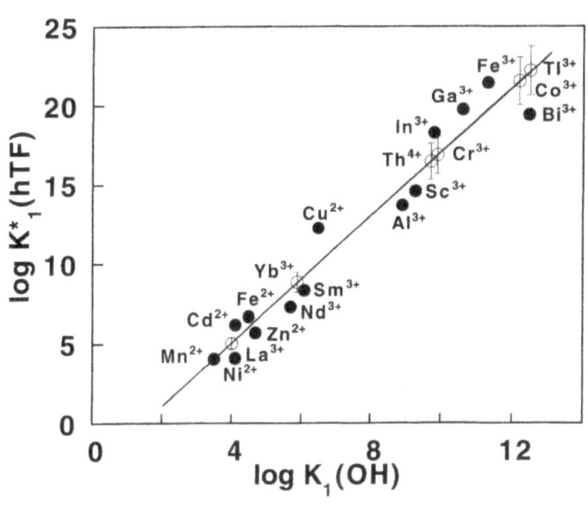

Fig. 6. Correlation of the strength of metal binding to human serum transferrin with metal ion acidity ($pK_a = 14 - \log K_{OH}$) [107]. The *open circles* are predicted binding constants

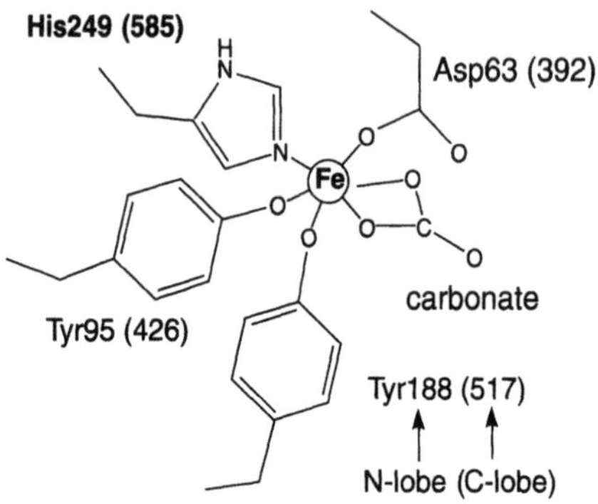

Scheme 8. Iron binding sites in human serum transferrin

Enzymes. Enzymes are thought to play a role in the mechanism of action of bismuth-containing drugs. The mode of action of colloidal bismuth subcitrate (CBS) in the treatment of gastroduodenal disorders may involve the prevention of adhesion of *H. pylori* to epithelial cells and inhibition of enzymes secreted by *H. pylori*, such as proteases, lipases, glycosidases and phospholipases [110]. It has also been found that CBS can induce a dose-dependent inhibition of phospholipase A (PLA2) and C activity of both *H. pylori* lysates and filtrates, and it has been suggested that bismuth binds to the calcium site on PLA2 [111]. Bismuth subcitrate inhibits *H. pylori* F_1-ATPase, an enzyme involved in bacterial energy metabolism. The inhibition can be prevented and reversed by the thiol glutathione. It is possible that bismuth binds to Cys thiol groups on the enzyme [112] but other sites may also be involved. It has been reported that CBS can inhibit the enzyme pepsin at pH 1.0 and 2.0, and it has been suggested that inhibition is due to negatively charged bismuth salts derived from CBS at pH 1.0 and 2.0 via ionic interactions with positively charged groups of pepsin, thereby inactivating the enzyme [113].

Drugs such as colloidal bismuth subcitrate and bismuth subsalicylate can inhibit alcohol dehydrogenase (ADH) from *H. pylori* and consequently suppress acetaldehyde (toxic to mucosal cells) production from endogenous or exogenous ethanol [Eq. (3) where NAD^+/NADH is the dehydrogenase coenzyme] [114,115]. ADH contains two Zn(II) ions which occupy two separate domains. One of them resides in the catalytic domain and is coordinated to two Cys thiolate groups, nitrogen from a His imidazole and to one water molecule. The other Zn(II) (structural) is bound to four thiolates of Cys residues [116]. The inhibition may be due to the displacement of zinc(II) by Bi(III), since Bi(III) has a higher affinity for thiolate groups than Zn(II).

$$R - CH_2 - OH + NAD^+ \xrightarrow{ADH} R - CHO + NADH + H^+ \tag{3}$$

The unusual feature of *H. pylori* is its growth under highly acidic conditions. For this it relies on the activity of the nickel-containing enzyme, urease. The recent X-ray crystal structure of urease from *Klebsiella aerogenes* shows the presence of a binuclear Ni(II) active site. The two Ni(II) ions are 3.5 Å apart and bridged by a carbamylated Lys residue [117]. One of the Ni(II) ions coordinates pesudo-tetrahedrally to two His nitrogens, one oxygen from bridging carbamylated Lys and one water oxygen. The second Ni(II) ion is five-coordinate and approximately trigonal-bipyramidal via one oxygen of the bridging carbamylated Lys, two His nitrogens, one carboxylate oxygen of Asp and one water oxygen. Bismuth complexes inhibit the *H. pylori*-produced urease, and the Bi-mercaptoethanol complex has ca. 1000 times higher activity than mercaptoethanol alone against urease [71]. Thiols may form disulfides with the active-site Cys residue on the protein and inactivate it. Bismuth(III) inhibition of urease may play an important role in the antibactericidal activity of bismuth-containing drugs [118].

7 Pharmacology and Toxicity of Bismuth Compounds

7.1 Absorption of Bismuth into the Blood

Bismuth may act both locally within the stomach and systematically as a result of gastrointestinal absorption. After oral intake of single doses of bismuth drugs such as CBS and BSS (bismuth subsalicylate), a protective coating (probably BiOCl and Bi citrate complexes) is thought to form on the ulcer crater, and there is also a significant increase in bismuth levels in the mucosa with a peak level of 30–60 μg/l (0.14 to 0.28 μM). Concentrations of bismuth in the blood increase by 51–1483-fold. The maximum level occurs unexpectedly rapidly (15–60 min) [119]. Intake of CBS swallowable tablets gave peak concentrations of Bi in blood plasma ranging from 25–300 μg/l (0.12 to 1.4 μM) after a mean time of 0.5 h [120]. Multiple dosing with CBS shows that an apparent steady state is reached after 3–4 weeks. Despite the extremely high solubility of ranitidine bismuth citrate in water, a study in human volunteers has shown that absorption of bismuth in the blood is significantly lower than that observed after dosage of De-Nol (colloidal bismuth subcitrate) [121]. The geometric mean maximum level of Bi in plasma of 12 ng/g after a 10-day course of RBC (1 g/kg body) can be compared with 21 ng/g attained after a lower dosage of De-Nol (0.24 g/kg body). Radio-bismuth studies using ^{206}Bi citrate suggest that Bi binds readily to serum components (83% after 1 h at biological pH) and that this binding is strong and not readily reversed by dialysis against phosphate buffer [122]. A gel filtration study of human blood after incubation with bismuth subgallate showed an association of bismuth with high molecular mass ligands [123], possibly transferrin or albumin.

7.2 Distribution of Bismuth in Tissues

In humans and other animals, ca. 70–90% of bismuth is excreted into the urine after dosing with bismuth complexes. The absorption of bismuth after oral intake of bismuth drugs is very low, normally less than 0.5% of the ingested bismuth drugs and depends mainly on the ligands bound to bismuth. The uptake is probably from the upper small intestine and is also into the gastric mucosa. Soluble bismuth drugs, such as CBS and ranitidine bismuth citrate, are better absorbed from the gastrointestinal tract than insoluble bismuth adducts [58a]. Although there are large variations in human and animal studies, the data on biodistribution of bismuth in tissues are similar; the organ with the highest content is always found to be the kidney. Dense intranuclear inclusion bodies, which probably contain bismuth-protein complexes, have been found in proximal renal tubular lining cells [124]. The retention time of bismuth in the kidney is always longer than that of any other organ. After 14 months of administration of CBS to rats, concentrations of bismuth have been reported to range from 13.9 μg/l wet weight in the kidney, up to 0.3 μg/l in muscle. In terms of bismuth concentration, the order of bismuth levels in organs ranges from (high to low): kidney > liver > bone > lung > spleen > brain > heart [125]. This order, however, may be influenced by the physicochemical form of the administered bismuth complex. After oral intake of trimethylbismuth, the bismuth concentration in the liver was found to be higher than that in the kidney, probably due to the organic character of this molecule [126].

The concentration of bismuth in brain tissue after administration of antiulcer drugs has been reported to be higher than in controls [127,128]. For patients who have died of bismuth encephalopathy, the concentration of bismuth in the gray matter was found to be about twice as high as that in white matter, with the highest concentration probably in the thalamus and cerebellar cortex [129]. It is not clear how bismuth is transferred to the brain, and the mechanism needs to be studied.

7.3 Human Toxicity

Many toxic effects in humans have been attributed to bismuth complexes: encephalopathy, nephropathy, osteoarthropathy, gingivitis, stomatitis, colitis and hepatitis. Different adverse effects on the various organ systems have been associated with different bismuth complexes. It seems that CBS and BSS are less toxic compared to previously used complexes. Signs and symptoms of bismuth encephalopathy have been described in detail by Slikkerveer and de Wolff [5], mainly from reported cases in France and Australia. The diagnosis is generally confirmed by the detection of bismuth in, for example, blood, plasma and serum. In patients with encephalopathy, the bismuth levels in blood usually exceed 100 μg/l (0.48 μM); most of these patients had blood levels of >500 μg/l (2.39 μM) at the time of presentation, but there is no clear correlation between clinical illness and bismuth concentration in the patient's blood. Thus, the interpretation of 'Hillemand safety levels' [130] (50–100 μg/l bismuth) as a warning sign of toxicity is not reliable.

8 Mechanism of Action of Bismuth Against *H. pylori*

The mechanism by which bismuth inhibits the growth of *H. pylori* is still not well understood. Bismuth appears to exert its bactericidal action by several mechanisms, including inhibition of cell wall synthesis, inhibition of cell membrane function, inhibition of enzymes and binding to metalloproteins (Scheme 9). After administration of bismuth citrate adducts, most of the Bi(III) may be deposited as polymeric $[Bi_2(cit_2)]_n^{2n-}$, $[Bi_6O_4(OH)_4]^{6+}$ and BiOCl on the ulcer crater to form a protective coating. A very small amount is absorbed, most probably in the small intestine. Bismuth may pass through the small intestinal mucosa or the membranes of bacteria (*H. pylori*) via some kind of endocytosis [131]. Since there is a good correlation between the strength of binding of Bi(III) and Fe(III) to O- and N-containing ligands, Bi(III) may block Fe(III) sites in bacteria (siderophores, lactoferrin and other non-heme iron binding proteins) [104]. Bi(III) has been shown to inhibit gram-negative bacteria and this inhibition is inversely related to iron concentration and also depends on the iron transport mechanism [132]. It inhibits a series of enzymes in bacteria, including urease, catalase, ADH and others. Acetaldehyde produced from ethanol via ADH catalysis inhibits gastric mucosal regeneration and forms stable adducts with mucosal proteins, both of which can cause gastric injury.

Once bismuth is adsorbed, it is distributed to different tissues, predominantly the kidney and the liver. GSH may transport Bi(III) from liver to bile; metallothionein could transfer it from liver to kidney, where it may be stored or excreted. In blood plasma, serum transferrin may be a target for Bi(III) even though the concentration of albumin is ca. 20 times higher [133]. It is still not clear why bismuth complexes can be neurotoxic (encephalopathy). The relationship between bismuth levels in blood and plasma, and the encephalopathy found by Hillemand et al. [130], indicates that proteins in blood plasma may play a role in this toxic side effect. There is also another hypothesis suggesting that increased bacterial colonization of the small intestine may lead to neurotoxic bismuth adducts [6].

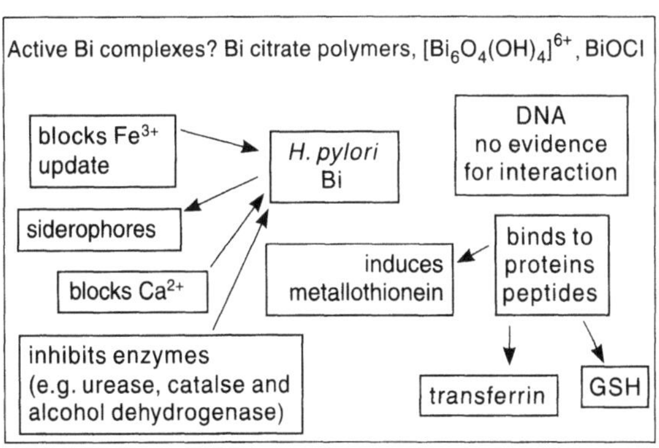

Scheme 9. Some events of interest in relation to the mechanism of action of Bi(III)

9
Future Outlook

Despite the widespread use of bismuth compounds in medicine for over two centuries, its chemistry and biochemistry are currently poorly understood and are much less advanced than the medical studies. Recent work has begun to elucidate the structures of Bi(III) thiolates, carboxylates and aminocarboxylate complexes in particular. The occurrence of a highly variable coordination number (3–9) and coordination geometry, together with the apparent activation of a strong lone-pair effect in certain complexes (e.g. those with alkoxide ligands), is highly characteristic of Bi(III). So too is the strong acidity of Bi(III) aqua complexes. Complexes crystallised from aqueous solution commonly contain bridging carboxylate ligands and so the chemistry of Bi(III) citrate antiulcer drugs appears to be dominated by polymeric species. The interaction of these polymers with membrane surfaces may be important to their bioactivity. The rates of ligand exchange on Bi(III) are highly variable and pH-dependent.

Little is known about the interaction of Bi(III) with proteins or enzymes, although this could be very important to the biological activity. It appears that Bi(III) can bind strongly to both Fe(III) sites (e.g. N and O ligands in transferrin) and Zn(II) sites (e.g. S ligands in metallothionein). Glutathione forms a strong complex $[Bi(H_{-1}gsh)_3]$ which may be involved in the transport of Bi(III) in cells or bacteria. Further exploration of the chemistry and biochemistry of bismuth is now warranted, particularly in view of current interest in the introduction of new Bi(III) antiulcer drugs, in the use of ^{212}Bi in radiotherapy, and the discovery of bismuth complexes with anticancer and anti-HIV activity. Such studies may lead to the design of new bioactive compounds and to a better understanding of their mechanism of action.

Acknowledgments. Work in our laboratory on bismuth has been supported by GlaxoWellcome plc., EPSRC, BBSRC and MRC. We thank GlaxoWellcome for permission to reproduce Fig. 3, and Dr. G. Klinkert for helpful comments. We acknowledge the use of the Cambridge Crystallographic Database.

References

1. Odier L (1786) J Med Chir Pharm (Fr) 68:49
2. Marcet A (1805) Memoirs Med Soc Lond 65:155
3. Balzer F (1889) C R Soc Biol 41:537
4. Morison R (1916) Lancet 2:268
5. Slikkerveer A, de Wolff FA (1989) Med Toxicol Adverse Drug Exp 4:303
6. Menge H, Gregor M, Brosius B, Hopert R, Lang A (1994) Eur J Gastroenterol Hepatol 4(S2):41
7. (a) Herrmann WA, Herdtweck E, Pajdla L (1991) Inorg Chem 30:2579; (b) Asato E, Driessen WL, de Graaf RAG, Hulsbergen FB, Reedijk J (1991) Inorg Chem 30:4210
8. (a) Asato E, Katsura K, Mikuriya M, Fujii T, Reedijk J (1993) Inorg Chem 32:5322; (b) Herrmann WA, Herdtweck E, Pajdla L (1992) Z Kristallogr 198:257
9. Lee SP (1981) Res Commun Chem Pathol 34:359
10. For example: (a) Marshall BJ, Goodwin CS, Warren RJ (1988) Lancet 2:1437; (b) Wagstaff AJ, Benfield P, Monk JP (1988) Drugs 36:132; (c) Marshall BJ (1986) J Infect Dis 153:650
11. (a) Lowe JD (1974) Med J Aust 2:664; (b) Martin-Bouyer G, Barin C, Beugnet A, Cordier J, Guerbois H (1978) Gastroenterol Clin Biol 2:349
12. Clitherow JW (1990) UK Patent GB2220937A
13. McColm AA, McLaren A, Klinkert G, Francis MR, Connolly PC, Grinham CJ, Campbell CJ, Selway S, Williamson R (1996) Aliment Pharmacol Therap 10:241
14. Kozak RW, Waldmann TA, Atcher RW, Gansow OA (1986) Trends Biotech 4:259

15. Kozak RW, Atcher RW, Gansow OA, Friedman AM, Hines JJ, Waldman TA (1986) Proc Natl Acad Sci USA 83:474
16. Krejcarek GE, Tucker KL (1977) Biochem Biophys Res Comm 77:581
17. Junghans RP, Dobbs D, Brechbid MW, Mirzadeh S, Raubitschek AA, Gansow OA, Waldmann TA (1993) Cancer Res 53:5683
18. Kirschner S, Wei Y, Francis D, Bergman JG (1966) J Med Chem 9:369
19. Skinner SM, Swatzell JM, Lewis RW (1978) Res Comm Chem Pathol Pharmacol 19:165
20. Köpf-Maier P, Klapötke T (1988) Inorg Chim Acta 152:49
21. Mahmood N, Burke A, Anner RM, Anner BM (1995) Antiviral Chem Chemother 6:187
22. Partington JR (1961) A history of chemistry. Macmillan, London
23. Dittes U, Keppler BK, Nuber B (1996) Angew Chem Int Ed Engl 35:67
24. (a) Baxter GF (1992) Chem Brit 28: 445; (b) Abrams MJ, Murrer BA (1993) Science 261:725
25. For example: (a) Jha AK, Mendiratta RG (1996) J Mater Sci 31:1735; (b) Lobo RPSM, Gervias F (1995) Phys Rev B - Condensed Mat 52:13294
26. (a) Pearson RG (1963) J Am Chem Soc 85:3533; (b) Pearson RG (1966) Science 151:172
27. Fraústo da Silva JJR, Williams RJP (1991) The biological chemistry of the elements: the inorganic chemistry of life. Clarendon Press, Oxford, pp 538
28. (a) Hancock RD, Cukrowski I, Baloyi J, Mashishi J (1993) J Chem Soc Dalton Trans 2895; (b) Hancock RD, Cukrowski I, Antunes I, Curkrowska E, Mashishi J, Brown K (1995) Polyhedron 14:1699; (c) Luckay R, Reibenspies JH, Hancock RD (1995) J Chem Soc Chem Comm 2365
29. Hancock RD, Martell AE (1989) Chem Rev 89:1875
30. Pettit G, Pettit LD (1993) IUPAC stability constants database. IUPAC and Academic Software, Otley, UK
31. (a) Cotton FA, Wilkinson G (1988) Advanced inorganic chemistry, 5th edn. John Wiley & Sons, New York; (b) Wilkinson G, Gillard RD, McCleverty JA (1987) Comprehensive coordination chemistry. Pergamon Press, Oxford
32. Frank W, Reiss GJ, Schneider J (1995) Angew Chem Int Ed Engl 34:2416
33. Lazarini F (1979) Cryst Struct Comm 8:69
34. Lazarini F (1981) Bull Bismuth Inst 32:3
35. (a) Rogers RD, Bond AH, Aguinaga S (1992) J Am Chem Soc 114:2960; (b) Rogers RD, Bond AH, Aguinaga S, Reyes A (1992) J Am Chem Soc 114:2967
36. Katayama S, Sekine M (1991) J Mater Res 6:36
37. Farrugia LJ, Lawlor FJ, Norman NC (1995) J Chem Soc Dalton Trans 1163
38. Atwood DA, Cowley AH, Hernandez RD, Jones RA, Rand LL, Bott SG, Atwood JL (1993) Inorg Chem 32:2972
39. Carmalt CJ, Farrugia LJ, Norman NC (1996) J Chem Soc Dalton Trans 455
40. (a) Farrugia LJ, Lawlor FJ, Norman NC (1995) Polyhedron 14:311; (b) Clegg W, Elsegood MRJ, Farrugia LJ, Lawlor FJ, Norman NC, Scott AJ (1995) J Chem Soc Dalton Trans 2129
41. Agocs L, Burford N, Cameron TS, Curtis JM, Richardson JF, Robertson KN, Yhard GB (1996) J Am Chem Soc 118:3225
42. Calderazzo F, Poli R, Pelizzi G (1984) J Chem Soc Dalton Trans 2365
43. Tokitoh N, Arai Y, Okazaki R, Nagase (1997) Science 277:78
44. Doak GO, Freedman LD (1995) J Organomet Chem 485:1; (b) Norman NC (1995) Adv Inorg Chem 41:233
45. Suzuki H, Ikegami T, Matano Y, Azuma N (1993) J Chem Soc Perkin Trans 1 2411
46. (a) Morgan K, Sayer BG, Schrobilgen GJ (1983) J Magn Reson 52:139; (b) Brevard C, Granger P (1981) Handbook of high resolution multinuclear NMR. Wiley-Interscience, New York, pp 210–211
47. Harvey I, Cheung C, Li H, Sadler PJ, Sun H, unpublished work
48. Li H, Keohane BM, Sun H, Sadler PJ (1997) J Anal At Spectrom 12:1111
49. Vanhoe H, Versieck J, Vanballenberghe LD, Dams R (1993) Clin Chem Acta 219:79
50. Warren JR, Marshall B (1983) Lancet 1:1273
51. Blaser M (1992) Gastroenterology 102:720
52. Mobley HLT, Island MD, Hausinger RP (1995) Microbiol Rev 59:451
53. Labigne A, de Reuse H (1996) Infect Agents Dis 5:191
54. Tomb JF, White O, Kerlavage AR, Clayton RA, Sutton GG, Fleischmann RD, Ketchum KA et al (1997) Nature (Lond) 388:539

55. Summers S, Abboud KA, Farrah SR, Palenik GJ (1994) Inorg Chem 33:88
56. (1989) Merck Index, 11th edn. Merck & Co Inc, Rahway, NJ, USA, p 197
57. (a) Asato E, Katsura K, Mikuriya M, Turpeinen U, Mutikainen I, Reedijk J (1995) Inorg Chem 34:2447; (b) Asato E, Katsura K, Mikuriya M, Fujii T, Reedijk J (1992) Chem Lett 1967
58. (a) Lambert JR, Midolo P (1997) Aliment Pharmacol Ther 11(Suppl):27; (b) Vogt K, Warrelmann M, Hahn H (1989) Zentralbl Bakt 271:304; (c) Stables R, Campbell CJ, Clayton NM et al (1993) Aliment Pharmacol Ther 7:237
59. Sadler PJ, Sun H (1995) J Chem Soc Dalton Trans 1395
60. Barrie PJ, Djuran MI, Mazid MA, McPartlin M, Sadler PJ, Scowen IJ, Sun H (1996) J Chem Soc Dalton Trans 2417
61. Sun H (1996) PhD Thesis, University of London
62. Parkinson JA, Sun H, Sadler PJ (1998) Chem Commun 881–882
63. Coghill SB (1985) Pathogenesis and the treatment of peptic ulcer disease. In: Axon ATR (ed) Proceedings of International Symposium, Cairo, Egypt, pp 7–12
64. Shweky I, Bino A, Goldberg DP, Lippard SJ (1994) Inorg Chem 33:5161
65. Diemer R, Keppler BK, Dittes U, Nuber B, Seifried V, Opferkuch W (1995) Chem Ber 128:335
66. Dittes U, Vogel E, Keppler BK (1997) Coord Chem Rev 163:345
67. Diemer R, Dittes U, Nuber B, Seifried V, Opferkuch W, Keppler BK (1995) Metal Based Drugs 2:271
68. Mobley HLT, Hausinger RP (1989) Microbiol Rev 53:85
69. (a) Clark PA, Wilcox DE, Scott RA (1990) Inorg Chem 29:579; (b) Clark PA, Dixon NE (1989) Inorg Chem 28:1326; (c) Todd MJ, Hausinger RP (1989) J Biol Chem 264:15835
70. Benini S, Rypniewski WR, Wilson KS, Ciurli S, Mangani S (1998) J Biol Inorg Chem 3:268
71. Asato E, Kamamuta K, Akamine Y, Fukami T, Nukada R, Mikuriya M, Deguchi S, Yokota Y (1997) Bull Chem Soc Jpn 70:639
72. Agocs L, Briand GG, Burford N, Cameron TS, Kwiatkowski W, Robertson KN (1997) Inorg Chem 36:2855
73. Domenico P, Salo RJ, Novick SG, Schoch PE, van Horn K, Cunha BA (1997) Antimicro Agents Chemother 41:1697
74. (a) von Itzstein M, Wu WY et al (1993) Nature (Lond) 363:418; (b) von Itzstein M, Dyason JC, Oliver SW, White HF, Wu WY, Kok GB, Pegg MS (1996) J Med Chem 39:388
75. (a) Holzer CT, von Itzstein M, Jin B, Pegg MS, Wu WY (1993) Glycoconjug J 10:40; (b) Hayden FG, Treanor JJ, Betts RF, Lobo M, Esinhart JD, Hueesy EK (1996) J Am Med Assoc 274:295
76. Midolo PD, Norton A, von Itzstein M, Lambert JR (1997) FEMS Microbio Lett 157:229
77. (a) Kiprof P, Scherer W, Pajdla L, Herdtweck E, Herrmann WA (1992) Chem Ber 125:43; (b) Herrmann WA, Herdtweck E, Scherer W, Pajdla L (1993) Chem Ber 126:51
78. Hancock RD, Martell AE (1995) Adv Inorg Chem 42:89
79. Williams DR (1977) J Inorg Nucl Chem 39:711
80. Chaleil D, Lefevre F, Allain P, Martin GJ (1981) J Inorg Biochem 15:213
81. Alonzo G, Bertazzi N, Consiglio M (1984) Inorg Chim Acta 85:L35
82. Herrmann WA, Herdtweck E, Pajdla L (1993) Chem Ber 126:895
83. Sadler PJ, Sun H, Li H (1996) Chem Eur J 2:701
84. (a) Rao N, Feldman S (1990) Pharmaceu Res 7:188; (b) Rao N, Feldman S (1990) Pharmaceu Res 7:237
85. Bernhard W, Good M, Vasák M, Kägi JHR (1983) Inorg Chem Acta 79:154
86. Paradies HH (1990) In: Malfertheiner P, Ditschuneit H (eds) *Helicobacter pylori*, gastritis and peptic ulcers. Springer, Berlin Heidelberg New York, p 409
87. Naganuma A, Satoh M, Imura N (1987) Cancer Res 47:983
88. Satoh M, Naganuma A, Imura N (1988) Toxicology 53:231
89. Boogaard PJ, Slikkerveer A, Nagelkerke JF, Mulder GJ (1991) Biochem Pharmacol 41:369
90. Shaw III CF, Savas MM (1992) In: Stillman MJ, Shaw CF III, Suzuki KT (eds) Metallothioneins: synthesis, structure and properties of metallothioneins, phytochelatins and metal-thiolate complexes. VCH, New York, pp 144–162
91. Szymanska JA, Zelazowski AJ (1979) Environ Res 19:121
92. Komsta-Szumska E, Chmielnicka J, Piotrowski JK (1976) Biochem Pharmacol 25:2539
93. Szymanska JA, Zychowicz M, Zelazowski AJ, Piotrowski JK (1978) Arch Toxocol 40:131
94. Sun H, Li H, Cheung K, Harvey I, Sadler PJ, unpublished results

95. Carter DC, He XM (1992) Nature (London) 358:209
96. Peters T Jr (1985) Adv Protein Chem 37:161
97. Carter DC, Ho JX (1994) Adv Protein Chem 45:153
98. Christodoulou J, Sadler PJ, Tucker A (1994) Eur J Biochem 225:363
99. Christodoulou J, Sadler PJ, Tucker A (1995) FEBS Lett 376:1
100. Sun H, Li H, Sadler PJ, unpublished work
101. Sadler PJ, Viles JH (1996) Inorg Chem 35:4490
102. Lambert JR (1991) Rev Infect Dis 13:S691
103. Harris DC, Aisen P (1989) Physical biochemistry of the transferrins. In: Loehr TM (ed) Iron carriers and iron proteins. VCH, Weinheim, pp 239–351
104. (a) Bruns CM, Nowalk AJ, Arvai AS, McTigue MA, Vaughan KG, Mietzner TA, McRee DE (1997) Nature Struct Biol 4:919; (b) Crosa JH (1984) Ann Rev Microbiol 38:69; (c) Mietzner TA, Tencza SB, Adhikari P, Vaughan KG, Nowalk AJ (1998) Curr Topics Microbiol Immunol 225:113
105. Li H, Sadler PJ, Sun H (1996) J Biol Chem 271:9483
106. Sun H, Cox MC, Li H, Mason AB, Woodworth RC, Sadler PJ (1998) FEBS Lett 422:315
107. Li H, Sadler PJ, Sun H (1996) Eur J Biochem 242:387
108. Sun H, Cox MC, Li H, Sadler PJ (1997) Struct Bonding (Berl) 88:71
109. Sun H, Li H, Weir RA, Sadler PJ (1998) Angew Chem Int Ed Engl 37:1577
110. Lee SP (1991) Scand J Gastroenterol 26:1
111. Ottlecz A, Romero JJ, Hazell SL, Graham DY, Lichtenberger LM (1993) Digestive Dis Sci 38:2071
112. Beil W, Birkholz C, Wagner S, Sewing KF (1995) Pharmacol 50:333
113. Beil W, Bierbaum S, Sewing KF (1993) Pharmacol 47:141
114. Roine RP, Salmela KS, Hooknikanne J, Kosunen TU, Salaspuro M (1992) Life Sci 51:PL195
115. Salmela KS, Roine RP, Hooknikanne J, Kosunen TU, Salaspuro M (1994) Scand J Gastroenterol 29:528
116. Zeppezauer E, Samama J-P, Eklund H (1982) Biochemistry 21:4895
117. Jabri E, Carr MB, Hausinger RP, Karplus PA (1995) Science 268:998
118. Asato E, Akamine Y, Nukada R, Mikuriya M, Deguchi S, Yokota Y (1997) J Inorg Biochem 67:146
119. Hespe W, Stall HJM, Hall DWR (1988) Lancet 2:1258
120. Nwokolo CU, Gavey CJ, Smith JTL, Pounder RE (1989) Aliment Pharm Ther 3:29
121. Lacey LF, Frazer NM, Keene ON, Smith JT (1994) Eur J Clin Pharmacol 47:177
122. Russ GA, Bigler RE, Tilbury RS, Woodard HQ, Laughlin JS (1975) Radiation Res 63:443
123. Thomas DW, Sobecki S, Hartley TF, Coyle P, Alp MH (1983) Variable absorption and excretion of bismuth and its potential toxicity. In: Brown SS, Savory J (eds) Chemical toxicology and clinical chemistry of metals. Academic Press, London, pp 391–394
124. Goyer RA, Cherian MG (1977) Tissue and cellular toxicology of metals. In: Clinical chemistry and chemical toxicology of metals. Elsevier/North-Holland Biomedical Press, pp 93–95
125. Luppino MA, McLean AJ (1995) Analyst 120:883
126. Sallmann T, Seifer J (1939) J Pharm Exp Therapeu 67:17
127. Lee SP, Lim TH, Pybus J, Clarke AC (1980) Clin Exp Pharmacol Physiol 7:319
128. Hamilton EI, Minski MJ, Cleary JJ (1972) Sci Total Environ 1:341
129. Escourolle R, Bourdon R, Galli A, Galle P, Jaudon MC, Hauw JJ, Gray F (1977) Rev Neurol (Paris) 133:153
130. Hillemand P, Palliere M, Laquais B, Bouvet P (1977) Sem Hop Paris 53:1663
131. Stiel D, Murray DJ, Peters TJ (1985) Gut 26:364
132. Domenico P, Reich J, Madonia W, Cunha BA (1996) J Antimicro Chemother 38:1031
133. Sun H, Li H, Mason AB, Woodworth RC, Sadler PJ, unpublished results

Chrysotherapy: Gold-Drug Metabolism and Immunochemistry

C. Frank Shaw III

Department of Chemistry, Eastern Kentucky University, Richmond, KY, 40475-3124, USA
E-mail: CHESHAW@eku.edu

Keywords. Gold, Metallodrugs, Auranofin, Myochrysine, Arthritis

1 Introduction

1.1 Chrysotherapy for Rheumatoid Arthritis

Chrysotherapy, the treatment of disease (principally rheumatoid arthritis) with gold-based drugs, is well established in current medical practice. The name is derived from Chryses, the golden-haired heroine of ancient Greek mythology. Figure 1 presents the formulations most commonly used for arthritis therapy. Bis(thiosulfato)gold(I) contains gold bound to the terminal sulfur atoms of the $S_2O_3^{2-}$ ligands. Auranofin, the most recent addition to the array of approved drugs, is also a discrete monomeric gold(I) complex, with tetraacetylthioglucose and triethylphosphine ligands. The crystal structures of both have been determined [1–3]. The thiomalate, thioglucose, and thiopropanolsulfonate complexes are oligomers with linear gold(I) centers linked by bridging thiolate ligands. Myochrysine and solgonal, the thiomalate- and thioglucose-containing formulations, respectively, are more complex than the ideal formulas, as they contain additional components and the gold and thiol are not in exact 1:1 stoichiometry [4,5].

Single crystals of $(Na_2Cs)_n[Au_2H(STm)_2]_n$, grown by vapor diffusion, allowed the first structure of a myochrysine analogue to be determined (Fig. 2). The structure is polymeric, as expected from the stoichiometry, with thiolates bridging linear gold(I) centers. It consists of spirals containing two interpenetrating gold–sulfur chains with approximately fourfold helical symmetry. The Au–S bond distances are normal, 228.3 and 228.6 pm. One S-Au-S angle is essentially linear, 179.0°, while the second is distorted, 169.8°, perhaps due to packing forces. The Au-S-Au angles are 99.4°. Thiomalate is racemic and the *S*-thiomalate forms exclusively left-handed helices, while the *R*-thiomalate forms right-handed helices. As a result the crystal is centrosymmetric. Interestingly, the related bis[(±)thiomalato-*S*)gold(I)] structure consists of [bis(*R*-thiomalato-*S*)gold(I)] and [bis(*S*-thiomalato-*S*)gold(I)] anions in equal proportions [7].

The bond lengths and angles found in the gold drugs can be compared with the data for related compounds of 2,4,6-triisopropylphenylthiolate [8–11] in Table 1. This is the only ligand for which structures with the stoichiometries $[AuSR]_n$, $Au(SR)_2^-$, and R'_3 PauSR have been determined. It is interesting to note that although the nature of the ligands in the gold drugs vary widely, the Au–S bond lengths span only a small range, regardless of whether the thiolate is bridging or terminal and independent of the second ligand bound to the linear gold(I) ion. In all the struc-

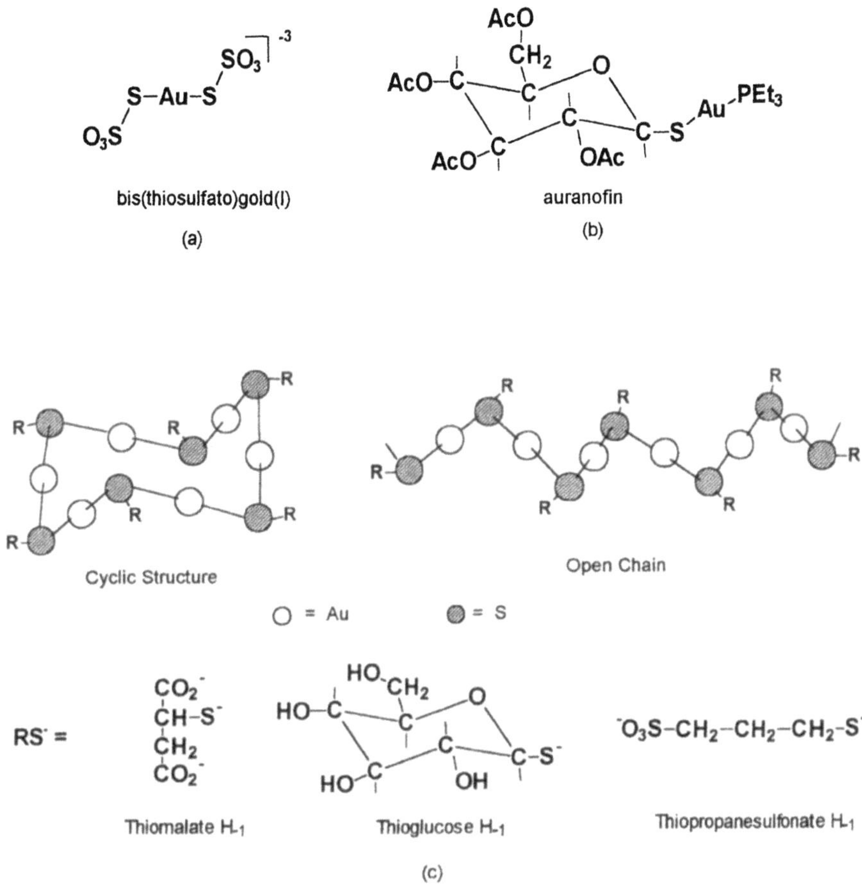

Fig. 1. Structures of antiarthritic gold complexes in clinical use. (**a**) sanocrysin $Na_3[Au(S_2O_3)_2]$; (**b**) auranofin, $Et_3PAuSAtg$; (**c**) the oligomeric gold(I) thiolates: myochrysine (RS^- = thiomalate); solganol (RS^- = thioglucose-H_{-1}); allochrysine (RS^- = thiolproanolsulfonate-H_{-1})

tures, gold(I) is linear and two-coordinate. A recent review of three- and four-coordinate gold(I) compounds reveals that to date no tris(thiolato)gold(I) species, $[Au(SR)_3]^{2-}$, have been prepared or characterized [12].

Chrysotherapy is effective for about 70% of the patients taking the treatment. The injectable gold thiolates exhibit superior activity to auranofin, the only orally administered gold drug [14–16]. Some patients suffer side effects and must cease the treatment. Many side effects are mild and some, like skin rashes, are primarily cosmetic. Only occasionally are there life-threatening consequences (1:10,000 patients). Patients are monitored by atomic absorption spectroscopy for blood gold levels, which are typically kept below 300 μg dl^{-1} (ca. 15 μM). This level minimizes the risk of side effects, but does not correlate with effectiveness. Interestingly, Panayi et al. observed a correlation between side effects to gold and patients' responses to the treatment [17] and suggest that both result from common mechanisms in the immune system for which there are genetic predispositions [17].

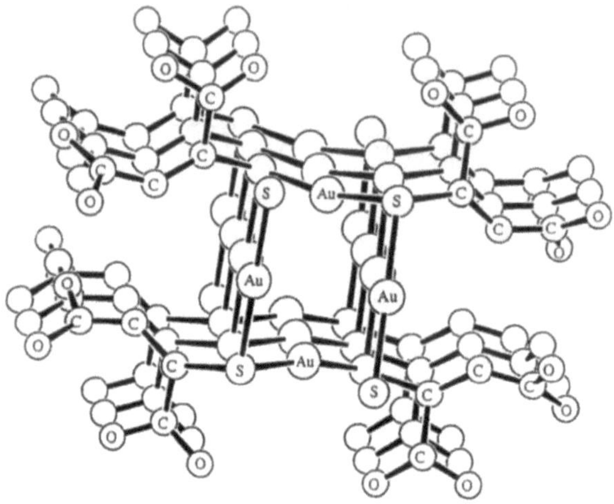

Fig. 2. Structure of $(CsNa_2)_n[Au_2(STm)(STmH)]_n$. Reprinted with permission from [6] 1998 American Chemical Society

Table 1. Bond lengths and angles of medicinal gold complexes and their 2,4,6-triisopropylphenylthiolatogold(I) analogues[a]

Compound	$d_{Au\text{-}S/P}$	S-Au-S/P angle	Au-S-C/S angle	$d_{Au\text{-}Au}$	Au-S-Au angle	Ref
Oligomeric gold(I)thiolate,-[-AuSR-]-						
$[Au(SC_6H_2Pr^i_3)]_6$	229 ± 1	176 ± 1	108 ± 2	356 ± 10	102 ± 4	8
$Au_n(STm)_m$	230	-	-	335	94	13
$[CsNa_2Au_2H(STm)_2]$	228 ± 1	170, 179		323, 348	99	6
Bis(thiolato)gold(I), $[Au(SR)_2^-]$						
$[Au((SC_6H_2Pr^i_3)_2^-]$	229	176	105 ± 3	-	-	11
$[Au(S_2O_3)_2]^{3-}$	228 ± 1	177	104 ± 1	-	-	2,3
$[Au(STm)_2]^{5-}$	226 ± 1	179	108 ± 1	-	-	7
Phosphinegold(I)thiolate, R'_3PAuSR						
$(C_6H_5)_3PAuSC_6H_2Pr^i_3$	S 229 P 226	176	105	-	-	9,11
$Et_3PAuSAtg$	S 229 P 226	174	106	-	-	1

[a] Error limits give the range of values for independent bond lengths or angles. Individual crystallographic standard deviations are less than 1 pm or 1° and are not given

1.2 Possible Mechanisms of Chrysotherapy

Rheumatoid arthritis is an inflammatory condition that leads to progressive erosion of the articular cartilage lining the joints [14]. The attack on the joints exhibits many characteristics of an autoimmune disease and, if not checked, leads to fusion of the two bones. The inflammation proceeds from the synovial membrane that surrounds the joints into the synovial cavity between the bones. Lysosomal enzymes, including collagenase and other proteases, are released as a result of the inflammation. They mediate the tissue damage, which releases cells and tissue fragments into the joint. In a cyclic process, these stimulate further inflammation and attract additional immune cells including macrophages into the joints. As a result, a cycle of degradation and release of destructive enzymes is established.

Gold drugs have well-documented anti-inflammatory activity, which they share with many organic medicines such as aspirin and ibuprofen. But simple anti-inflammatory agents lack the more significant ability of gold to inhibit the underlying tissue destruction. A small but significant fraction of patients enjoy long term remission of the disease state. Such patients are given maintenance therapy consisting of less frequent treatment with gold in order to avoid accumulating too much gold in tissues.

Table 2 lists proposed mechanisms of chrysotherapy. They range from very specific biochemical systems that are inhibited to very general, systemic effects such as immunomodulatory action. These mechanisms have plausible connections to the therapeutic benefits that ensue, but none has been unambiguously established as the dominant effect. A critical analysis, however, is beyond the scope of this chapter and the interested reader is referred to the medical and pharmacological literature [4, 14–16,18].

One difficulty encountered in assessing various mechanisms of action is that gold drugs are pro-drugs that undergo ligand exchange reactions in vivo [5,19–21]. Many of the protein complexes formed by direct reaction of the drugs with serum albumin are known [a68, bl] and should better model the in vivo interactions of cells and enzymes with circulating gold. However, these albumin complexes may not be the ultimate circulating metabolites. Efflux of gold from red cells treated with auranofin generates an albumin-gold-glutathione complex [22], consistent with findings that both the phosphine and tetraacetylthioglucose ligands are displaced within 24 hours [23]. $[Au(CN)_2]^-$ has been identified as a metabolite common to the injectable gold drugs and auranofin [24], and is known to be formed from cyanide generated by certain immune cells [18]. It is taken up by cells more readily than gold thiolates and their metabolites [18,25,26]. Finally, evidence in mice [27,28] and humans [17]

Table 2. Possible mechanisms of chrysotherapy

Anti-Inflammatory Activity
Altered Prostaglandin Biosynthesis
Immunomodulatory Activity
Altered Peptide Presentation
Inhibited Macrophage Phagocytosis
Inhibited Histamine Release by Mast Cells
Inhibited Complement Activity
Inhibited Monocyte Migration
Inhibited T-Cell Stimulation and Proliferation
Reduced Immunoglobin (esp. Rheumatoid Factor) Titres
Inhibition of Lysosomal Enzymes Activity and/or Release
Altered Copper Homeostasis
Altered Metabolism of Reactive Oxygen Species
Catalysis of Singlet-to-Triplet Oxygen Conversion
Glutathione Peroxidase Inhibition
Modulation of the Oxidative Burst
Perturbation of Zn-Fingers by Au(I) Thiol Competition
Inhibition of Jun-Fos binding to AP-1

suggests that the immunogenesis of gold(III) may be central to the interaction of chrysotherapy agents with the immune system. These findings suggest that direct interactions of gold drugs with cells and enzymes should be interpreted cautiously.

Gold drugs and their metabolites behave promiscuously. They distribute themselves widely among tissues, cells and proteins of the body. Thus, they are unlike many drugs that bind tightly and specifically to receptors at their sites of action and thereby facilitate the identification of their mechanisms of action. Marginally higher concentrations are found in the inflamed joints than in the blood, but only by a factor of two- to threefold. Once definitive mechanisms for the action of gold are identified, the rational design of new gold drugs which can target the correct tissues and receptors will be possible.

1.3 Use of Gold for Other Diseases

1.3.1 *Anti-HIV Activity*

Investigations of the anti-HIV activity of gold complexes were stimulated by reports that AuSTg is a reverse transcriptase (RTase) inhibitor [29]. This enzyme, which converts viral RNA into DNA in the host cell, is a key therapeutic target. AuSTg is an effective inhibitor in the cell-free extracts as studied [29], but it is unable to enter the cells, which is where RTase acts [30]. $[Au(STg)_2]^-$, which can be generated in situ from AuSTg and TgSH, has a different mechanism of action [30]. It inhibits the infection of MT-4 cells by HIV strain HL4-3 without inhibiting the RTase activity in the intact virons. The critical target site has been tentatively identified as a thiol group, Cys-532 on gp160, which is a glycoprotein of the viral envelope [30]. Interestingly, $Au(STm)_2^{5-}$ is also active, but the oligomeric 1:1 thiolates, [AuSTm] and [AuSTg], are not active. Auranofin and two analogues ($Et_3PAuSTg$ and Et_3PAuCl), that are able to enter cells, are inactive against the HL4-3 strain. Several trialkylphosphinegold(I) cyanides (R_3PAuCN) were inactive below the onset of cytotoxicity at ca. 1 μM concentrations [31].

Tepperman and colleagues [25] have found that $[Au(CN)_2]^-$ is taken up into H9 cells, a continuous T-cell line that is susceptible to HIV infection. At concentrations as low as 20 ppb, it retards the proliferation of HIV in these cells. The concentration used is well tolerated in patients with arthritis, which suggests that $[Au(CN)_2]^-$ may have promise as a complement for existing HIV treatments [25].

1.3.2 *Anti-Tumor Activity*

Extensive research has found many gold drugs with anti-tumor activity, although none have as yet proceeded to clinical testing [32–35]. Several rationales have driven the search for active compounds. One has been the analogy between square planar metal centers of gold(III) and Pt(II). Early examinations of organogold(III) compounds and gold(III) halides were not promising [32–35], but two compounds of the damp ligand ($2\text{-}Me_2NCH_2C_6H_4$) exhibited promising activity, as discussed below.

A second is that the immunomodulatory activity of gold in treating rheumatoid arthritis may be beneficial in cancer chemotherapy. Thus, a burst of activity followed reports that auranofin was active against HeLa cells in culture and P388 leukemia in mice, but screening of large numbers of analogues against a variety of tumors showed that the activity of these agents was limited to the P388 tumor line, despite their potent cytotoxicity in cell culture. Interest shifted to complexes with chelating diphosphine ligands such as *cis*-$Ph_2PCH{=}CHPPh_2$. Promising results in cell culture and in mouse models were obtained, but further testing revealed cytotoxic effects on the cardiovascular system which result from actions in the mitochondria. The gold may act as a carrier of the phosphine ligands which are the active agents but are oxidized to phosphine oxides if they are not protected by chelation. Further testing with other metals as anti-tumor agents and for non-human use of these complexes as biocides is continuing. A third approach is to complex gold(I) or gold(III) to known anti-tumor agents such as steptonigrin, 5-fluorodeoxyuridine, and tegafur. These approaches have been examined in three reviews [32–35].

The damp ligand (Fig. 3) is the basis for a promising new class of organogold anti-tumor agents. It forms *cis*-chelates with N,C coordination that stabilize gold(III) against reduction to gold(I) or elemental gold and leaves two-coordination sites open for additional ligands which are cis to one another, as are the chloride ligands in cisplatin [36]. The chloride and acetate compounds have been extensively studied. The acetate complex is more soluble and, in general, shows better biological activity. NMR studies indicate that the acetate ligands are subject to stepwise hydrolysis in aqueous solution. $[Au(OAc)_2(damp)]$ is effective against bacteria such as *Staphylococus aureus* and *Enterococus faecalis*, more cytotoxic to CHO cells than cisplatin, and equally as effective against HT 1376 bladder tumor and SK-OV-3 ovarian tumor in vitro. Despite similarities in structure and reactivity with cisplatin, its mechanism of action is different from cisplatin, since it is not cell-cycle specific in its mode of action and does not induce intra- and inter-strand crosslinks [37].

1.4
Protein Complexes of Gold Drugs and Metabolites

Gold(I) has a high affinity for soft ligands such as thiolates, thioethers, phosphorus and carbon. As a result, it tends to react in vivo with low-molecular-weight thiols, cyanide, and solvent accessible cysteine and methionine side chains of proteins. Protein complexes were thoroughly reviewed in 1989 [19] and have since been updated in other reviews [20,21,38]. In this section the recent progress in understanding the reactions of albumin and the regulatory proteins JunC and Fos in chrysotherapy are presented briefly. Section 2.3 examines the reactions of aurocyanide, $[Au(CN)_2]^-$, with proteins.

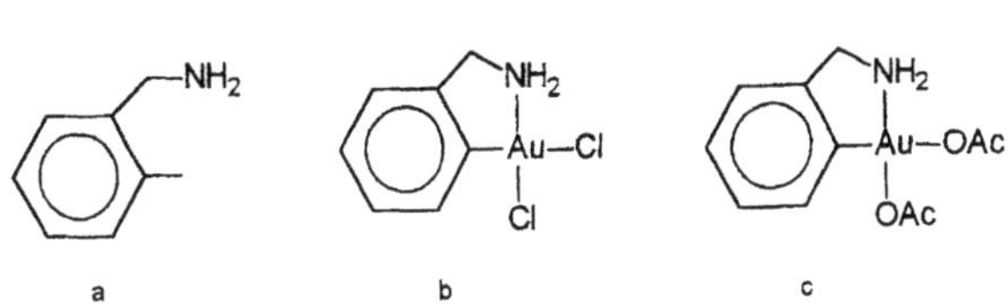

Fig. 3. Structures of (**a**) the damp ligand, 2-[(dimethylamino)methyl]phenyl, which forms a C,N chelate with metal ions; (**b**) $Au(damp)Cl_2$; (**c**) $Au(damp)(OAc)_2$

1.4.1
Serum Albumin

Serum albumin carries 80 to 95% of the gold that circulates in the blood serum. Cysteine 34 has been identified as the critical binding site for the reactions of gold drugs with albumin. Both human albumin (HSA) and the highly homologous bovine albumin (BSA) are microheterogeneous. The principle form is mercaptalbumin ($AlbS^-$) in which cysteine 34 is in reduced form. The thiol of cysteine 34 is unusually acidic and exists in deprotonated form at biological pH. Also in circulation are species in which cysteine 34 is present as mixed disulfides with predominantly cysteine and smaller amounts of glutathione present. Cysteine 34 is the only cysteine among 35 which is not engaged in a disulfide bond. The recently determined crystal structure [39] shows that it is located in a turn between helices h2 and h3 in domain IA.

The reactions of AuSTm proceed through association of the oligomer adducts from which gold appears to redistribute progressively to other albumin molecules so that a single gold thiomalate moiety is eventually associated with each cysteine 34 [40,41].

$$AlbS^- + [AuSTm]_n \rightarrow AlbS\text{-}Au\text{-}STm\text{-}[Au\text{-}STm]_{n-1} \qquad (1)$$

$$AlbS\text{-}Au\text{-}STm\text{-}[Au\text{-}STm]_n + n\ AlbS^- \rightarrow AlbS\text{-}Au\text{-}STm \qquad (2)$$

It is presumed that AuSTg and other oligomeric gold complexes react similarly.

Auranofin and analogues in which Et_3P is replaced by other neutral phosphines and/or the tetraacetylthioglucose is replaced by other thiolates, chloride or cyanide, etc., have been extensively studied. Figure 4 presents the reactions of auranofin ($Et_3PAuSAtg$) and its triisopropylphosphine analogue with serum albumin. All of the phosphine species are denoted by their ^{31}P chemical shifts, which are spectroscopic signatures that facilitate rapid progress in exploring their bioinorganic chemistry. The phosphine introduces additional complexity into the metabolic reactions since it binds more tightly than the thiolate or other anion ligands opposite it and is oxidized to $Et_3P{=}O$ in animal models and patients. Figure 4 presents the mechanism for the oxidation of the phosphine via an unfavorable ligand exchange reaction that displaces trace amounts of free phosphine that are irreversibly oxidized by reactions with disulfide bonds of the albumin. The thiolatophosphonium ion of triisopropylphosphine accumulates in detectable amounts because its hydrolysis is slow due to steric factors that inhibit the attack of water at the phosphorus atom. In contrast, the same reaction at the unhindered Et_3P analogue is rapid and the intermediate does not accumulate.

Examination of the auranofin-albumin reaction by kinetic methods [42] and by ^1H NMR studies [43] probing a histidine residue sensitive to the state of cysteine 34 has led to the conclusion that a rearrangement of the albumin is the first step of the reaction. Kinetic studies revealed a multistep mechanism in which $^*AlbS^-$ represents the rearranged protein with an accessible cysteine 34 residue [42]:

$$Alb\text{-}S^- \underset{k_{-5}}{\overset{k_5}{\rightleftharpoons}} {}^*Alb\text{-}S^- \qquad (3)$$

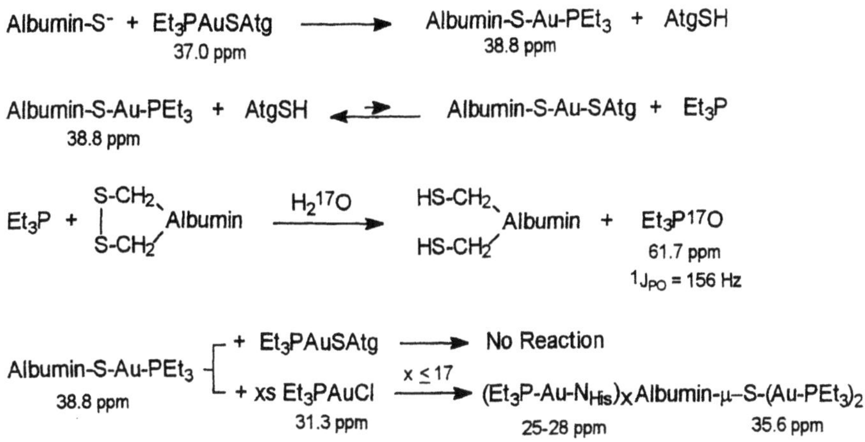

(a)

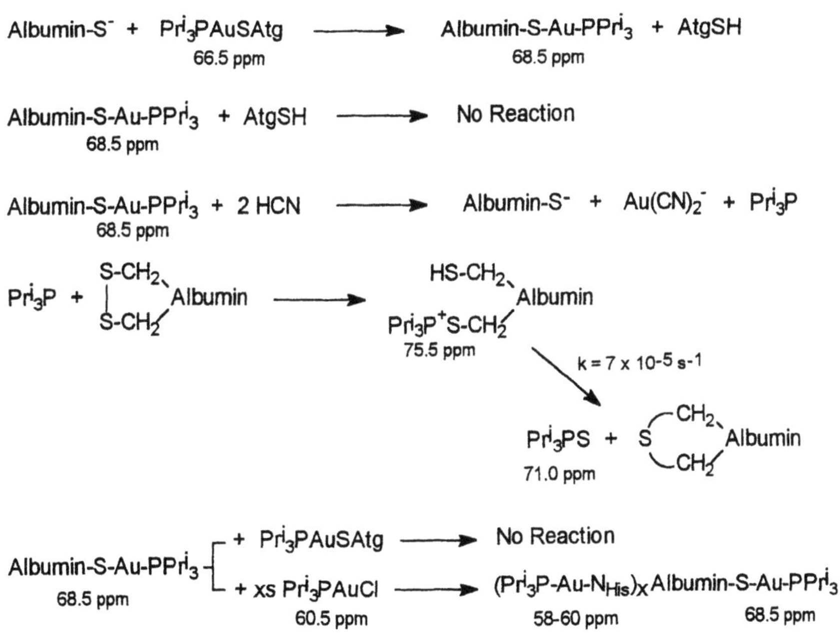

(b)

Fig. 4. Reactions of (**a**) auranofin, $Et_3PAuSAtg$, and (**b**) its triisopropylphosphine analogue, $Pr^i_3PAuSAtg$, with serum albumin. ^{31}P NMR chemical shifts are indicated for all relevant species

$$^{*}\text{Alb-S}^{-} + \text{Et}_3\text{PAuSAtg} \underset{-\text{H}^+,k_{-6}}{\overset{\text{H}^+,k_6}{\rightleftharpoons}} {}^{*}\text{Alb-S-AuPEt}_3 + \text{AtgSH} \quad (4)$$

$$^{*}\text{Alb-S-AuPEt}_3 \underset{k_{-7}}{\overset{k_7}{\rightleftharpoons}} \text{Alb-S-AuPEt}_3 \quad (5)$$

The ^{1}H resonances of the His 3 residue near the N-terminal end of the protein are sensitive to the position of cysteine 34 [43]. Figure 5 shows the rearrangement of cysteine 34 from its protected environment to a solvent-exposed condition where it can react with auranofin or other species in solution [43]. The exposed position of the albumin-bound gold may have important implications for the transport and bioavailability of gold-drug metabolites.

Under in vivo conditions where the concentration of mercaptalbumin is ca. 400 μM and gold is 5–15 μM, the mechanism above reduces to a very rapid first-order process in which $AlbSAuPEt_3$ is formed with a rate constant of 2 s^{-1}. Thus, the release of AtgSH into the blood is faster than the rate of clearance of AtgSH from the blood, as would be expected when the release is extremely rapid.

1.4.2 *Jun and Fos*

The proteins Jun and Fos, which are products of the proto-oncogenes *c-jun* and *c-fos*, bind to the activator protein-1 (AP-1) DNA regulatory element as the Jun-Fos heterodimer and the Jun-Jun homodimer. Dimerization occurs through the interaction of leucine zipper sequences on the monomers. Each monomer contains two conserved cysteine residues. One is located in the positively charged DNA-binding region (Cys272 in Jun and Cys 154 in Fos) and the other on the N-terminal side of the leucine zipper regions (Cys 323 and Cys 204, respectively). The cysteines in the DNA-binding region occur as a highly conserved Lys-Cys-Arg sequence and are essential for DNA binding [44]. If they are oxidized or alkylated, DNA binding is inhibited. Cysteines 323 and 204, on the other hand, can be linked as disulfide bonds

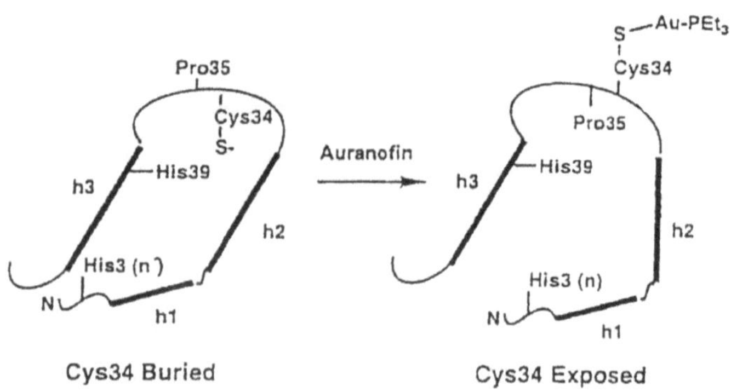

Fig. 5. Rearrangement of cysteine 34 of serum albumin upon gold binding. Used with permission from [21]

in the dimers, without loss of binding [44]. The reduced state of the cysteines in the DNA-binding domain, and hence the DNA-binding activity, is regulated by a cysteine-containing protein known as redox factor-1 (ref 1) [45,46]. The positive charges of the Lys-Cys-Arg environment enhance the acidity of the cysteine (lower pK_{SH}) which, in turn, increases the propensity for oxidation, allowing redox control of the activity. Lower pK_{SH} values also correlate with greater affinity of thiols for gold(I) [47].

Handel and co-workers [48] postulated that gold can inhibit DNA binding by dimers of Jun and Fos which may, in turn, decrease the expression of important pro-inflammatory genes. They determined that AuSTg and AuSTm react with Jun and Fos and inhibit their homo- and heterodimers from binding to the AP-1 site on DNA. Both compounds were active inhibitors at gold concentrations that prevail in vivo. The IC_{50} concentration for AuSTm is 5 μM. Thiomalate alone had no effect on AP-1 binding, but 10 mM glutathione or 1 mM DTT reversed the inhibition by AuSTm. The Cys272Ser and Cys154Ser double mutant of the Jun-Fos dimer, which binds the AP-1 regulatory element more tightly than the native sequence [44], is not affected by AuSTm [48]. These results are in accord with a mechanism in which the gold binds to Cys 272 of Jun and Cys 154 of Fos and prevents their interaction with DNA at the AP-1 biding site. This in turn may be an important mechanism by which the beneficial effects of chrysotherapy are mediated.

2
Cyanide Metabolites of Gold

The emerging role of cyanide in the metabolism of gold-based antiarthritic drugs and the ability of $[Au(CN)_2]^-$ to inhibit HIV infection of T cells have given new importance to the study of gold cyanides [18,25,26,49]. These findings have a pleasant irony since the rational use of gold-based medicinal agents can be traced back to the German physician and bacteriologist Robert Koch, whose investigations of the bacteriostatic properties of aurocyanide led to the first rational use of gold compounds - albeit ineffectively - for treating tuberculosis [50]. This section begins with the inorganic chemistry of gold cyanides, then discusses the accumulation of aurocyanide in red blood cells, the evidence for its formation at inflamed sites, its protein binding ability, and finally its activity against HIV.

2.1
Gold(I) and Gold(III) Cyanide Complexes

Gold forms two homoleptic cyanide complexes, $[Au(CN)_2]^-$ and $[Au(CN)_4]^-$ in addition to AuCN, which is a solid consisting of linear -Au-CN-Au-CN- polymeric chains [51]. Crystal structure determinations on several salts of $[Au(CN)_2]^-$ have been reported. Three independent ions are linear with d(Au-C) = 197.1(1) pm and d(CN) = 146(1) pm in the isostructural Cs and Tl salts [52]. The equilibrium binding constant for the cyanide ion has been reported as $\log b_2 = 10^{36.6}$ [53] and 10^{39} [54]. It should be noted that these are for binding of cyanide ion, and that protonation at physiological pH moderates the binding affinity, albeit only slightly. Structures of the potassium and hydronium ion salts, $K[Au(CN)_4] \cdot H_2O$ and $H[Au(CN)_4] \cdot 2H_2O$, have been determined [55,56]. In each case, the expected square planar coordination of cyanide to gold(III) with Au—CN bond lengths of 197–198 pm is observed. In both salts, the waters of crystallization are hydrogen bonded to the cyanide ions [55,56].

The finding that gold metabolites readily enter the red blood cells of smokers, who inhale cyanide in the smoke, stimulated studies of reactions between cyanide and gold drugs and their analogues. The scheme in Fig. 6 summarizes the major equilibria that have been determined for gold(I) thiolates. It applies broadly to glutathione [26,57], thiomalate [26,57,58], thioglucose [59,60], captopril [61], cysteine [26,62] and thiosulfate [63]. The equilibria qualitatively favor cyanide over the thiols [26,57–62]. One exception is that $[AuSCy]_n$ precipitates readily and thereby drives the upper pathway in Fig. 6 in reverse of its usual direction [57], unless a large excess of cysteine is present [26]. The lower pathway in Fig. 6 is a ligand scrambling (or redistribution) reaction which generates the homoleptic cyano and thiolato complexes from the mixed-ligand species. The scrambling reaction explains the observation of two cyanide signals by NMR, even when the $Au/CN/RS^-$ ratio is 1:1:1.

Elder and co-workers measured apparent equilibrium constants for reactions of cyanide with gold(I) thiolates at pH 7.4 [Eqs. (6) and (7)] [26]. The values confirm the preference of gold(I) for cyanide ligands, but also indicate that under biological conditions where [RSH] ≫ [HCN], substantial displacement of cyanide may take place. This may facilitate the accumulation of gold by red cells when the extracellular $[Au(CN)_2]^-$ concentration is 10–20 μM and GSH is 1–2 mM [27,64]. On the other hand, in serum where the free thiol concentration is low, aurocyanide would be favored.

$$1/n[AuSTm]_n + 2\,HCN \xrightleftharpoons{K_{app}^{7.4}=6\times 10^2 M^{-1}} Au(CN)_2^- + TmSH + H^+ \qquad (6)$$

$$Au(SCy)_2^- + 2\,HCN \xrightleftharpoons{K_{app}^{7.4}=1.7\times 10^2} Au(CN)_2^- + 2\,CySH \qquad (7)$$

Auranofin also reacts with cyanide under conditions that suggest that the reaction could occur in vivo. The tetraacetylthioglucose ligand is displaced in preference to the phosphine ligand. Thus, when auranofin is in excess, the first reaction would

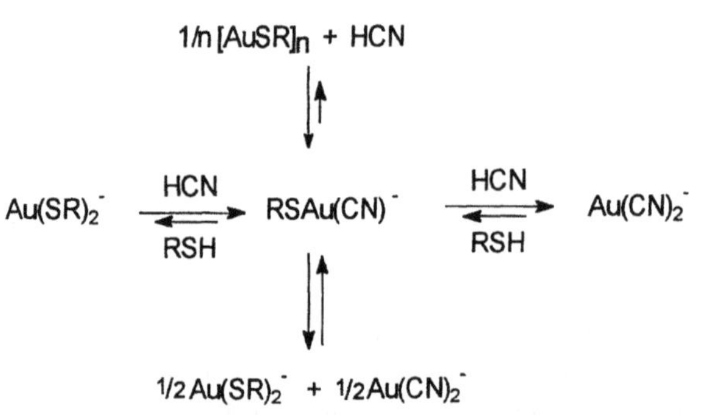

Fig. 6. Reaction scheme for cyanide and gold(I) thiolates. Adapted from [57] copyright 1986 American Chemical Society

predominate but if cyanide is in excess the second reaction, which generates aurocyanide, would take place [Eqs. (8) and (9)].

$$Et_3PAuSAtg + HCN \rightarrow Et_3PAuCN + AtgSH \quad (8)$$

$$Et_3PAuCN + HCN \rightarrow Et_3P + Au(CN)_2^- + H^+ \quad (9)$$

Phosphine gold(I) cyanides, such as Et_3PauCN and Ph_3PAuCN, undergo a ligand scrambling reaction analogous to that of $[RSAuCN]^-$ complexes [65,66]:

$$R_3PAuCN \leftrightarrow 1/2[(R_3P)_2Au]^+ + 1/2[Au(CN)_2]^- \quad (10)$$

When R = Ph, the redistribution is obscured by rapid ligand exchange which gives rise to a single ^{31}P resonance at room temperature (Fig. 7A), but is clearly evident from analysis of the ^{13}C and ^{31}P spectra obtained at 200 K with 30% enriched $^{13}CN^-$ (Fig. 7B). The resonances with $^2J_{PC}$ = 126 Hz in each spectrum are unambiguously assigned to the neutral compound. The additional singlets in ^{31}P and ^{13}C spectra correspond to the chemical shift of the ionic, homoleptic products, $[(Ph_3P)_2Au]^+$ and $[Au(CN)_2]^-$, respectively [Eq. (10)]. Investigation of complexes with other aliphatic phosphines showed that the redistribution is a general phenomenon and allowed equilibrium constants to be calculated for the reactions [65,66]. Interestingly, Et_3PAuCN and Ph_3PAuCN exist as such in the solid form and scramble in solution [65,66], while the tris(cyanoethyl)phosphine analogue crystallizes as the ionic species $[\{(NCCH_2CH_2)_3P\}_2Au][Au(CN)_2]$ and equilibrates to form the neutral species, $[(NCCH_2CH_2)_3PAu(CN)]$, in solution [67].

2.2
^{13}C and ^{15}N NMR Studies of Gold Cyanides

^{13}C NMR spectroscopy is used frequently to characterize the chemistry of gold cyanides in aqueous media. Measurements at natural abundance (1.11%) require moderately high concentrations, but gold cyanides can be prepared using enriched $K^{13}CN$, a readily available starting material, which confers up to a 99-fold increase in sensitivity. The chemical shifts of diamagnetic metal cyanide complexes typically are upfield of CN^- (d_C = 166.2) and a few are even upfield of HCN (d_C = 114.7 ppm) (Table 3). The upfield shifts are greater for the higher oxidation states of a given metal [68]. Thus, in aqueous solution, $Au(CN)_2^-$ and $Au(CN)_4^-$ have shifts of 154 and 104 ppm, respectively [68]. There are extensive studies, typically using enriched $^{13}CN^-$, of substituted gold(I) cyanides, but relatively few of gold(III) cyanides. The chemical shift ranges of neutral and anionic gold(I) cyanides, LAuCN and $XAuCN^-$, respectively, are rather small (Table 3). Thiolate ligands generate a small upfield shift, typically less than 2 ppm from $Au(CN)_2$. A single exception is the assignment of a broad resonance at 143.8 ppm to the *cis*-captopril complex. The phosphine ligands in R_3PAuCN, on the other hand, cause a downfield shift, which increases with the basicity of the phosphine as measured by the Tolman ν_{CO} parameter [69] with Pr^i_3P (160.9 ppm) > Et_3P (160.4 ppm) > Me_3P (158.3 ppm) > Ph_3P (156.2 ppm) appearing at 2059, 2062, 2064, and 2069 cm^{-1}, respectively.

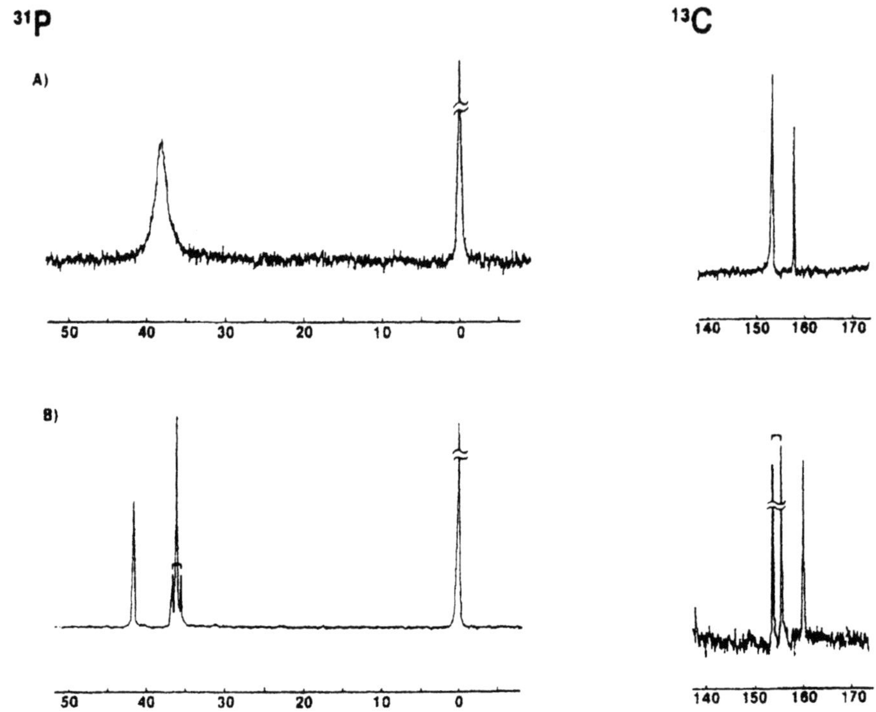

Fig. 7. $^{31}P\{^1H\}$ (101.3 MHz) and $^{13}C\{^1H\}$ (62.9 MHz) NMR spectra of $Ph_3PAu^{13}CN$ (26% enriched in CH_3OD at (*A*) room temperature and (*B*) 200 K. (**A**) The single ^{31}P resonance at room temperature results from rapid exchange according to Eq. (10). The room temperature ^{13}C spectrum exhibits two signals but no coupling of ^{13}C to ^{31}P. (**B**) The 200 K spectra each consist of a singlet due to the homoleptic complexes formed by scrambling reactions and a doublet with $^2J_{PC}$ = 126 Hz. In the ^{31}P spectrum the coupling generates a pseudo-triplet due to partial (26%) enrichment of the $^{13}CN^-$ ligand. Used with permission from [65] 1986 American Chemical Society

The spectrum of [*cis*-Au(damp)$(CN)_2$], an analogue of the anti-tumor agent [*cis*-Au(damp)Cl_2], exhibits cyanide signals at 109.8 and 144.2 ppm. The first is close to the resonance of $[Au(CN)_4]^-$ (104.5 ppm) and is assigned to the cyanide trans to the amino group, while the resonance at 144.2 ppm is assigned to the cyanide opposite the phenyl group [36].

2.3 Aurocyanide Complexes of Albumin and Other Proteins

Since aurocyanide can be generated by the reactions of polymorphonuclear monocytes (PMN) at inflammed sites and is found in urine, there must be a transport mechanism. A logical candidate is serum albumin, which transports many gold species by forming adducts at cysteine 34 (cf. Sect. 1.4.1). Several possible modes of binding can be considered: (a) ligand exchange at cysteine 34 to form albumin-S-AuCN$^-$; (b) formation at cysteine 34 of a three-coordinate gold(I) species, albumin-S-$Au(CN)_2^{2-}$; or (c) adducts of intact $Au(CN)_{2^-}$ associated through hydrogen bonding, van der Waals and other non-covalent forces. Extensive chromato-

Table 3. $^{13}CN^-$ and $C^{15}N^-$ NMR parameters for gold cyanide complexes

Compound	Conditions	δ_C (ppm)	δ_N (ppm)[a]	Coupling constants (Hz)	Ref
KCN	(D_2O, RT)	166.2			68
			274.7[b]	$^1J_{CN}$ = 6.1	70
HCN	(D_2O, RT)	114.7[c]			57
$[Au(CN)_2]^-$	(D_2O, RT)	154.1			68
	(D_2O, 300 K)		265.9		62
			266.5[b]	$^1J_{CN}$ = 11.1	70
$[Me_3PAuCN]$	(CH_3OD, 298 K)	156.0[c]		$^2J_{PC}$ = 129.4	66
			263.0	$^3J_{PN}$ = 3.6	62
$[Et_3PAuCN]$	(CH_3OD, 298 K)	158.1[c]		$^2J_{PC}$ = 122.4	66
			262.5	$^3J_{PN}$ = 2.9	62
$[Pr^i_3PAuCN]$	(CH_3OD, 298 K)	158.6[c]		$^2J_{PC}$ = 116.4	66
			262.2		62
$[Cy_3PAuCN]$	(CH_3OD, 298 K)	158.6[c]		$^2J_{PC}$ = 116.4	66
$[Ph_3PAuCN]$	(CH_3OD, 298 K)	153.8[c]		$^2J_{PC}$ = 126	65
			265.5	$^3J_{PN}$ = 4.0	62
$[Au(CH_2PPh_3)CN]$	($CDCl_3$, 198 K)	153.8			71
$[Au(\text{thioglucose-}H_{-1})CN]^-$	(D_2O, RT)	153.2			59
			260.3		60
$[Au(\textit{cis}\text{-captopril-}H_{-1})CN]^-$	(D_2O, RT)	143.8			61
$[Au(\textit{trans}\text{-captopril-}H_{-1})CN]^-$	(D_2O, RT)	152.2			61
$[Au(\text{thiomalate-}H_{-1})CN]^{3-}$		~153.4			58
		~153.6[c]			57
			260.3		60
$Au(\text{glutathione-}H_{-1})CN]^{2-}$		153.1[c]			57
Albumin $[Au(CN)_2]^-{}_{1\text{-}7}$		152.5[c]			67
Albumin $[Au(CN)_4]^-$		103.2			d
$[Au^{III}(CN)_4]^-$	(D_2O, RT)	104.5			68
$[Au^{III}(CN)_3Cl]^-$	10 M HCl/D_2O,				d
trans Cl		94.7 t		$^2J_{CC}$	
trans CN		111.1 d			
$[Au^{III}(CN)_2Cl_2]^-$	(D_2O, RT)	119			d
$[Au^{III}(CN)_2Cl(OH)]^-$	(D_2O, RT)	121			d
$[Au^{III}(CN)_2(OH)_2]^-$	(D_2O, RT)	121			d
cis-$[Au^{III}(Me_2NCH_2C_6H_4)$	(DMSO, RT),				
$(CN)_2]$	*trans* N	109.8			36
	trans C	144.2			

[a] Relative to external $NH_4{}^{15}NO_3$
[b] Downfield relative to external $Na^{15}NO_3$
[c] Corrected from internal TSP to TMS [δ_C ($Au(CN)_2$) = 154.1 ppm]
[d] A. Canumalla, PhD thesis, University of Wisconsin-Milwaukee (1998)

graphic, spectroscopic and radiolabelling studies produced apparently discrepant results: techniques in which the aurocyanide and albumin are at equilibrium show robust interactions, while methods that disturb the equilibrium depict very weak interactions [49,72]. Labile, non-covalent binding of intact aurocyanide ions at multiple binding sites resolves the discrepancies [49,72]. Thermodynamic binding constants were determined for a tight binding site (≈0.8 per albumin) and three weaker sites:

$$\text{Albumin} + [\text{Au(CN)}_2]^- \underset{n=0.8}{\overset{K_1=5.5\times10^4\,M^{-1}}{\rightleftharpoons}} \text{Albumin} \cdot [\text{Au(CN)}_2]_t^- \tag{11}$$

$$\text{Albumin} \cdot [\text{Au(CN)}_2]_t^- + 3[\text{Au(CN)}_2]^- \underset{n=3}{\overset{K_2=7.0\times10^3\,M^{-1}}{\rightleftharpoons}} \text{Albumin} \cdot [\text{Au(CN)}_2]_t^- \cdot [\text{Au(CN)}_2]_w^- \tag{12}$$

Approximately half of the bound gold dissociates during size-exclusion chromatography over small, rapidly eluted Penefsky columns and most of it is lost when using larger conventional columns that are eluted over longer times. The facile dissociation allows the low-molecular-weight $[Au(CN)_2]^-$ ion to be trapped in the resin pores, despite the strength of its binding to albumin [72].

The sharp ^{13}C resonance of aqueous aurocyanide (154.1 ppm vs. TMS) broadens and shifts to 152.5 ppm when it reacts with albumin. At least seven $[Au(CN)_2]^-$ ions [72], and perhaps as many as ten (A. Canumalla and C.F. Shaw III, unpublished), bind with essentially the same chemical shift. No displaced HCN or free $[Au(CN)_2]^-$ was detected. The direction and magnitude of the chemical shift can be accounted for by solvation changes as the aurocyanide ions are incorporated into the less polar albumin structure with reduced hydrogen bonding [72]. Mössbauer parameters for complexes with one and three bound aurocyanide ions indicate that they bind primarily as intact molecules and do not form three-coordinate structures with or undergo ligand exchange reactions at cysteine 34 [73]. Radiolabelling with $[Au(^{14}CN)_2]^-$ revealed that a small, spectroscopically silent fraction of the bound gold (<11%) is associated with cysteine 34 following a ligand exchange reaction and does not dissociate during chromatography, but it was not observed in NMR or Mössbauer spectra.

$[Au(CN)_2]^-$ is widely used as a heavy atom probe for protein crystallography, including inter alia the structures of haloalkane dehydrogenase (HD) [74], human carbonic anhydrase I (CA) [75], horse liver alcohol dehydrogenase [76], and human liver interleukin-1 receptor antagonist protein [77]. In haloalkane dehydrogenase, the intact aurocyanide ligand sits between two tryptophan nitrogens and is stabilized by the partially positive ring nitrogens at one end and by phenylalanine ring hydrogens, also slightly positive, at the other [74]. The aurocyanide in carbonic anhydrase binds at the active site and is oriented so that one cyanide nitrogen atom is 340 pm from the catalytic zinc atom and within hydrogen-bonding distance of the coordinated water [75]. Aurocyanide binds at the active site of HD in the same location as the halide ion substrate [74]. It is interesting that the enzymatically determined constants (K_i) for inhibition alcohol dehydrogenase by $[Au(CN)_2]^-$ [78] are somewhat weaker in magnitude ($K_1 = 1.2 \times 10^3$ and $K_2 = 1.3 \times 10^2\ M^{-1}$) than the equilibrium binding constants for albumin and $[Au(CN)_2]^-$ [Eqs. (11) and (12)] [72]. The crystal structures and inhibition constants support the proposed model of non-covalent binding of $[Au(CN)_2]^-$ to serum albumin.

Multiple albumin binding sites for anions such as Cl^-, Br^-, and SCN^- have been documented [79–81], consistent with the strong and weak aurocyanide binding sites identified by thermodynamic binding studies [72] and NMR titrations showing that at least 7–10 aurocyanide ions bind to albumin [72, Canumalla and CF Shaw III, unpublished studies]. The NMR chemical shifts [66] and the Mössbaur parameters [73] do not change significantly as the additional sites are populated, suggesting that only subtle differences distinguish the weak and strong binding sites. The free energies of binding calculated from the aurocyanide-albumin binding constants −6.5 and −5.3 kcal/mol [Eqs. (11) and (12)] are also in agreement with the non-covalent bonding model.

The significance of albumin–$[Au(CN)_2]^-$ adduct formation is twofold. First, the $[Au(CN)_2]^-$ measured in the ultrafiltrates of blood and urine from chrysotherapy patients accounts for about 0.1–1.0% of the total gold present in these fluids [24]. The ability of albumin, and perhaps other serum proteins, to bind intact aurocyanide implies that additional quantities may be associated with the high-molecular-weight species in the retentate, so that the total aurocyanide concentrations in circulation are greater. Second, the ability of albumin to function as a transport mechanism for aurocyanide is firmly established by the evidence that it binds predominantly as intact ions that are demonstrably labile to dissociation [49,72,73].

2.4 Cyanide and Cellular Accumulation of Gold

Evidence accumulated over the past decade has established the importance of cyanide complexes, principally $[Au(CN)_2]^-$, in the metabolism of gold drugs. $Au(CN)_{2^-}$ also has anti-HIV activity (see Sect. 1.3.1). The first evidence that $[Au(CN)_2]^-$ is a metabolite was the report that cyanide facilitates the accumulation of gold metabolites in patients' red blood cells (RBCs), although the red cells of laboratory animals typically do not accumulate them [82,83]. The smoke inhaled from tobacco products contains up to 1700 ppm of HCN [84], which is absorbed through the lungs. It can react with gold and facilitate the transfer of gold into various cells [83]. (Of course, this phenomenon was not observed in laboratory animals, whose life spans do not extend to the minimum age for cigarette purchase.) Red cells exposed to 4.5 μM (0.9 ppm) aurocyanide absorb about 95% of the gold [64]. The uptake is not affected by the anion channel blocker DIDS (diisothiocyanatostilbene-2,2′-disulfonic acid), which rules out the possibility that linear aurocyanide ion enters RBCs through the anion channel [26,64]. Preincubation of cells with NaCN does limit the uptake, as would be expected if the aurocyanide had to react with a thiol or other ligand that mediates its transport into cells:

$$[Au(CN)_2]^- + RSH \rightleftharpoons [RSAuCN]^- + HCN \tag{13}$$

In the red cell lysates, the majority of the gold is protein bound. Although the largest pool of available thiols are the Cys β93 residues of hemoglobin, relatively little gold is hemoglobin bound. The largest fraction is associated with an unidentified protein of molecular weight 330,000 [64].

The ability of cyanide to alter the metabolism of gold also explains the higher incidence of side effects from gold drugs in patients who smoke [18,83].

2.5
Immunogenesis of Cyanide and Aurocyanide

Graham and co-workers [18,85,86] have assembled considerable evidence that aurocyanide is formed at the sites of inflammation and may mediate the cellular effects of gold during chrysotherapy. During the oxidative burst carried out by polymorphonuclear monocytes (PMNs) and other cells, oxygen is reduced in sequential reactions that release superoxide ($O_2^{-\bullet}$), hydrogen peroxide (H_2O_2), hydroxyl radical ($OH^{\bullet}$) and hypochlorite (OCl^-). The cyanide is generated from thiocyanate (SCN^-), a component of many extracellular fluids in the body, under the action of the enzyme myleoperoxidase [87,88]. The reaction is complex and also generates $OSCN^-$ and ClCN as products [87,88]. Evidence for formation of cyanide from glycine has also been reported [89]. The cyanide reacts, as described in Sect. 2.1, with gold drugs and their metabolites to form aurocyanide. The aurocyanide can inhibit the oxidative burst of the PMNs. At aurocyanide concentrations of 1, 5 and 10 μM, the production of superoxide (measured as lucigenin-dependent chemiluminescence from stimulated PMNs suspended in 50% synovial fluid) is decreased in intensity and duration. These gold concentrations fall within the range observed in patients' blood during chrysotherapy [85]. Greater inhibition is achieved at 50 μM aurocyanide.

The formation of aurocyanide from thiocyanate and AuSTm can be demonstrated in vitro in the presence of polymorphonuclear monocytes stimulated to undergo the oxidative burst [86]. Inhibition of the oxidative burst although less effectively under these conditions, nontheless demonstrates that aurocyanide can be formed simultaneously with the inhibitory effect. Azide, an inhibitor of myleoperoxidase, prevents the inhibitory effect of AuSTm and thiocyanate, but not that of aurocyanide [18], demonstrating that the enzyme is active in the cellular generation of aurocyanide.

3
Biological Oxidation of Gold

3.1
Redox Properties of Gold

The aquated ions $[Au(OH_2)_n]^+$ and $[Au(OH_2)_4]^{3+}$, unlike those of most other transition metals, are thermodynamically unstable and can be readily reduced to elemental gold. While the reduction potentials are sensitive to the choice of ligand, the large, positive E_o values in Table 4 indicate that mild reducing agents are capable of reducing gold(III) and gold(I) halides. For example, $[AuCl_4]^-$ reacts slowly with water to release elemental oxygen. While the E_o value for $[AuBr_4]^-$ is significantly less than that for free Au^{3+}, it is still sufficiently positive for $[AuBr_4]^-$ to act as a powerful oxidant. In contrast, the negative potential of $[Au(SCy)_2]^-$ reflects stabilization of the +1 oxidation state by the coordinated thiolate ligands. Significantly, gold(I) and gold(III) can be stabilized as their cyanide complexes. Thus, gold(III) tetrahalide complexes are powerful oxidizing agents, but gold(I) can be stabilized by

Table 4. Reduction potentials for selected gold complexes

Couple		E_o (V)	Ref
$Au^{+1} + e^-$	$= Au(0)$	+1.68	89
$AuCl_{2^-} + 1e^-$	$= Au(0) + 2Cl^-$	+1.15	89
$AuBr_{2^-} + 1e$	$= Au(0) + 2Br^-$	+0.959	89
$Au(SCy)_{2^-} + 1e^-$	$= Au(0) + 2CyS^-$	−0.14	90
$Au(CN)_{2^-} + 1e^-$	$= Au(0) + 2CN^-$	−0.48	89
$Au^{+3} + 3e^-$	$= Au(0)$	+1.42	89
$AuCl_{4^-} + 3e^-$	$= Au(0) + 4Cl^-$	+1.00	89
$AuBr_{4^-} + 3e^-$	$= Au(0) + 4Br^-$	+0.854	89
$Au(CN)_{4^-} + 3e^-$	$= Au(0) + 4CN^-$	−0.10	89

thiolate ligands and both gold(I) and gold(III) by cyanide. This is consistent with the metabolic properties of gold as described in the following sections.

Reduction of gold(III) to gold(I) or gold(0) is often observed in biological milieu. The reaction can be driven by naturally occurring reductants such as thioethers, thiols or even disulfides:

$$Au^{III}L_4 + RSR + H_2O \rightarrow Au^IL_2 + R_2S{=}O + 2L + 2H^+ \tag{14}$$

$$Au^{III}L_4 + 2RSH \rightarrow Au^IL_2 + RSSR + 2L + 2H^+ \tag{15}$$

$$mAu^{III}L_4 + \tfrac{1}{2}RSSR + nH_2O \rightarrow mAu^0 + RSO_nH + 4mL + (2n-2)H^+ \tag{16}$$

In agreement with the chemistry of Eqs. (14–16), Sadler has demonstrated that the reduction of gold(III) can occur in cell culture media [91].

3.2 Oxidation States In Vivo

Chemical reactions of gold drugs exposed to body fluids and proteins are predominantly ligand exchange reactions that preserve the gold(I) oxidation state [19,21]. They are exemplified by the protein reactions described in Sect. 1.4. Considerable evidence suggests that gold exists as, and is expected to remain, primarily gold(I) in vivo. Aurosomes (lysosomes that accumulate large amounts of gold and undergo morphological changes) isolated from gold(III)-treated rats, as well as those treated with gold(I), contain predominantly gold(I) [92,93]. Peptide and protein methionine residues, as well as other thioethers, also reduce gold(III) to gold(I) [94]. Even disulfide bonds react rapidly to reduce gold(III) [95,96]. Thus, the bulk of gold present in vivo is likely to be gold(I). Nonetheless, the *potential* for oxidizing gold(I) to gold(III) in vivo has long been recognized.

Immunological assays provided the first concrete evidence for gold(III) formation in vivo. Gleichmann and co-workers observed that gold drugs can be activated in vivo to a gold(III) metabolite that is responsible for many of the immunological reactions to gold drugs [27,28]. This finding is based on the observation that, after treating mice with AuSTm for several weeks, gold(III) elicits a response in the popliteal lymph node assay (PLNA) but AuSTm does not. The PLNA is important because it discriminates between the effects of a drug and its metabolites, in order to determine which is immunogenic. Subsequent research in an independent laboratory confirmed that T cells from human chrysotherapy patients are sensitized against gold(III) but not gold(I) [16,17]. Thus, there is clear biological evidence for the formation of gold(III), and it is demonstrably relevant to human therapy.

Beverly and Couri reported briefly that hypochlorous acid (HOCl), which is generated by the enzyme myleoperoxidase during the oxidative burst, can oxidize the gold in AuSTm to gold(III) [97]. This has been confirmed and extended to additional gold compounds [49,72]. For example, AuSTg and auranofin are oxidized to gold(III), and the ligands are oxidized preceeding or in concert with the gold oxidation.

$$\tfrac{1}{n}[AuSR]_n + 4OCl^- \rightarrow AuCl_4^- + RSO_3^- \tag{17}$$

$$Et_3PAuSAtg + 5OCl^- \rightarrow AuCl_4^- + AtgSO_3^- + Et_3PO \tag{18}$$

$[Au(CN)_2]^-$ is oxidized to a *trans*-dicyanogold(III) species, $[Au(CN)_2X_2]^-$, in which the remaining coordination positions are occupied by chloride or hydroxide ions in a pH-dependent equilibrium [A. Canumalla and C.F. Shaw III, unpublished studies]. Thus, the chemical feasibility of gold oxidation by hypochlorite is clearly established [49] and provides a plausible mechanism for the immunological findings of Gleichmann [27,28] and Verwilghen [16,17].

3.3 Redox Cycling of Gold In Vivo

The apparent dichotomy between the observation that gold is present primarily as gold(I) in vivo and the findings that the oxidative burst can generate gold(III) and that T cells are sensitized to gold(III) not the gold(I)-containing drugs is easily resolved if a redox cycle is operative [49] (Fig. 8). The role of hypochlorite in oxidizing gold is described in Sect. 3.2 and examples of reduction of gold(III) by sulfur-containing amino acids and related molecules are given in Sect. 3.1.

This cycle is consistent with observations that relatively low concentrations of gold are present during chrysotherapy (10–25 μM Au), yet the changes in tissue levels of metals, thiols, proteins, etc., in responding patients are much larger than can be accounted for on a stoichiometric basis.

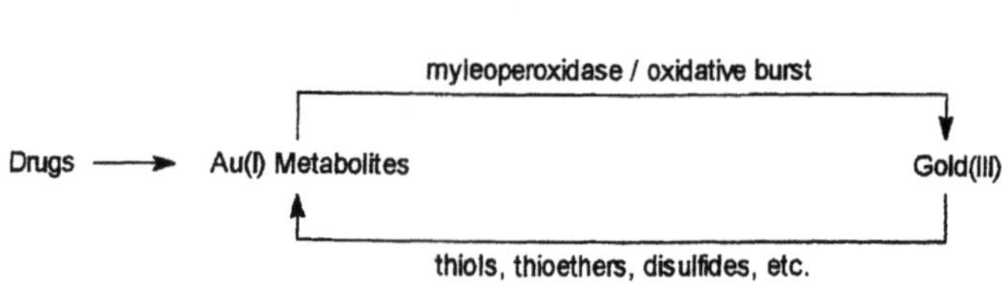

Fig. 8. Redox cycle for the immunogenesis of Au(III) and its subsequent reduction to Au(I)

3.4 Gold(III) Peptide Complexes

Several peptide complexes of gold(III) have been characterized by Lippert and co-workers [98]. At low pH glycyl-L-histidine binds to Au(III) as a tridentate ligand

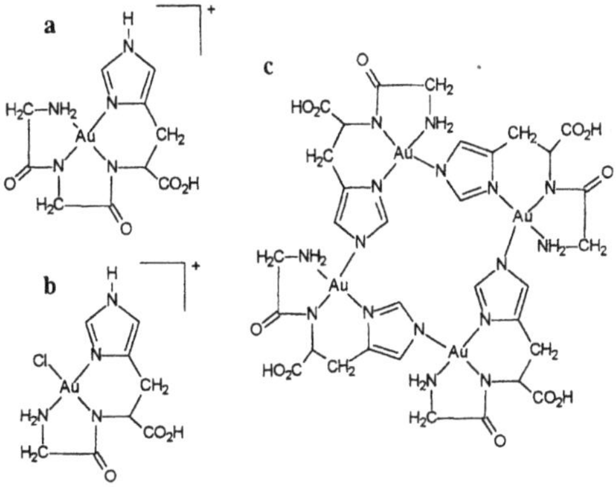

Fig. 9. Structures of gold(III) peptide complexes. Adapted from [21]

(Fig. 9b). The donor atoms are the *N*-terminal amino nitrogen, the amide nitrogen and N_δ in the imidazole ring of histidine. The fourth coordination site of gold is occupied by a chloride ion. N_ε of the ring is protonated. At higher pH the imidazole ring deprotonates and bridges to another gold ion, which leads to the formation of a tetrameric structure (Fig. 9a). A peptide which contains an additional glycine on the *N*-terminal end can acts as a tetradentate ligand (Fig. 9c). The glycyl-glycyl-L-histidine–gold(III) complex forms slowly at pH 1.5–2 [99]. The fourth donor atom is the additional amide nitrogen, which is also deprotonated. The imidazole N_δ is again the donor atom. The motif of coordination by the histidine imidazole N_δ and deprotonated amide nitrogens is frequently observed with Ni(II) and Cu(II) peptide complexes, which can sometimes be oxidized to form stable M(III) complexes analogous to glycyl-glycyl-L-histidine–gold(III).

4
Gold and the Immune System

4.1
Hypersensitivity Reactions

Human immunological reactions to metal ions have enormous social and economic consequences in industrial and environmental toxicology and medicine. Metal ions are significant inducers of hypersensitivity reactions [100] but can also exhibit immunosuppressive effects. The types of hypersensitivity reactions induced by gold and other immunologically active metals are given in Table 5. The type I, II and III hypersensitivity reactions are humoral and require reactive antibodies. Type IV or delayed hypersensitivity requires reactive T cells and the presentation of an antigen by antigen-presenting cells such as macrophages. For example, contact dermatitis is frequently induced after exposure to nickel in jewelry alloys or industrial settings [101]. Following the use of skin ointments or occupational exposure, mercury can induce immunological manifestations in humans [102,103]. Interestingly, rats (Brown Norway and Lewis strains) and mice (A.SW and DBA2 strains) generate similar autoimmune responses following gold or mercury exposure [102].

Table 5. Hypersensitivity reactions and coordination chemistry of selected metal ions

Ion	Hypersensitivity[a]	Coordination modes[b]	Acidity[c]	Amino acid binding	Notes
Au(I)	Types II, III & IV	dig > trig > tet	soft	cys > met > his > asp	
Au(III)		sq pl	border line	variable	oxidant
Be(II)	Type IV	tet	hard	asp > his > met > cys	
Cr(III)	Types I & IV	oct	hard	asp > his > met > asp >	inert
Cr(VI)		tet > oct	hard	[$HCrO_4^-$ until reduced]	oxidant
Hg(II)	Types III & IV	dig > trig > tet	soft	cys > his > met > asp	
Ni(II)	Types I & IV	oct, sq pl, tet > spy,tbp	borderln	variable	
Pt(II)	Type I	sq pl	soft	cys > met > his > asp	inert
Pt(IV)		oct	soft	cys > met > his > asp	inert

[a] Hypersensitivity reaction types [100]

[b] dig: 2-coord, linear; trig: 3-coord, planar; tet: 4-coord, tetrahedral; sq pl: 4-coord, planar; spy: 5-coord, square pyramidal; tbp: 5-coord, trigonal bipyramidal; oct: 6-coord, octahedral; 7- and 8-coord: variable [104]

[c] Lewis acidity classified as hard or soft acid [105]

As part of the discrimination between "self" and "non-self" by the immune system, proteins are proteolytically digested by various cells, including macrophages. Peptide fragments generated during the digestion are bound to the multihistone compatibility complexes (MHC molecules) and transferred to the surface of the cells, where T cells react to non-self peptides. Two independent systems are present and functional [106,107]. In the first, peptides generated from a wide variety of cells are presented on class I major histocompatibility complexes and recognized by cytotoxic T cells (Tc) which contain $CD8^+$ markers. The second system, which is relevant to this discussion, requires processing of the peptides by antigen-presenting cells (APCs) followed by presentation on class II MHCs and recognition by $CD4^+$-bearing T-helper cells.

Drugs, metal ions, natural and man-made environmental toxins [e.g. procainamide, Pt(II) and Au(I), poison ivy and cosmetic components] can generate immunotoxic reactions when they interact in antigen-presenting cells to alter self-proteins and cause the presentation of cryptic peptides. They are designated cryptic peptides, because they are typically neither presented nor recognized, but are part of the original self-proteins. In some cases the toxin itself is not the etiologic agent, but is first modified in vivo to a more reactive species. Figure 10 [108] shows the process schematically. In step I, a reactive compound, or its metabolite, which has entered the body reacts with a self-protein. The protein adduct is processed in the APCs to generate a cryptic peptide which is presented on an MHC. In step II, a specific $CD4^+$ T cell reacts with the MHC–peptide complex and stimulates the release of signals 1 and 2. In step III, the cytokines that are necessary for activation of further immune response are released thereby initiating allergic reactions.

Chrysotherapy for rheumatoid arthritis, which provides limited relief for many patients and spectacular remissions for some [15,109], requires immunosuppressive action. Yet, immunotoxic side effects to gold frequently require patients to withdraw from chrysotherapy before it can be established whether the treatment is effective

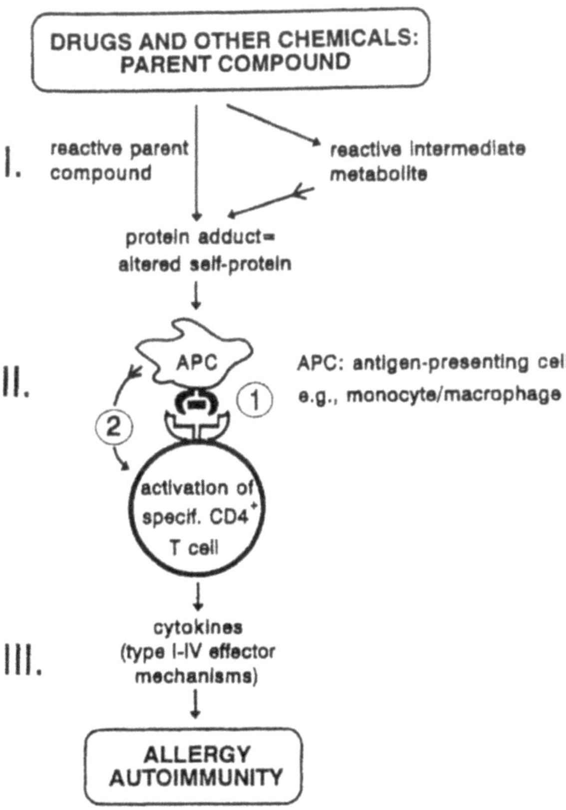

Fig. 10. The induction of allergy and autoimmunity by metals and other xenobiotic species. Used with permission from [106] 1997 Pergamon Press

[16,17]. These patients often exhibit immunogenetic markers (either HLA-DR3 or HLA-DR2 positive) and typically exhibit poor sulfoxidation [16]. Fewer than 15% of the patients treated with chrysotherapy remain on gold for longer than 5 years. In autoimmune diseases, T cells may react to cryptic self-peptides that are not normally presented, thereby generating a reaction against a patient's own tissues, such as the joints.

Antigenic peptides may be generated (a) from foreign proteins, (b) from haptenized (i.e. chemically modified) self-proteins, and (c) as cryptic peptides, which are sequences of self-proteins that are normally not presented but which will stimulate T cells if their presentation takes place (Fig. 11). Enzymatic protein processing, which generates the peptides that bind to the class II MHC molecules and are subsequently presented, occurs in the endosomes of antigen-presenting cells (APCs) such as macrophages. The lowered pH within the endosomal vesicles, ca. 5–5.5 [107], facilitates partial unfolding of the proteins and allows greater access of the proteases to the protein chain. Reduction of thiol groups during the processing also leads to enhanced peptide presentation [110,111].

The class II MHC molecules are membrane-imbedded heterodimers, which are stabilized by a so-called invariant chain which is replaced when a peptide binds to

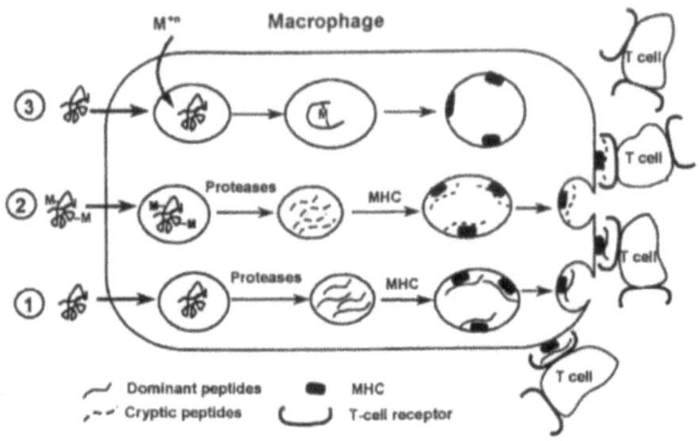

Fig. 11. Effects of metal ions on the presentation of dominant and cryptic peptides by antigen-presenting cells such as macrophages. (*1*) a dominant peptide formed from an antigenic protein and presented on a class II MHC can generate a T-cell response; (*2*) metal ion complexation or oxidation of a protein can generate cryptic peptides from self proteins to which T cells react; and (*3*) metal complexation of a dominant peptide may inhibit its presentation by the MHC molecule and prevent an immune response

the MHC. The peptides that bind to the class II MHCs may range from 12 to 20 amino acid residues. A core of 7–9 residues, some of which are anchored in the MHC, provides much of the specificity of interaction. The crystal structure (2.5 Å resolution) of an antigen–MCH complex, HLA-DR1 associated with the antigenic Haemagglutinin 306–318 peptide ($HA_{306-318}$), revealed that five residues (Tyr 308, Gln 311, Thr 313, Leu 314 and Leu 316) are accommodated in pockets of the peptide-binding surface while 35% of the peptide surface is accessible to solvent and hence to the T-cell receptor (TCR) [112]. The anchored peptide conformation exhibits a left-handed twist (−130° ± 15° with $\phi = -82° \pm 15°$ and $\theta = 139° \pm 16°$) and resembles the polyproline type II helix. In contrast, the strands of parallel and non-parallel β-sheets have slight right-handed twists [112].

4.2 Model Systems

The application of inorganic coordination chemistry to understanding the mechanisms of metal-induced immunotoxicity can best be achieved when the questions are posed within the context of well-defined immunological models. Two important model systems that are sensitive to gold-induced alterations in the process are briefly described below. The first uses insulin as an antigen to model the possible immunosuppressive effects of gold. The second uses ribonuclease to model the immunotoxic effects of gold.

4.2.1 *Gold(I) Inhibition of Insulin A1–14 Presentation*

Bovine insulin is used widely in murine immunological model systems and many properties of reactions against it are well known [110,111]. T cells react to the peptide generated from A-chain residues 1–14 (Fig. 12a) when they are presented on class II MHC molecules A^d and A^b [111,113]. Griem, Gleichmann and co-workers observed that the antigenicity of bovine insulin injected into mice is modulated by coinjection of antiarthritic gold(I) complexes and have developed an excellent model for studying metal-altered peptide presentation [113].

Presentation of the dominant peptide, A1–14, which binds to the class II MHC molecules during processing, is recognized by several specific T-cell clones [114] and can be studied in vitro. When gold is added during processing, presentation is blocked for the native peptide (CCC) which has cysteines at positions 6, 7 and 11 (see pathway 3, Fig. 11). To distinguish whether merely binding gold to the peptide suffices to inhibit presentation or whether chelation is required, single and double Cys-to-Ser mutants were examined. The single mutants which retain two of the three cysteines, Cys6Ser = CCS (Fig. 12b), Cys7Ser = CSC and Cys11Ser = SCC, are inhibited by gold, but the double mutant SCS = Cys6,11Ser was not [113]. Thus, they postulated that chelation of gold(I) by two of the three cysteine residues distorts the peptide structure so it is unable to bind to the MHC.

Gold binding to Ins_{A1-14}(SCC), in which the Cys6Ser substitution reduces the coordination possibilities, is presently under study [115]. The peptide reacts with $Au(STm)_2^{5-}$ to form complexes with one or two Au(I) ions (A. Muñoz and C.F. Shaw III, unpublished studies). The 1:1 complex is formed with loss of the thiomalate ligands (Fig. 12c). In contrast, excess $Au(STm)_2^{5-}$ generates a digold complex which retains two thiomalate ligands (Fig. 12d). These structures are consistent with the tendency of gold to form two-coordinate linear structures with thiolate ligands. The ^{1}H NMR spectrum of Au-Ins_{A1-14}(SCC) exhibits dispersion of the peptide and cysteinyl Hα and Hβ resonances characteristic of metal-altered peptides (A. Muñoz and C.F. Shaw III, unpublished studies). The gold adducts will have electronic and steric properties very different to the apo-peptide. In particular, the steric constraints which bring cysteines 7 and 11 into position to form the mono-gold complex will preclude forming the extended polyproline-like structure necessary for binding of the peptide to the class II MHC molecules. Hence, the effect of gold(I) thiomalate in this immune model system is presumably to alter the peptide structure and prevent its binding to the MHC molecule. This, in turn, suggests that the mechanism

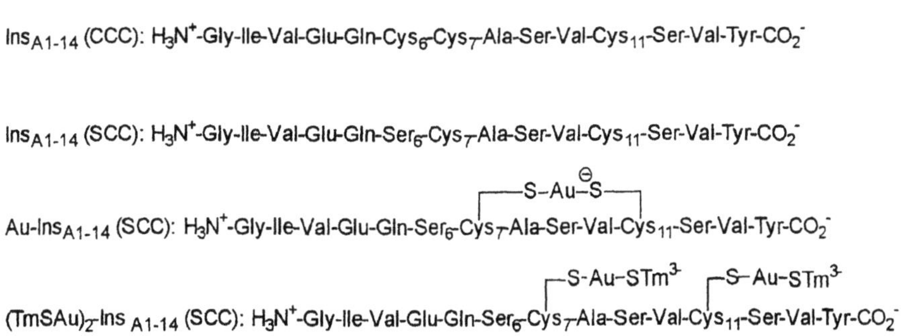

Fig. 12. Insulin peptides: (*a*) Ins_{A1-14}(CCC), the dominant peptide which can reduce cysteines at positions 6, 7 and 11; (*b*) Ins_{A1-14}(SCC), the Cys6Ser mutant; (*c*) Au-Ins_{A1-14}(SCC), with gold chelated by cysteines 7 and 11; (*d*) $(TmSAu)_2$-Ins_{A1-14}(SCC), with two thiomalatogold(I) moieties bound in monodentate fashion to cysteines 7 and 11

by which gold therapy causes remission of rheumatoid arthritis might be to prevent the presentation of a cysteine-containing immunogenic self-peptide to which the patients' T cells react.

The insulin model has been extended to investigate other metal ions [114]. The response of two clones to an array of heavy metal ions showed that Hg(II), Ag(I), Cu(II), Pd(II), Pd(IV), Pt(IV) are equally as effective as Au(I).

4.2.2
Gold(I)-Stimulated Cryptic Peptides from Ribonuclease

Gleichmann, Griem and co-workers [28,116,117] have developed a second model system using ribonuclease A as an antigen from which cryptic peptides can be generated by pretreatment with gold(III) and other metal ions. When mice are immunized with native bovine RNase, the dominant peptide is segment 74–88 (RN_{74-88}) of the sequence (Fig. 13). After challenging the same strain of mouse with Au(III)-treated RNase, several $CD4^+$ T-cell hybridoma lines responsive to Au(III)-treated RNase but not the native protein were established. The treatment at low pH with gold(III) – a powerful oxidant – caused two cryptic peptides to be presented and recognized by T cells: RN_{7-21} and RN_{94-108} [28]. A plausible hypothesis based on previous studies by Sadler and Isab [94] is that gold(III) oxidizes the methionines of ribonuclease and that the structural changes give rise to altered processing and peptide presentation. Changes in the CD spectra of gold(III)-treated ribonuclease are consistent with this view [117].

This RNase model clearly shows that an oxidizing metal species can generate cryptic peptides (pathway 2, Fig. 12). Thus, it mimics the response of mice which have been immunized with gold(I) compounds, but responds to gold(III) in the secondary PLNA [27,118], and also mimics the immunotoxic reactions of human

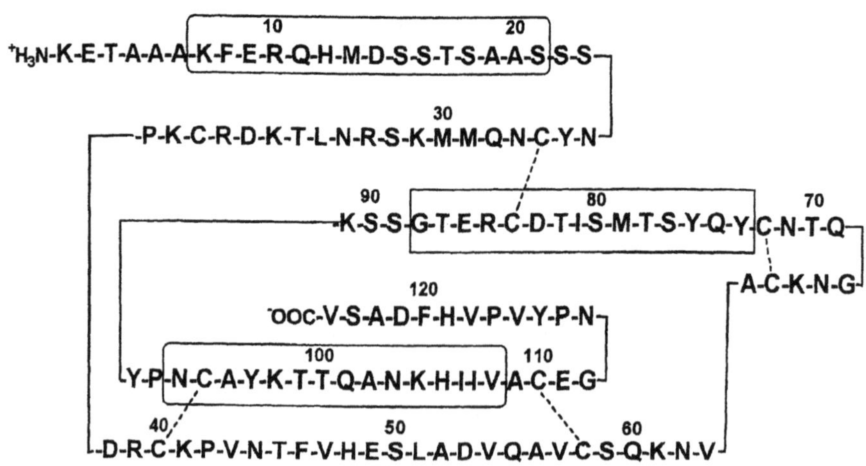

Fig. 13. The amino acid sequence of bovine ribonuclease showing the dominant peptide 73–98 [*square box*], and the two cryptic peptides 7–21 and 94–108 (*rounded boxes*) that are generated by pretreating the RNase with gold(III) at low pH

patients to chrysotherapy, in which T cells respond anamnestically to gold(III) in vitro [16,17]. The actual responses are not to the gold(III), which cannot function directly as an antigen, but to unknown self-proteins in the mice and humans that are modified by gold(III) and subsequently activate T cells.

The mouse T-cell clones which are specific to gold(III)-altered proteins were also screened against ribonuclease treated with Au(I), Au(III), Ni(II), Ni(IV), Pt(II), Pt(IV), Pd(II) and Pd(IV) [117]. Three lines of T cells reacted in every case to Au(III), Pd(II) and Pd(IV); two lines also reacted to Pt(II) and Pt(IV); all reacted weakly or not at all to Ni(II) or Au(I) [117]. These findings demonstrate that diverse metals may exert common effects on the immune system and suggest that the ribonuclease model can contribute to an understanding of immunological cross-reactivity to metal ions.

The RNase model also reveals a mechanism by which gold(III) generated in vivo may exert an immunosuppressive effect. The action of gold(III) causes cryptic peptides, but not the dominant peptide of the antigenic protein to be presented. Thus a plausible model for the action of gold(III) generated in vivo may be to prevent the presentation of the dominant (arthritogenic) peptide from a self-protein that is responsible for autoimmune response in rheumatoid arthritis [118].

5
Conclusions

Our understanding of gold complexes at the biochemical and cellular levels has evolved rapidly in the last two decades. Among the most significant findings are: (1) the roles of protein-gold complexes; (2) the metabolic generation of $Au(CN)_2^-$ and its ability to inhibit polymorphonuclear leukocytes (PMN); (3) the conclusion that the medicinal agents are actually pro-drugs; (4) the immunogenesis of gold(III); and (5) the emerging mechanisms for the effects of gold on T-cell activation. The first three findings are consistent with an earlier prediction that the parent drugs and auranofin should form common metabolites [19]. The results from many laboratories and disciplines described in this chapter are consistent with this view. Similar conclusions have been reached from consideration of the pharmacology of gold complexes [119]. The absence of a widely accepted mechanism of action for chrysotherapy, despite six decades of use and the tremendous advances in our understanding of the bioinorganic chemistry of gold drugs and metabolites, clearly points to the need for new experimental designs involving these metabolites, because they either: (1) constitute the active species, or (2) generate it (them) at the elusive sites of action.

The development of anti-tumor agents based on gold and the use of gold to modulate organic chemotherapy drugs through ligation remain tantalizing possibilities for research. The recent reports of anti-HIV activity for cyanide and thioglucose derivatives of gold may be even more significant and clearly merit active pursuit of their mechanism(s) of action.

References

1. Hill DT, Sutton BM (1980) Cryst Struct Commun 9:679
2. Baggio RG, Baggio S (1973) J Inorg Nucl Chem 35:3191
3. Reuben Z, Zalken A, Faltenns MO, Templeton DH (1974) Inorg Chem 13:1836
4. Kean WF, Lock CJL, Howard-Lock H (1991) Inflammopharmacology 1:103
5. Grootveld MC, Razi, MT Sadler PJ (1984) Clin Rheumatol 3:5
6. Bau R (1998) J Am Chem Soc 120:9380
7. LeBlanc DJ, Smith RW, Wang Z, Howard-Lock HE, Lock CJL (1997) J Chem Soc Chem Commun 3263
8. Schröter I, Strähle J (1991) Chem Ber 24:2161
9. Tong YY, Pombiero AJL, Hughes DL, Richards RL (1995) Transition Met Chem 20:372
10. Bates PA, Waters JM (1985) Acta Crystallogr C41:862
11. Nakamoto M, Hiller W, Schmidbaur H (1993) Chem Ber 126:605
12. Gimeno MC, Laguna A (1997) Chem Rev 97:511
13. Elder RC, Ludwig K, Cooper JN, Eidsness MK (1985) J Am Chem Soc 107:5024
14. Jaffe I (1989) Arthritis. In: Tapley DF, Morris TQ, Rowland LP, Wiess, RJ (eds) The Columbia University complete home medical guide. pp 602–625
15. Lipsky PE (1985) Disease modifying drugs. In: Utsinger PD, Zvaifler NJ, Ehrlich GE (eds) Arthritis: etiology, diagnosis and management. Lippincott, Philadelphia, pp 602–634
16. Panayi GS (1990) Immunological and immune-mediated toxic effects of gold compounds used in the treatment of rheumatoid arthritis. In: Dayan AD, Hertel RF, Heseltine E et al (eds) Immunotoxicity of metals and immunotoxicology. Plenum, New York, pp 155–160
17. Verwilghen J, Kinglsley GH, Gambling L, Panayi GS (1992) Arthr Rheum 35:1413
18. Graham GG, Campion GD, Ziegler JB (1994) Metal Based Drugs 1:395
19. Shaw CF III (1989) Comments Inorg Chem 8:233
20. Shaw CF III (1998) The biochemistry of gold. In: Schmidbaur (ed) Gold: progress in chemistry, biochemistry and technology. Wiley, New York, pp 259–308
21. Best SL, Sadler PJ (1996) Gold Bulletin 29:87
22. Shaw CF III, Isab AA, Coffer MT, Mirabelli CK (1990) Biochem Pharmacol 40:1227
23. Intoccia AP (1983) Animal pharmacokinetics and metabolism of ridaura. In: Sutton BM, Franz RG (eds) Bioinorganic chemistry of gold coordination compounds. Smith Kline & French, Philadelphia, pp 21–33
24. Elder RC, Zhao Z, Zhang Y, Dorsey JG, Hess EV, Tepperman K (1993) J Rheumatol 20:268
25. Tepperman K, Zhang Y, Roy PW, Floyd R, Dorsey JG, Elder RC (1994) Metal Based Drugs 1:433
26. Elder RC, Jones WB, Zhao Z, Dorsey JG, Tepperman K (1994) Metal Based Drugs 1:363
27. Schuhmann D, Kubicka-Muranyi, Mirtcheva J, Günther J, Kind P, Gleichmann E (1990) J Immunol 145:2132
28. Takahashi K, Griem P, Goebel C, Gonzalez J, Gleichmann E (1994) Metal Based Drugs 1:483
29. Blough H (1990) Abstr 3rd Int Conf on Gold and Silver in Medicine, p 14
30. Okada T, Patterson BK, Ye S-Q, Gurney ME (1993) Virology 192:631
31. Okada T, Gurney ME, Arendt AL, Shaw CF III, unpublished data
32. Shaw CF III (1994) Gold. In: Fricker SP (ed) Metal compounds in cancer therapy, Chapman and Hall, pp 46–64
33. Haiduc I, Silvestru C (1989) In vivo 3:363
34. Berners-Price SJ, Sadler PJ (1988) Struct Bonding 70:27–102
35. Sadler PJ, Nasr M, Narayanan VL (1984) The design of metal complexes as anti-tumor agents. In: Hacker MP, Douple EB, Krakoff, IH (eds) Platinum coordination complexes in cancer chemotherapy. Martinus Nijhoff, Boston, pp 290–304
36. Parish RV, Howe BP, Wright JP, Mack J, Pritchard R, Buckley RG, Elsome A, Fricker SP (1996) Inorg Chem 35:1659
37. Buckley RG, Elsome AM, Fricker SP, Henderson GR, Theobald BRC, Parish RV, Howe BP, Kelland, LR (1996) J Med Chem 39:5208
38. Smith WE, Reglinski J (1994) Metal Based Drugs 1:497
39. Carter D, Ho JX (1994) Adv Protein Chemistry 45:153
40. Shaw CF III (1983) In: Sutton BM, Franz RG (eds) Bioinorganic chemistry of gold coordination compounds. Smith Kline & French Laboratories, Philadelphia, pp 98–123

41. Shaw CF III, Schaeffer NA, Elder RC, Eidsness MK, Trooster JM, Calis GHM (1984) J Am Chem Soc 106:3511
42. Roberts JR, Xiao J, Schliesman B, Parsons DJ, Shaw CF III (1996) Inorg Chem 35:424
43. Christodoulou J, Sadler PJ, Tucker A (1994) Eur J Biochem 225:363
44 Abate C, Patel L, Rauscher FJ III, Curran T (1990) Science 249:1157
45. Xanthoudakis S, Curran T (1992) EMBO J 11:653
46. Xanthoudakis S, Curran T (1992) EMBO J 11:3323
47. Isab AA, Sadler PJ (1982) J Chem Soc Dalton Trans 135
48. Handel ML, Watts CKW, DeFazio A, Day RO, Sutherland RL (1995) Proc Natl Acad Sci USA 92:4497
49. Shaw CF III, Schraa S, Gleichmann E, Grover YP, Dunemann L, Jagarlamudi A (1994) Metal Based Drugs 1:351
50. Koch R (1890) Dtsche med Wochenschr 16:756
51. Bowmaker GH, Kennedy BJ, Reid JC (1998) Inorg Chem 37:3968
52. Blom N, Ludi A, Bürgi H-B, Tichy K (1984) Acta Crystallogr C40:1767
53. Skibsted LH, Bjerrum (1977) Acta Chem Scand A31:155
54. Hancock RD, Finkelstein NP, Evers AJ (1972) J Inorg Nucl Chem 34:3747
55. Bertinotti C, Bertinotti A (1970) Acta Crystallogr B26:422
56. Penneman RA, Ryan RR (1972) Acta Crystallogr B28:1629
57. Lewis G, Shaw CF III (1986) Inorg Chem 25:58
58. Graham GG, Bales JR, Grootveld MC, Sadler PJ (1985) J Bioinorg Chem 25:163
59. Isab AA (1992) J Inorg Biochem 46:145
60. Isab AA, Ghazi IH, Wazeer IM, Perzanowski HP (1993) J Inorg Biochem 50:299
61. Isab AA, Ghazi I, Al-Arfaj AR (1993) J Chem Soc Dalton Trans 841
62. Akhtar MN, Gazi IH, Isab AA, Al-Arfaj AR, Wazeer MIM, Hussain MS (1995) J Coord Chem 36:149
63. El-Hinnawi MA, Peter L, Meyer B (1984) J Raman Spectrosc 16:371
64. Zhang Y, Hess EV, Pryhuber KG, Dorsey JG, Tepperman K, Elder RC (1995) Inorg Chim Acta 229:271
65. Hormann AL, Shaw CF III, Bennett DW, Reiff WM (1986) Inorg Chem 25:3953
66. Hormann-Arendt AL, Shaw CF III (1990) Inorg Chem 29:4863
67. Hussain MS, Al-Arfaj AR, Akhtar MN, Isab AA (1996) Polyhedron 15:2781
68. Pesek JJ, Mason RW (1979) Inorg Chem 18:924
69. Tolman CA (1977) Chem Reviews 77:313
70. Sano M, Yosikawa Y, Hideo H (1982) Inorg Chem 21:2521
71. Bowmaker GA, Schmidbaur H (1984) J Chem Soc Dalton Trans 2859
72. Canumalla A, Schraa S, Isab AA, Shaw CF III, Gleichmann E, Dunemann L, Turfeld F (1998) J Biol Inorg Chem 3:9
73. Canumalla A, Shaw CF III, Wagner FE (1999) Inorg Chem 38: in press
74. Verschueren KHG, Franken SM, Rozeboom HJ, Kalk KH, Dijkstra BW (1993) FEBS 323:267
75. Kumar V, Kannan KK, Sathyamurthi P (1994) Acta Crystallogr D50:731
76. Söderberg B-O, Zeppezauer E, Boive T, Nordström B, Brändén C-I (1970) Acta Chem Scand 24:3567
77. Clancy LL, Finzel BC, Yem AW, Deibel MR Jr, Strakalaitis NA, Brunner DP, Sweet RM, Einspahr HM (1994) Acta Crystallogr D50:197
78. Gunnarsson P-O, Pettersson C, Zeppezauer M (1974) Eur J Biochem 43:479
79. Norne J-E, Lilja H, Lindman B, Einarsson R, Zeppezauer M (1975) Eur J Biochem 59:463
80. Scatchard G, Scheinberg IH, Armstrong SH Jr (1950) J Am Chem Soc 72:540
81. Scatchard G, Yap WT (1964) J Am Chem Soc 86:3434
82. James DW, Ludvigsen NW, Cleland LG, Milazzo SC (1982) J Rheumatol 9:532
83. Graham GG, Haavisto, TM, Jones HM, Champion GD (1984) Biochem Pharmacol 33:1257
84. Smoking and Health, Report of the Advisory Committee to the Surgeon General of the Public Health Service (1964) Report 1103, p 53
85. Rudkowski R, Graham GG, Champion GD, Ziegler JB (1990) Biochem Pharmacol 39:1687
86. Graham GG, Dale MM (1990) Biochem Pharmacol 39:1697
87. Stelmaszynska T (1986) Int J Biochem 18:1107
88. Thomas EL, Fishman M (1986) J Biol Chem 261:9694

89. Stelmaszynska T (1985) Int J Biochem 17:373
90. Chernyak AS, Shestopalova LF (1976) Russ J Inorg Chem (Engl Transl) 21:464
91. Bell DJ, Norman RE, Sadler PJ (1989) J Inorg Biochem 31:241
92 Elder RC, Tepperman KG, Eidsness MK, Heeg MJ, Shaw CF III, Schaeffer NA (1983) ACS Symp Ser 209:385
93. Eidsness MK, Elder RC (1987) Chem Rev 87:1027
94. Isab AA, Sadler PJ (1977) Biochim Biophys Acta 492:322
95. Shaw CF III, Cancro MP, Witkiewicz PL, Eldridge J (1980) Inorg Chem 19:3198
96. Witkiewicz PL, Shaw CF III (1981) J Chem Soc Chem Comm 1111
97. Beverly, Couri D (1987) Fed Proc 46:854
98 Wienken M, Lippert B, Zangrando E, Randaccio L (1992) Inorg Chem 31:1983
99. Best SL, Chattopadhyay TK, Durjan MI, Mazid MA, Palmer RA, Sadler PJ (1994) Metal Based Drugs 1:510
100. Kazantzis G (1990) Hypersensitivity: clinical aspects. In: Dayan AD, Hertel RF, Heseltine E et al (eds) Immunotoxicity of metals and immunotoxicology. Plenum, New York, pp 67–74
101. Avstorp C, Menne T, Maibach H (1990) Contact allergy to chromium and nickel. In: Dayan AD, Hertel RF, Heseltine E et al (eds) Immunotoxicity of metals and immunotoxicology. Plenum, New York
102. Pelletier H, Tounade H, Druet P (1990) Immunologically mediated manifestations of metals. In: Dayan AD, Hertel RF, Heseltine E et al (eds) Immunotoxicity of metals and immunotoxicology. Plenum, New York, pp 121–134
103. Friberg L, Eneström S (1990) Toxicology of inorganic mercury. In: Dayan AD, Hertel RF, Heseltine E et al (eds) Immunotoxicity of metals and immunotoxicology. Plenum, New York, pp 163–173
104. Cotton FA, Wilkinson G (1988) Advanced inorganic chemistry, 5th edn. Wiley, New York, chaps 1 & 2
105. Douglas B, McDaniel D, Alexander J (1993) Concepts and models of inorganic chemistry, 3rd edn. Wiley, New York, pp 337–343
106. Schafer PH, Pierce SK (1990) Antigen processing and presentation. In: Lawrence DA (ed) Comprehensive toxicology, vol 5. Pergamon, pp 215–233
107. Germain RN, Margulies DH (1993) Ann Rev Immunol 11:403
108. Griem P, Gleichmann E, Shaw C F III (1997) Chemically induced allergy and autoimmunity: what do T cells react against? In: Lawrence DA (ed) Toxicology of the immune system. Pergamon, New York, pp 323–338
109. Gordon DA (1989) Gold compounds. In: Kelly WN, Harris ED, Ruddy S, Sledge CB (eds) Textbook of rheumatology, WB Saunders, New York, pp 804–823
110. Collins DS, Unanue ER, Harding CV (1991) J Immunol 147:4054
111. Hampl J, Gandehandt G, Kalbacher H, Rüde E (1992) J Immunol 148:2664
112. Stern LJ Brown JH, Jardetzky TS, Strominger JL et al (1993) Nature 368:215
113. Griem P, Takahashi K, Kalbacher H, Gleichmann E (1995) J Immunol 155:1575
114. Griem P, Gleichmann E (1998) personal communication
115. Muñoz A, Canumalla, A, Singer M, Shaw C. F.III (1998) Abstract 124, Great Lakes Regional Meeting. June 1–3 (1998)
116. Griem P, Pantel K, Kalbacher H et al (1996) Eur J Immunol 26:279
117. Griem P, von Vultée C, Pantel K, Best SL, Sadler PJ, Shaw CF III (1998) Eur J Immunol 28:1941
118. Goebel C, Kubicka-Muranyi G, Tonn T, Gonzalez J, Gleichmann E (1995) Arch Toxicol 69:450
119. Graham GG, Champion GD, Ziegler JB (1991) Inflammopharmacology 1:99; Gordon, DA In: Kelly WW, Harris ED, Ruddy S, Sledge CB (1989) Textbook of rheumatology. WB Saunders, New York, pp 804–823

Springer and the environment

At Springer we firmly believe that an international science publisher has a special obligation to the environment, and our corporate policies consistently reflect this conviction.

We also expect our business partners – paper mills, printers, packaging manufacturers, etc. – to commit themselves to using materials and production processes that do not harm the environment. The paper in this book is made from low- or no-chlorine pulp and is acid free, in conformance with international standards for paper permanency.

MIX
Papier aus verantwortungsvollen Quellen
Paper from responsible sources
FSC® C105338

If you have any concerns about our products,
you can contact us on
ProductSafety@springernature.com

In case Publisher is established outside the EU,
the EU authorized representative is:
Springer Nature Customer Service Center GmbH
Europaplatz 3, 69115 Heidelberg, Germany

Printed by Libri Plureos GmbH
in Hamburg, Germany